# CLIMATE CHANGE AS A CRISIS IN WORLD CIVILIZATION

## Why We Must Totally Transform How We Live

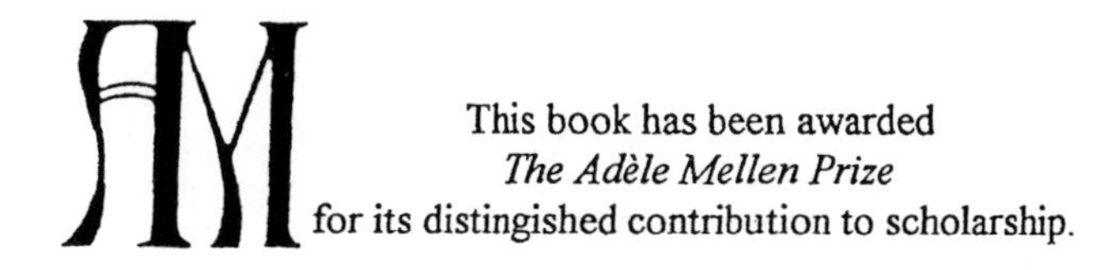

This book has been awarded
*The Adèle Mellen Prize*
for its distinguished contribution to scholarship.

# Climate Change as a Crisis in World Civilization

## Why We Must Totally Transform How We Live

Joseph Wayne Smith
David Shearman
Sandro Positano

With a Foreword by
Graham Lyons

The Edwin Mellen Press
Lewiston•Queenston•Lampeter

**Library of Congress Cataloging-in-Publication Data**

Smith, Joseph Wayne.
Climate change as a crisis in world civilization : why we must totally transform how we live / Joseph Wayne Smith, David Shearman, Sandro Positano ; with a foreword by Graham Lyons.
p. cm.
Includes bibliographical references and index.
ISBN-13: 978-0-7734-5162-9
ISBN-10: 0-7734-5162-5
I. Title.

*hors série.*

A CIP catalog record for this book is available from the British Library.

Front cover: Painting courtesy of David Shearman



The Edwin Mellen Press
Box 450
Lewiston, New York
USA 14092-0450

The Edwin Mellen Press
Box 67
Queenston, Ontario
CANADA L0S 1L0

The Edwin Mellen Press, Ltd.
Lampeter, Ceredigion, Wales
UNITED KINGDOM SA48 8LT

Printed in the United States of America

## Contents

## Foreword

Once again Dr Joseph Wayne Smith brings his formidable intellect to bear upon a compelling issue. And this may be the biggest one yet. And once again he has risen to the challenge and rigorously examined what appear to be all the major issues at the core of this subject, which is the most important challenge facing humanity at this time. He draws on his previous works published by Macmillan, Praeger *et al* on subjects including population growth, ecology, global health and economics to bolster his arguments in the current work. And the co-authors bring their own specialized knowledge to add further weight to the arguments. I enjoyed the analogy of the need to examine this issue as the physician (e.g. Professor Shearman) examines the plight of a patient. There needs to be a blend of holism with specialized knowledge to achieve a positive outcome, that is, (ideally) a cure.

The aim of the book is stated by the authors to be to understand the threat of global climate change. I would suggest that they have clearly achieved this and more. This work is not some unbalanced polemic. The authors point out that the West has produced the medical and agricultural technology which has allowed the populations of the poor countries to rapidly expand over the last 200 years. However, globalization has clearly been a disaster for the poorer nations. From a *limits to growth* perspective (the core theme of this work), both the developed and developing world have unsustainable social systems.

Climate change skeptics like Professor Ian Plimer of the University of Adelaide, who have debunked some of the more outlandish findings from some climate change models, are dealt with in a reasoned manner. The book is not for the fainthearted. A compelling case is built that the risks and effects of global climatic change may be even greater than the IPCC consensus on climate change. An important point is that the climate crisis is one crisis occurring with a number of other converging crises, including the peak oil issue and the population problem. None of these can be solved in isolation...and they are all examined in this book. The dangers of the *business as usual* scenario are clearly delineated.

The book is very thoroughly referenced and annotated, the information is up to date, and the text is eminently readable, even entertaining, which is not easy to achieve with such a complex and at times depressing subject. What really stands out to me with this work is its broad scope, the fearlessness of the authors in tackling ALL of the big issues involved/associated with global climate change, the rigour of their examination of these issues, the strength of the arguments and the honesty of their recommendations for/solutions to this mess that we are in. There is no false optimism as the book concludes (unlike in many recent works that conclude with a plaintive cry that humankind's ingenuity will undoubtedly save the day) that somehow we will be saved by a *technological fix*. Indeed, this is one of the many falsehoods that the authors debunk. Their recommendations will dismay those who expect that growth economies can continue "forever". It is argued that individual societies driven by economic growth are unsustainable, being out of balance with nature. This is perceived by many (perhaps even most) people in our society, but this book demonstrates it, clearly articulates it and offers realistic (and perhaps the only plausible) alternatives.

A quotation from the introduction asks humanity to "look at life with respect and wonder". Can this be achieved within the framework of the "received religions"? Even this subject is dealt with in the book, including a masterly examination of the historicity of the Old Testament and the inherent danger for the environment in adherence to belief in *Rapture*, the time when the righteous are taken directly to heaven, leaving the ungodly to deal with hell on earth! I am reminded of a quote from E.O. Wilson, the noted sociobiologist:

> Material reality discovered by science possesses more content and grandeur than all religious cosmologies combined.

Even Richard Dawkins is taken to task, for failing to discuss the issue of religion and the environment, along with the irrationality of his key premise for writing *The God Delusion*:

> To suppose as Richard Dawkins does that the weight of scientific and rational argument will change people's core religious beliefs, is in our opinion, not only unsubstantiated by empirical and scientific evidence...but is contradictory with his basic argument that religious beliefs are *irrational.* (Smith et. al.,p.196)

Moral and ethical issues are frequently neglected by climate change commentators, but not here. The pivotal *conserver society* work of Ted Trainer is discussed, including the importance of social responsibility. The essence of the dilemma is captured in a quote from Stephen Gardiner:

> Future generations are not here to assert claims and have no bargaining power. The present generation, despite some altruistic actions, is still dominated by self-interest. This same egoistic inertia grips each generation...resulting in continuing degradation of the environment.

(Smith et. al., p.211)

Dr Smith's strong background in philosophy is brought to bear, especially in chapter 6. Surprisingly little has been written about how humans can make a peaceful transition to a sustainable society. Without a solution to this problem humans, if they survive, will still make a transition to sustainable lifestyles, but they will do so because of *the remorseless working of things*. This is the title of a previous book by Dr Smith and is drawn from a quote from the great philosopher and mathematician, A.N. Whitehead. It goes to the very core of human and societal behaviour and what shapes history.

In chapter 5 the authors address the vexed question of the relevance of the modern university in contributing meaningfully to "saving the world" while hamstrung by rationalist economics, managerialism, compartmentalization, corporatization and (ironically) suppression of free speech. The demise of university publishing houses is covered in some detail. As noted, *reform*, *the global market* and *research excellence* are terms which are bandied about freely. However, relatively quiet are the students, who understandably worry about future jobs in a competitive world with high unemployment, and the academics who fear for their jobs and are too busy to say much anyway. I can certainly relate to this! Dr Smith's own case bolsters the authors' argument: his astonishing intellect and outstanding academic achievements (especially in publishing terms) in several disciplines over a long period have rarely been recognized within the (less than) hallowed cloisters of modern academe, due in part to his fearless nature and willingness to seek and state the truth. However, this is also a reflection of the prevailing intellectual mediocrity.

Also examined in chapter 5 is science, how it is part of the problem yet, if handled properly, can be a major part of the solution. I can attest from an insider perspective that the analysis and conclusions provided in this book are accurate and profound:

> The dream of the Enlightenment is compromised by elitism, self-interest and individualism...The problem now is that science proceeds at such a pace that society, its ethical and legal framework trail behind. Why does it proceed at this rapid rate? Mainly because of competition between egos of scientists who cannot resist the challenge and because science is a competitive economic force. There is an urgent need to revise and control this situation. (Smith et. al., p.p.137-138)

The fragmentation or compartmentalization of science, perhaps an inevitable consequence of reductionism, continues apace. Expertise becomes compartmentalized and the scientist is often incapable of taking a broad view. The recent imposition of the so-called *Research Quality Framework* (RQF) in Australian universities, which favours grants directed towards scientists who publish in *high impact factor* (IF) journals, is the latest manifestation. In my own area of plant science and agronomy, this inevitably means preference is given to prestigious molecular journals and not agronomic/soil science journals, despite the importance of many of the articles in the latter, especially from an environmental perspective. The RQF has resulted, in the UK, in the demise of basic/generic science departments, such as chemistry (and the same would occur in Australia, but most of these departments have already gone due to the influences mentioned above). Hence few new science graduates have a solid grounding in chemistry, but there are many who can follow recipes to analyse DNA and mRNA and characterise genes and perform transformations and discuss phenomics and genomics and even metabolomics. The discipline of molecular science/genetics requires intimate knowledge of a particularly esoteric language.

This area, which attracts perhaps 90 per cent of funding in plant science now, and has produced numerous articles in high IF journals (made so by this clique's avid citing of its own articles!), has achieved little of practical value to farmers and there is little visible on the horizon.

To conclude, this is a comprehensive, detailed, scholarly work in which the authors have fearlessly tackled all of the major issues involved in global climatic change. They point the way forward, again fearlessly, and of course it is not a *business as usual* scenario. Those whose living depends on the continuation of the growth economy - and perhaps this includes most of us - will find some painful reading here. Many will prefer to ignore the findings - which are not optimistic and not glossed-over - and recommendations of this book, with its essential *limits to growth* theme. These findings require radical changes in the way humans interact with the world. This may be the most important book yet to be published on this compelling issue, and deserves the widest possible readership.

Graham Lyons B Agric Sci M Public Health PhD
Research Fellow
School of Agriculture, Food & Wine,
University of Adelaide,
Waite Campus, Glen Osmond, SA 5064.

# Preface

Many scientists and intellectuals believe that the threat posed by global climatic change to human welfare and health, may be greater than the position expressed by the Intergovernmental Panel on Climate Change, which represents the scientific consensus on climate change. This book examines the scientific evidence on "abrupt" or "dangerous" climate change, as well as the health effects. The work attempts to locate the "climate crisis" within the context of a wider "crisis of civilization," which includes a series of "converging" threats to human survival. These threats are well known and have been discussed in a large number of books and reports, and include matters of environmental change such as the degradation of land resources, forests, biodiversity, fresh and marine waters, and the coming collision between the world's ever-growing demand for oil and other resources and the finitude of this resource and others. Few environmental books attempt a broad-picture analysis of the human predictament, and fewer yet of these books attempt to locate the "climate crisis" within the context of these converging threats. The present book will make such an attempt.

Further to this, it will be argued that the climate crisis cannot be resolved by a technological fix. It will require a fundamental change to the economy and also present patterns of living and thinking. The fundamental conflict which the modern world faces is between the promotion of economic growth as the ideal by which society functions, on the on hand, and on the other, the need to conserve nature through building ecologically sustainable societies. There is now a detailed literature arguing that these ideals are incompatible and that societies founded upon unending economic growth and the associated ideologies of consumerism and materialist hedonism are not sustainable. This book will apply arguments found in the limits to growth tradition to the issue of global climatic change. However, the work will go further than most limits to growth books in advocating the need for an intellectual revolution to deal with the crisis of civilization. Thus, there is a need for academic disciplines such as philosophy and psychology, to abandon research concerns with many tradition problems, and for academics to focus cognitive energies upon problems associated with the threats to civilization, particularly the environmental crisis. Beyond this there is an urgent need for universities to rise to the challenge of the times and encourage such a research focus, instead of pursuing a corporate model which puts the quest for profits above the values of ecological sustainability and integrity.

## Acknowledgments

We are grateful to a number of scientist and scholars, who helped us in various ways with this book. Of course they are not responsible for the work as presented here. We thank: Dr Paul Babie, University of Adelaide; Dr Catherine Bennett, Centre for Molecular, Environmental, Genetic and Analytic Epidemiology, University of Melbourne; Associate Professor Malcolm Bond, Flinders University of South Australia; Professor Ove Hoegh-Guldberg, University of Queensland; Professor Graeme Hugo, University of Adelaide; Professor Barry Jones, University of Melbourne; Professor Sue Koger, Willamette University; Dr R. A. D. Kooswinarsinindyah, Bundung Institute of Technology, Indonesia; Dr Graham Lyons, University of Adelaide; Professor Brian Moss, University of Liverpool; Dr Barrie Pittock, CSIRO Marine and Atmospheric Research, Australia; Professor Graham Priest, University of Melbourne; Professor Nigel Stocks, University of Adelaide; Dr Haryono Tandra, Parahyangan Catholic University, Indonesia; Professor Sir Crispin Tickell, University of Oxford and Dr Kevin White, Australian National University.

As this book was being completed, the mother or Joseph Wayne Smith, Vera Smith (June 5, 1917- September 9, 2007), and the father of Sandro Positano, Giuseppe Positano (March 28, 1928- June 17, 2007), died. This book is dedicated to their memory. Their love for nature was conveyed to us in many ways throughout their lives. They taught us many lessons: to struggle against adversity,

**patience, persistence, but above all, they inspired us to work to preserve this planet for our children and for future generations.**

# Chapter 1

## Introduction: The Threat of Global Climatic Change

> The ideology of industrial society, driven by nations about economic growth, ever-rising standards of living, and faith in the technological fix, is in the long run unworkable. In changing our ideas, we have to look forward towards the eventual target of a human society in which population, use of resources, disposal of waste, and environment are generally in healthy balance.
> Above all we have to look at life with respect and wonder. We need an ethical system in which the natural world has value not just for human welfare but for and in itself. The universe is something internal as well as external.
>
> - Sir Crispin Tickell[1]
> Chairman Emeritus
> Climate Institute, Washington DC

**Global Climatic Change: The Danger**

The aim of this book is to understand the threat of global climate change, with particular reference to human health and the sustainability of human civilisation and to evaluate the various strategies and proposals to prevent or mitigate this threat in the form of a type of "meta-plan", an overview and synthesis of plans. Our aim is to examine the plight of the planet as a physician would examine the plight of a patient who has a mental or physical illness. It is necessary for a satisfactory diagnosis to understand the full medical and/ or psychological facts of the case. To cure a disease, one first needs to know the nature and full extent of the disease. Thus, first, by way of introduction, what is the threat of global climatic change?

Sir David King, Britain's chief scientific advisor, has said that global warming is a greater threat than global terrorism.[2] He believes that the world in this century with experience a temperature rise greater than 3 C threatening millions of lives.[3] James Lovelock, who has been described by *The Observer* newspaper as "one of the environmental movement's most influential figures"[4] has written: "Before this century is over billions of us will die and the few breeding pairs that survive will be in the Arctic, where the climate will remain tolerable."[5] The view expressed by this distinguished environmentalist is that the climate change is "already insoluble" and that the planet has passed the point of no return so that human civilization is unlikely to survive.[6] Lovelock does not believe that the big greenhouse gas processing nations - the United States, China and India - will cut emissions in time, so that wildlife and whole ecosystems will go extinct.[7] Thus: "[t]he worse will happen and survivors will have to adapt to a hell of a climate."[8]

James Hansen, head of NASA's Goddard Institute for Space Studies, and considered by many to be the doyen of American climate research[9] said in his keynote address to the Third Annual Climate Change Research Conference in Sacramento, California on September 13 2006: "I think we have a very brief window of opportunity to deal with climate change... no longer than a decade at most."[10] There must be a limit of the increase in global temperature to 1 C because on the "business-as-usual" scenario, where greenhouse gas emissions continue at the present rate, the increase in the global temperature of the Earth will be from 2 to 3 C which will product a "different planet." This "different planet", Hansen envisages, will be one with melting ice sheets, a rise in sea levels which will submerge Manhattan, droughts, hurricanes striking new areas, heat waves and the extinction of 50 per cent of species.[11] In a paper published in the Proceedings *of the National Academy of Sciences*,[12] Hansen (*et al*) state that the Earth's temperature is within 1 C of its highest temperature levels in the past million years. Over the past 30 years (relative to 2005), global surface temperatures have increased by approximately 0.2 C per decade, with the warming being larger in the Western Equatorial Pacific than in the Eastern Equatorial Pacific.

Hansen (*et al*) note that the "conclusion that global warming is a real climate change, not an artifact due to measurements in urban areas, is confirmed by surface temperature change inferred from borehole temperature profiles at remote locations, the rate of retreat of alpine glaciers around the world, and progressively earlier break up of ice on rivers and lakes."[13] On the "business-as-usual" (BAU) scenario, carbon dioxide emissions grow at the rate of approximately two per cent per year in the first half of the 21st century and other greenhouse gas emissions, such as methane and nitrogen dioxide, as well as black carbon and aerosols also grow. This will result in global warming by 2100 of at least 2-3 C. The last time the Earth was 2-3 C warmer than at present was three million years ago during the Middle Pliocene and sea levels were 25-35 meters higher than today.[14] A warming of 2-3 C "would bathe most of Greenland and West Antarctica in melt-water during lengthened melt seasons."[15]

The scientists associated with the journal, *Bulletin of the Atomic Scientists* have moved the "Doomsday Clock" forward two minutes, to now stand at five minutes to midnight or global catastrophe.[16] Cosmologist, Professor Stephen Hawking of the University of Cambridge said at the time (January, 2007), that climate change is as great a threat to the world as terrorism and nuclear war.[17] The *Bulletin of the Atomic Scientists* said in a press release:

> The dangers posed by climate change are nearly as dire as those posed by nuclear weapons. The effects may be less dramatic in the short term than the destruction that could be wrought by nuclear explosions, but over the next three or four decades climate change could cause irremediable harm to the habitats upon which human societies depend for survival.[18]

The Intergovernmental Panel on Climate Change (IPCC) in its assessment reports has been viewed as representing the scientific consensus on climate change,[19] since the IPCC reports are a peer-reviewed study of published scientific research on climate change.[20] The Fourth Assessment Report proposes that by 2050 there is predicted to be a doubling of carbon dioxide levels from the pre-industrial level of 270 parts per million. This doubling of carbon dioxide would bring a temperature increase of 2 - 4.5 C, with a less probability of higher temperatures of up to 5.8 C, and with a 3 C rise being the most likely figure if no action is taken to cut carbon dioxide greenhouse gas emissions. If carbon dioxide emissions are held at present levels, then the temperature increase could be 2 C by 2100, which involves stabilizing carbon dioxide levels at 400 ppm and for global emissions to be 50 per cent below 1990 levels by 2050.[21] The Third Assessment Report of the IPCC said that greenhouse gas emissions were "probably" responsible for sea level changes, glacial melting, droughts, floods and ocean acidification; the Fourth Assessment Report gives a higher probability that this cause-effect relationship exists. For example, a rise in sea levels of between 14 cm to 43 cms, with further rises in the 22nd century as polar ice melts, are predicted for the 21st century.[22]

The oceans have been absorbing over 80 per cent of the added heat and consequently the sea water has expanded. Thermal expansion has resulted in a global sea level rise between 1993 - 2003 of 1.6 ± 0.5 mm per year.[23] According to the IPCC it is very likely that melts from the ice sheets of Greenland and Antarctica have also added to the rise in sea level over the 1993-2003 period: glaciers and ice caps producing a rise of 0.77 ± 0.22 mm per year; the Greenland ice sheet 0.21 ± 0.77 mm per year and the Antarctic ice sheet 0.21 ± 0.35 mm per year, giving a total observed sea level rise of 3.1 ± 0.7mm per year.[24] However some scientists in more recent publications argue that sea levels are rising faster than predicted by IPCC computer projects, or at least at the upper end of the IPCC projections.[25]

The IPCC has said:

> At continental, regional and ocean basin scales, numerous long-term changes in climate have been observed. These include changes in Arctic temperatures and ice, widespread changes in precipitation amounts, ocean salinity, wind patterns and aspects of extreme weather, including droughts, heavy precipitation, heat waves and the intensity of tropical cyclones.[26]

Further:

> Paleoeclimate information supports the interpretation that the warmth of the last half century is unusual in at least the previous 1300 years. The last time the polar regions were significantly warmer than [at] present for an extended period (about 125,000 years ago), reductions in polar ice volume led to 4 to 6 metres of sea level rise.[27]

Presumably, as a worst case scenario, this could happen again. Indeed, the IPCC observes, the situation could be presently worse, but volcanic matter and aerosols generated by human activity have offset some warming, so increases in greenhouse gas concentrations are likely to have caused more warming than observed.[28] On a business-as-usual scenario, the world faces an ecological disaster, with 20-30 per cent of plant and animal species placed at an increased risk of extinction, if global average temperatures exceed 1.5-2.5 C, with higher risks of extinction above 2.5 C.[29] The risks that this poses for the sustainability of human civilization are considerable.

Some newspaper headlines give the impression that the new IPCC projections "temper" some of the "more alarmist" climate change scenarios.[30] The IPCC's position on the question of the sensitivity of climate to a doubling of carbon dioxide levels, is that the most likely temperature increase is 3 C, which is a figure now widely supported by climate scientists.[31] Sir David King, Britain's chief scientific advisor, said that a 3 C rise in global temperature over the 21st century would lead to many ecosystems being unable to adapt, an increase in desertification and would place 400 million people at risk of starvation.[32] However, Sir David King believes that the IPCC prediction is optimistic in seeing carbon dioxide levels of 500 ppm in the 21st century as many scientists (whose work is examined in chapter 2 of this book) believe that the carbon dioxide emissions could reach 550 ppm or more, giving a temperature rise greater than 3 C. He said on BBC Radio: "If we go beyond 500 parts per million we reach levels of temperature increase and sea level rise in terms of the coming century which would be extremely difficult for world populations to manage."[33]

**The Kyoto Protocol: Its Limits**

At the present time, the Kyoto Protocol is developing problems. The Kyoto Protocol to the United Nations Framework Convention on Climate Change, aims to stabilize atmospheric greenhouse gas concentrations that would cause dangerous anthropogenic interference with the climate system. The Canadian Environment Minister, for example, wants the Federal Canadian government to abandon its Kyoto plan and adopt another strategy, as Canada's greenhouse gas emissions continue to rise.[34] To take another example: Germany. Germany is a leader in Europe in greenhouse gas emission reductions, having over the past decade reduced its emissions by 17 per cent from the 1990 baseline year for emission targets under the Kyoto Protocol.[35] However, in July 2006, the government of Angela Merkel said that it intended to issue almost the same number of carbon dioxide emission permits for industry in the 2008-2012 period as it did for the 2005-2007 period. This makes it doubtful whether Germany can reach its Kyoto target of a 21 per cent reduction in greenhouse gas emissions for the 2008-2012 period, although the government optimistically hopes that the reductions can be made from vehicle emissions.[36] Power companies in Germany are refusing to invest in new generation capacity without first being given exemptions from stringent emissions caps.[37]

There are other difficulties facing the Kyoto Protocol approach to controlling greenhouse gas emissions. One such difficulty relates to the issue of policing the limits of greenhouse gas emissions. At present governments declare their emissions of carbon dioxide, methane and for other greenhouse gases through "bottom-up" estimates of emission sources. Environment audits, independently conducted by researchers of the European Commission Joint Research Centre at Ispra, Italy and the Royal Holloway University of London, have found that the UK and France, among other nations, are under-reporting methane emissions.[38] The UK for example may be emitting 92 per cent more of methane (a gas which is weight-for-weigh, 23 times more effective at warming the atmosphere than carbon dioxide, that being the climate forcing potential of methane for a 100 year time horizon) than it reports under the Kyoto Protocol.[39] For Kyoto to work, there will need to be a global system of auditing greenhouse gas emission reporting.[40] In principle this can be achieved, but there needs to be the political will to implement this.

The Global Commons Institute, an independent group based in the UK, is highly critical of the Kyoto Protocol mechanisms as a way of dealing with global climatic change.[41] Kyoto Protocol rules allow tradable emissions entitlements that permit the developed-high-greenhouse gas emitting nations to effectively avoid making real and substantial greenhouse gas emission reductions, they maintain. This Institute believes that the Kyoto reductions over the long term will not result in the level of per capita reductions needed; the Global Commons Institute puts this figure at around 0.3 tonnes per capita by 2050.[42] To put this figure in perspective, the Oxford-based company, Climate Care, estimates that a global flight from London, to Toronto, Christchurch, Adelaide, then back to London, would emit six tonnes of carbon dioxide.[43] Further, even if all nations did faithfully remain on their Kyoto targets, Kyoto will only reduce projected warming in 2050 by about 1/20 of a degree C, whereas, a stabilization of atmospheric greenhouse gases at a level that is likely to result in a global temperature increase of no more than 2-3 C from pre-industrial levels would require, according to the UK Meteorological Office, a global greenhouse gas emission reduction of about 60-70 per cent by 2050.[44]

Frances Cairncross, president of the British Association for the Advancement of Science and chairwoman of the Economic and Social Research Council believes that the Kyoto Protocol is ineffectual. Although she believes that humanity should not abandon attempts to slow down the pace of global warming, in her opinion climate change would occur, and that attention should also be addressed to mitigation questions: adapting to climate change. This is also the position held by Australia's former Environment Minister, Ian Campbell.[45] According to Campbell, there is no point to Australia signing the Kyoto Protocol: "Signing Kyoto is like catching the 3 pm train from [Sydney's] central station when it's five o'clock."[46] Senator Campbell has said that even with the Kyoto Protocol "global gas emissions will actually go up by 40 per cent." Australia wants a "new Kyoto deal" for according to Senator Campbell: "even the existing Kyoto signatories know it's not working and we need a better agreement."[47] However, figures released by the secretariat for the United Nations Framework Convention on Climate Change on October 30 2006 indicate that Australia's greenhouse gas emissions rose by 25.1 per cent between 1990 and 2004.[48] In 2004, Australia's greenhouse gas emissions were already 15.8 per cent above what the Australian government promised they would be by 2012.[49] The US was 21.1 per cent above the bench mark year of 1990, with an emissions growth of 1.3

per cent from 2000 to 2004.[50]

Former British Prime Minister Tony Blair said in his closing speech at the World Economic Forum (WEF) in January 2007, that he looked forward to a more radical and comprehensive climate agreement after the Kyoto Protocol, which runs to 2012. The present Kyoto Protocol *reductions* do not include the major greenhouse gas producing developing nations of China and India, who are not required to reduce emissions yet as developing countries under the present agreement, and if "Britain shut down our emissions entirely...the growth in China's emissions would make up the difference within just two years."[51]

**Addressing the Climate Change Challenge**

A brief overview of the argument of this book will now be given. The starting point for the considerations in this book is that there exists substantial scientific evidence for the view that the risks of global climatic change may be even greater than the IPCC consensus view on climate change. For example, at the time of preparation of the camera – ready copy of the book (August 2007), the Met Office, Hadley Centre in the UK, released decadal climate predictions.[52]The model gives predictions for the annual global temperature to 2014. Over this 10-year period, the climate continues to warm, with 2014 probably being 0.3 C warmer than 2004. It is predicted on the basis of their model that at least half of the years after 2009 will be warmer than the warmest year currently on record.

Chapter 2 will also summarize evidence of the possibility of dangerous and abrupt climate change and positive feedback mechanisms that may accelerate global warming. For example, the chapter will examine evidence that sea levels are rising faster than predicted by the IPCC. As well, climate change itself may trigger events leading to additional rises in greenhouse gas emissions, in turn amplifying global warming. However, most importantly, the climate crisis is one crisis occurring with a number of other converging crises, including the peak oil issue and the population problem.[53] These problems and others will make it increasingly difficult to tackle the climate change challenge, for reasons that will be detailed.[54]

Chapter 3 examines legal, economic and technological responses to the climate crisis and the problem of fuelling the global consumer society. Although the answer to the climate crisis inevitably lies in utilizing technology and other human resources and institutions, such as the legal institution, to build a sustainable world, there are limits to legal, economic and technological solutions. In particular, these responses are generally not coordinated into any sort of a master plan which integrates the elements into a coherent whole. Further, while there is among many environmentalists and public policy makers, great enthusiasm about the prospects of renewable energy powering the globalized consumer society, it is argued here that this is most unlikely in a future world of around nine billion people, all seeking to attain not only the present day per capita resource consumption of the developed world, but a higher one from continuous economic growth. Ecological systems are already cracking under the strain of present economic growth.

There is an emerging consensus among environmentalists that a transition must begin now toward the creation of sustainable conserver, rather than consumer societies. In the light of evidence and argument that the supply side technological fixes to environmental problems are limited[55] public and academic attention must be increasingly directed towards demand side lifestyle changes. The transformation of an entire culture, or indeed, civilization, is an enormous task, even to conceptualized, and the present book is not bold enough to recommend any sort of blueprint for survival. However, there are important changes which should be made to the present intellectual culture which educates people to address the environmental crisis and the bulk of this book will deal with the changes which the present writers see as necessary for humanity to rise to the challenge of the times. We believe that the universities need to be radically reformed – or if this is impossible because they have gone too far down the corporate path – then replaced by Real Universities which are structured to focus on the environment as the key issue for study and analysis. In this book we discuss how psychology, religion and philosophy can be harnessed to aid in the quest for human survival. In the case of philosophy, for example, we see a very important role, but only if this discipline ceases to attempt to model itself upon mathematics and physics as much of Anglo-American philosophy continues to do.

Philosophy, like other disciplines, should be concerned with aiding people solve the great problems of living of our times.[56] This will involve something of an intellectual revolution, as researchers simply opt to abandon research on many age-old problems, and other problems of a highly abstract nature, and focus their attention upon the real world dangers that are rapidly approaching.

# Chapter 2

# Global Climatic Change, Human Health and the Crisis of Civilization

> [W]e are colliding with planet's ecological system, and its most vulnerable components are crumbling as a result.
>
> - Al Gore[1]

> Climate Change is the world's greatest environmental challenge. It is now plain that the emission of greenhouse gases, associated with industrialisation and economic growth from a world population that has increased six fold in 200 years, is causing global warming at a rate that is unsustainable... the risks of climate change may well be greater than we thought.
>
> - Tony Blair, Former UK Prime Minister[2]

**Introduction: The Crisis of Civilization**

The aim of this chapter is to summarize the specific evidence which indicates as, former UK Prime Minister Tony Blair has stated (as quoted above) that "the risks of climate change may well be greater than we thought" and that the risks of "abrupt" or "dangerous" climate change may be higher than represented by the IPCC consensus on global climate change.[3] Frequently discussions of global climatic change, especially relating to proposed mitigation strategies, ignore the impact of other ecological and socio-political phenomena, which will interact and impact upon human society. In the short term, some of these crises could have a substantial impact upon our world and some pessimists predict that a collapse of civilized order may occur. We will discuss these issues in the introduction to give a general context and backdrop for viewing the problem of global climatic change.

After outlining the argument that the risks of global climatic change are greater than usually thought, we turn to examine the harm arising from global climatic change to the environment and threats posed to national security. Further, in this book we will be focusing particularly upon the human health ramifications of global climatic change because we regard climatic change as ultimately a human health issue. At this point it is logically appropriate to introduce the broad conceptual framework underlying this work. Here we view human health from the framework of "conservation medicine" or ecological health. According to one researcher in this field, Gary M. Tabor, conservation medicine by uniting and unifying the disciplines of health and ecology:

> [R]epresents an attempt to examine the world in an inclusive way. Health effects ripple throughout the web of life. Health connects all species. The interaction of species is inextricably linked to the ecological processes that govern life[4]

The principle goals of the field of conservation medicine, according to Osfeld (*et al*) is "to develop a scientific understanding of the relationship between the environmental crisis and both human and nonhuman animal health and to develop solutions to problems at the interface between environmental and health sciences."[5] With respect to the issue of global climatic change its human health and other civilization-disrupting possibilities, that is also the goal of this book.

An ecological approach to understanding human health and well being is important for any comprehensive scientific understanding of the environmental crisis that humanity faces. A number of important publications have attempted to give an overview of humanity's plight, and although we cannot go anyway into doing full justice to these works, we shall mention them briefly to give a background to the seriousness of humanity's ecological problems.

*Global Environmental Outlook 3* (GEO3),[6] is the latest comprehensive health check on the planet, published by the United Nations Environment

Programme, with *Global Environmental Outlook 4* to be published in September 2007.[7] The GEO3 report found that human groups across the world are becoming particularly vulnerable to environmental change, especially in developing countries.[8] The degradation of land resources, forests, biodiversity, and fresh and marine waters often leads to the impairment of the "sink" capacities of the environment (involving natural processes of purification, filtering of air and water and nutrient recycling) which in turn can impair human health through contaminated groundwater supplies leading to diarrhea diseases and air pollution leading to respiratory disease.[9]

The GEO3 report was optimistic about progress made in addressing stratospheric ozone depletion, although four years after the publication of *Global Environmental Outlook 3*, on September 25 2006, the World Meteorological Organization (WMO) using NASA data, reported that the hole in the ozone layer above Antarctica was the largest that it has been in six years: 29.5 million square kilometers.[10] Worse, the European Space Agency reported, in October 2006 that a loss of 39.8 million tonnes of stratospheric ozone had occurred, which was greater than the previous loss of 39.6 million tonnes which had occurred in 2000.[11] The ozone layer has been damaged by human-made chemicals such as chlorofluorocarbons (CFCs) which were once used as aerosol gases in spray cans and for refrigerants, but which were banned by the Montreal Protocol in 1987.[12] Nevertheless, large ozone holes will continue to form for a number of decades due to the presence of ozone-depleting pollutants already in the atmosphere: chlorine and bromide remain for about 50 years in the stratosphere.

The consensus scientific view is that there are less ozone-depleting chemicals present in the stratosphere, but these chemicals are more efficient at ozone breakdown when cold and 2006 had been for the Antarctic stratosphere one of the coldest on record. As an example of the ecological principle that one can never do just one thing (i.e., that there are unanticipated consequences of actions) the chemical HFC134a which manufacturers used in air conditioning systems after the bans on CFCs, has been detected at "alarming levels" at the Arctic. The Norwegian Institute for Air Research, from research and testing at the Mount Zepplin station, has found that the HFC134a concentrations have doubled between 2001 and 2004.[13] One molecule of HFC134a has 1,300 times the warming effect of a molecule of carbon dioxide. According to Chris Rose of the Multisectoral Initiative on Potent Industrial Greenhouse Gases, London, air conditioning systems, although said by manufacturers to be designed to prevent HFC134a leaks, are in fact leaking,[14] This of course does not explain the 2006 Antarctic stratospheric depletion of ozone but it does indicate that there may be future surprise much like those which initially arose with CFCs. Obviously further scientific investigation and monitoring of HFC134a emissions is required.

*Global Environmental Outlook 3* outlined the existence of a number of environmental challenges and threats. Biodiversity decline was of particular concern, with the extinction rate of species, accelerating. The overuse of surface water was also of concern with 1.2 billion people lacking access to clean drinking water and 2.4 billion, lacking access to basic sanitation with over half of them being in China and India.

Such concerns are also reflected in the more recent Millennium Ecosystem Assessment Synthesis Report,[15] an assessment which was called for by the United Nations Secretary-General Kofi Annan in his 2000 report to the UN General Assembly, *We the Peoples: The Role of the United Nations in the 21st Century.*[16] The multiscale Millennium Ecosystem Assessment found that 60 per cent (that is 15 out of 24) ecosystem services examined are being degraded or used unsustainably, including freshwater resources, capture fisheries, air and water purification, the regulation of regional and local climate and biodiversity. Thus at present 10-30 per cent of mammal, bird and amphibian species face the threat of extinction; humans are thought to have increased the rate of global extinction by 1, 600 times as much as the typical "natural" background rate over the Earth's history.[17] Supporting this thesis, more recent opinion by 50 international amphibian experts have reported a rapid decline and extinction of amphibian species with extinctions of entire clades of species.[18] Approximately a quarter of amphibian species have been classified as "data deficient" in a global assessment of their conservation status.[19] This is an indication that something "very wrong" is happening to the ecological health of the planet because amphibians are considered "canaries in nature's coalmine," being more susceptible to changes in the environment than mammals and reptiles because they have a permeable skin which absorbs water and oxygen.[20] Climate change is, and will be, a leading cause of biodiversity loss, especially for changing disease dynamics arising from global warming.[21]

The World Water Council believes that the world is facing a water crisis as the withdrawals of water from groundwater sources for agriculture, industrial and household use has doubled in the last 40 years, resulting in humans using between 40-50 per cent of all fresh water from the land.[22] In North America and the Middle East up to 120 per cent of this freshwater is used, drawing on groundwater that is not recharged.[23] Increased usage, along with the impacts of global climatic change, has led to the world's mightiest rivers, such as the Yellow River in China, the Indus in Pakistan and the Nile in Egypt, regularly running dry in recent times, at least in the area where the river reaches the sea.

According to the United Nations World Water Development Report 2, *Water: A Shared Responsibility*,[24] by 2030 two thirds of the world's population will live in cities and towns, with most of the increasing concentration of population being in the urban areas of Africa, Asia and Latin America, increasing urban water demands, which are likely to be met by drawing water from increasing distant watersheds.[25] In many low-income and middle-income countries, urban population expansion has outgrown the provision of water sanitation.[26] Many such areas lack adequate capital to meet the substantial costs of maintaining and expanding household water and sewer connections.[27] Access to safe water and adequate sanitation are a major environmental cause of water-related diseases, including diarrhea, which leads to the deaths of some 3,800 children every day.[28] Water is also the breeding site of many disease vectors, such as mosquitoes, which spread diseases such as malaria, Japanese encephalitis, schistosomiasis and filariasis, to name but a few.

The "water crisis" is an important issue for first world nations, as a recent report prepared for the WWF (formerly known as the World Wildlife Fund) Freshwater Program, *Rich Countries, Poor Water*, details.[29] For example, with respect to the United State's water resources, the report concludes that more water in many areas is being used than can be naturally replenished. US agricultural products have a substantial "water footprint" or "virtual water" content, the amount of water usage embodied in products. Thus one 150g hamburger has a virtual water content of 2,400 liters, a 200ml glass of milk, 200 liters, a 500g cotton T-shirt medium size, 4,100 liters and a pair of leather shoes 8,000 liters.[30] With respect to the US water situation the report says: "This situation [of water overuse] will only be further exacerbated by climate change scenarios of lower rainfall, increased evaporation and changed snowmelt patterns."[31] The same problems confront Australia.[32] The European Environment Protection Agency said in 2005, the following about Europe's water crisis:

> [Europe has] also a wide variety of water uses, pressures and management approaches. A succession of floods and droughts in recent years has illustrated Europe's vulnerability to hydrological extremes. However, there are many other water-related pressures on Europe's environment. River systems and wetlands are increasingly at risk. The quality of Europe's rivers, lakes and groundwater is being threatened by the discharge of sewage and industrial waste and by excessive application of pesticides and fertilizers. Climate change and sea level rise add other potential pressures on European water resources and management.[33]

Developed countries over the last decades have faced "emerging water related health problems including dangerous pollutants in urban drinking water, nitrate contamination, cyanabacterial toxins and infections with dangerous pathogens.[34] For example, in 1993 the largest documented outbreak of a waterborne disease occurred in the United States when 403,000 people become ill and 69 died through the Wisconsin water supply being infected by the pathogen cryptosporidium. The research team of Corso (*et al*) estimate that the total cost of this illness was $US 96.3 million, with $US 31.7 million in medical costs and $US64.6 million in lost productivity.[35]

The final area of concern that we shall briefly mention in this introductory section is the question of emerging resource shortages and the interaction of this problem with other environmental crises. Lester Brown in *Plan B2.0: Rescuing a Planet Under Stress and Civilization in Trouble*[36] gives a clear and concise chapter by chapter examination of all (and more) of the environmental problems mentioned here. Brown points out in the preface to this book that an important factor which needs to be considered by environmental crisis theorists is that "China has now overtaken the United States in the consumption of most basic resources. Among the leading commodities in the food sector (grain and meat), in the energy sector (oil and coal), and in the industrial economy (steel), China now leads the United States in consumption of all except oil".[37] China is predicted, at its growth rate of 8 per cent per year, to surpass the US economy in 2031. If Chinese consumption levels are the same per capita as US consumption levels then this population of 1.45 billion "would consume an amount of grain equal to two thirds of the current world grain harvest, and it would use 99 million barrels of oil per day - well above current world production of 84 million barrels."[38]

This last projection - of a direct collision between the world's ever growing demand for oil - and the finitude of this resource is of particular significance. "Peak oil" theorists have argued that the world's oil production has reached or will soon reach its "Hubbert peak," after which world oil production will be in decline due to increased marginal costs of extraction.[39] The "peak oil" theory is that the rate of oil production follows a bell shape curve; before the peak there are increases in production with the addition of technical infrastructure, but after the peak resource depletion leads to a decline in production due to increasing difficulties (and hence costs) in getting extra increments of oil. The peak oil theorists believe that we cannot know the exact date of the Hubbert peak and that we will only know when it occurred, retrospectively. One such peak oil theorist, Chris Skrebowski of the Oil Depletion Analysis Centre and editor of *Petroleum Review* said in August 2006 in an address to the Committee for Economic Development of Australia (CEDA), that the Hubbert peak is likely to be reached by 2010. He also said: "Peak oil is when flows can't meet the required demand. This will cause an economic tsunami."[40] Skrebowski said that the rate of discovery of major new oil fields is decreasing and that of the world's 18 largest oil fields, 12 were in production decline. Oil supply will peak at 92 to 94 million barrels a day and then decline.[41]

Some disagree with this pessimistic assessment.[42] For example, Exxon Mobile Australia chief executive Mark Nolan has said at the Asia Pacific Oil and

Gas Conference, Adelaide September 11 2006 in rejecting the peak oil view: "The fact is that the world has an abundance of oil and there is little question scientifically that abundant energy resources exist."[43] Mr. Nolan quotes the US Geological Survey that the Earth has over three trillion barrels of conventional recoverable oil of which one trillion has been produced. He believes that when alternatives oil resources such as heavy oil and shale oil are added to this figure, the total recoverable reserves are over four trillion barrels. As well, technology enables twice as much oil to be recovered today than 100 years ago: "When you consider that a further 10 per cent increase in recoverability will deliver an extra 800 billion barrels of oil to our recoverable total, we have reason to be sure that the end of the oil is nowhere in sight."[44]

Peak oil theorists generally believe that the US Geological Survey overstates undiscovered oil and prospective oil reserves.[45] For example, oil expert Matthew R. Simmons in *Twilight in the Desert: The Coming Oil Shock and the World Economy* [46] argues that Saudi oil reserves are smaller than Saudi oil officials claim. The principal Saudi fields are decades old and oil fields, he argues; yield around 75 per cent of their oil during the first half of their productive life before going into decline. In Simmon's opinion it is remote that there will be giant fields like the Saudi Ghawar discovered.

Jeremy Leggett is a leading environmentalist, but has also expertise in the oil industry being a geologist and former faculty member of the Royal School of Mines in London. In his latest book is *The Empty Tank: Oil, Gas, Hot Air and the Coming Global Financial Catastrophe*.[47] Leggett argues against the oil optimists who believe that there are two trillion barrels or more of oil left, which would put the Hubbert peak in the 2030s.[48] He points out that Shell had overestimated its reserves by more than 20 per cent.[49] OPEC has vested interest in "massaging the data" as OPEC quotas for each country in the cartel are allocated according to the size of national reserves. It is in the interests of countries to over-estimate reserves. As a matter of fact when OPEC decided in 1982 to allocate a quota to each oil-producing nation based upon estimations of the size of national reserves, the estimation of reserves went up overnight. The estimated reserves of the Middle East have been approximately constant since then- for both the total oil

survival of human civilization in a hypothetical world where the Earth was experiencing rapid global cooling, especially if an ice age was approaching. Unfortunately for us, humanity faces all of the environmental problems mentioned in this chapter and many others, all at once. Consequently environmental problems cannot be satisfactorily understood as being isolated phenomena and their understanding and solution must be in a holistic and systematic way. This point may seem obvious, yet the sheer technical sophistication of modern science tends to lead to specialization rather than global and holistic approaches to knowledge,[64] and much research on global climatic change fails to take into account the holistic, chaotic and non-linear nature of ecological systems.[65]

**Climate Change Skepticism**

Before advancing our argument that risks of climate change may be greater than thought to be so by the consensus view of climate change, we will in this section address the issue of climate change skepticism. This position doubts whether observed trends in the rise of the global average temperature are due to human causes or whether there is as a matter of fact a rise in the global average temperature at all. As the Australian archaeologist Ian Lilley puts it: "Climate change is natural… and humans can adapt to deal with it… people manage the change in their environments and life goes on."[66] Whatever happens, we'll manage. In general, climate change skeptics believe that even if there are small changes in global average temperatures, this is not a problem because the Earth's climate has varied greatly over historical and certainly geological time.[67] After all it may be argued, only 20,000 years ago much of Europe and North America was covered by vast sheets of ice, yet humankind survived.

As an example of climate change skepticism we will consider an article published by scientist Professor Ian Plimer of the University of Adelaide School of Earth and Environmental Sciences.[68] He believes that the establishment or consensus view on climate change is short on evidence and big on politics, ideology and "hysteria". Climatic data, he argues, needs to be understood over "meaningful periods of observation" and the science of geology allows this. This means an examination over long periods of time measured in millions of years.

Professor Plimer argues that the data used by organizations such as the Intergovernmental Panel on Climate Change (IPPC), does not pass a simple scientific validation test. He says: "A body of data needs explanation. This is called a theory. A scientific theory needs testing and if it fails a test such as data repeatability or agreement with previously validated theories, then the new theory is discarded. The theory of human-induced climate change again fails this test."[69]

The IPCC itself is a scientific review mechanism and merely summarizes the research results of peer-reviewed scientific papers. Consider, for example, one of the many hundreds of such papers supporting the theory of human-induced climate change. D.J. Karoly (*et al*) investigated surface temperature variations in North American climate data over the 20$^{th}$ century.[70] This research team used various indices of large-scale temperature variation such as the area-mean surface temperature over land, and the mean land-ocean temperature contrast (this being area-mean temperature over land minus the mean sea surface temperature of the surrounding region) as well as other indices. These indices represented various climatic features of the modeled surface temperature - such as faster warming over the land than ocean and faster warming in winter than in summer. It was found that these results were consistent with various computer climatic model simulations. Karoly (*et al*) concluded from their analysis that human-induced warming is detectable in North American climatic data from a period from 1949 onwards. Returning now to Professor Plimer: for his climate change skepticism to be scientifically accepted, he would need to methodologically invalidate or refute by contrary evidence, many hundred scientific papers like the one reviewed here. This has not been done.

Professor Plimer also argues that there are many uncertainties associated with General Circulation Models - the computer models used in climate modeling. He rightly notes that the very complexity of the climatic system means that there are limits placed upon the capacity to understand and analyze the model processes and interactions. The role of aerosols, liquid or solid particles suspended in the air, and their interactions with the clouds, is for example, but one active field of research. Even the understanding of clouds on a regional scale is a challenging problem.[71] Physical models are idealizations of reality. But it does not follow from this that models in climatology are therefore useless in investigating climatic phenomena. To suppose so because of uncertainty - which is present in every field of science including pure mathematics itself[72] - is to deny that a scientific study of climate is possible at all. That is a "religious" attitude towards climate, not a scientific one.

Professor Plimer does seem to believe that a scientific study of climate change, at least by computer-based models, is impossible. He argues that since we cannot predict the weather for years in advance, how can one hope to predict such longer-term phenomena such as climate? However this argument is fallacious. Day-to-day weather patterns are seemingly "chaotic" and prediction is difficult because of the existence of many fine-scale or so called "sub-grid" phenomena. These are individual physical events, such as thunderstorms that are often impossible to precisely model for various technical reasons.[73] It is actually mathematically easier to deal with longer-time scale phenomena such as climate, than short-scale phenomena such as weather, where chaotic effects can make prediction difficult.

In his article Professor Plimer gives considerable significance to the argument that the theory of human-induced climate change is incorrect because viewed over a 590 million year period, the average temperature of the Earth has increased *before* carbon dioxide levels have risen.[74] Even if this is true, it is irrelevant. Over such vast periods of time the temperature of the Earth has risen, and fallen, for a variety of geophysical and even cosmological reasons, such as changes in solar activity, which have altered the influx of solar radiation. All of this is logically consistent with the proposition that present day global warming is occurring within a geological era where long term cooling is observed over periods of millions of years. It is clear that there can be multiple causes for temperature rises on Earth.[75] At present, over time scales of interest to most humans, the rise in mean global temperature arises from rising carbon dioxide and other greenhouse gases such as methane, rather than from say variations in solar radiation. However as will be seen from our discussion of the issue of "global brightening" in this chapter, there are other important relationships between solar radiation levels and greenhouse gas emission levels.

Professor Plimer says that by looking at the climatic change debate from the perspective of the earth sciences, balance and perspective can be obtained in the climate change debate. In particular he has said that in "the history of science, predictions of catastrophes have all been wrong."[76] But Professor Plimer is simply wrong on this point. Extinction is common.[77] For example, about 250 million years ago the Permian extinction occurred, where 90 per cent of marine life and 70 per cent of all land families disappeared. Even dead vegetation did not completely decompose, making the Earth a foul place indeed.[78] There are a number of theories about how this mass extinction occurred, including a meteorite strike. Some theorists believe that radical climate change occurred caused by glacial events at the north and south poles producing severe climatic fluctuations, including substantial temperature rises in some parts of the world.[79] Still other theorists posit global cooling leading to this mass extinction as volcanic eruptions in Siberia produced clouds of ash in the atmosphere which lowered global temperatures.[80] Other theorists hypothesize that a 2-3 C rise in global mean temperatures destabilized deposits of gas hydrated of methane, which by methane release and positive feedback mechanisms, resulted in a compounding effect leading to a 6 C temperature rise.[81] When cool, methane hydrate is a crystalline material. Warming releases the methane molecules. Large amounts of methane hydrate exists today in slush mixed in mud on ocean beds Methane is a greenhouse gas many times more effective at heating the atmosphere than carbon dioxide. This type of temperature change is at the upper end of the range predicted by some climate scientists who believe that the IPCC's models do not adequately represent the climate forcing and compounding effects of phenomena such as gas hydrate release. The release of last quantities of methane would accelerate global warming and also lead to some other spectacular problems: oceans become frothy reducing the buoyancy of water so that ships could suddenly sink.

In summary, the lessons which we can learn from examining geological history should give us pause and cause for concern. Climate change is "natural" - but so is death and destruction. Even if this theory of human-induced climate change was false as the climate change skeptics believe, it hardly follows that *naturally-caused* climatic change is of no concern. Indeed, William Kininmonth, a critic of the consensus view of climate change, still sees climate change as a natural hazard and a matter of concern.[82] Natural climate change could be less susceptible to human intervention than human-induced climate change.

Recently there have been some books published which challenge the consensus view of climate change by means of alternative cosmological or geo-historical theories. For example, Henrik Svenmark, a researcher at the Danish National Space Center and science writer Nigel Calder, have published *The Chilling Stars: A New Theory of Climate Change.*[83] The book advances the idea that cosmic rays (high energy particles falling on earth from space) are substantially involved in cloud formation. Due to unusually violent solar activity, less cosmic rays have hit the Earth in the last 100 years. Fewer clouds means a warmer world as there is less cloud mass reflecting solar radiation back into space. These authors believe that cosmic rays and fluctuations in the intensity of solar activities have a greater impact on the Earth's climate than anthropogenically produced carbon dioxide.

S. Fred Singer and Dennis T. Avery have written a book, *Unstoppable Global Warming Every 1,500 years* [84] which also offers a challenge to the consensus theory of climate change. They agree that the Earth is warming but argue that anthropogenic carbon dioxide plays only a minor role. The real cause is a 1,500-year climate cycle (plus or minus 500 years), discovered by Willi Dansgaard of Denmark and Hans Oeschger of Switzerland, based upon the analysis of oxygen isotopes from Greenland ice cores.[85] Further analysis of scientific data, such as from seabed sediment cores, cave stalagmites and fossilized pollen, supports the idea of this climatic cycle, Singer and Avery argue.[86] Thus, the Earth is presently in a warming phase, but it will ultimately go into a cooling phase, and that, not warming, Singer and Avery Argue, should be a matter of greater concern.

Technical issues about the relative significance of cosmic rays upon the climatic system must be evaluated by scientists in the relevant field. Likewise, the idea that the Earth has a 1,500 year cycle of warming and cooling could well prove to be correct. Here, the point which we wish to make, which is not addressed by these various authors, since they are critics of the theory of human-caused climate change, is that all of these theories could be mutually correct. There may well be a 1,500- year cycle of warming and cooling, which has *also* been augmented by anthropogenic carbon dioxide increases. This could mean that the net temperature increases - at least in the present warming phase, may be greater than either conventional or unconventional scientists expect.[87] The present debate about global climatic change is concerned with possible climatic changes which are, as these more cosmologically-inclined writers and researchers, rightly point out, *relatively* insignificant compared to the vast changes which have occurred on Earth over millions of years. However, over such time scales human existence becomes a blip in history. Over such time scales all of human civilization is relatively insignificant. However, *our* concern is with *human* survival rather than cosmological issues, and climate changes which may not be substantial over the expanse of millions of years are quite important, looked at from the relativistic perspective of human survival and welfare.

Another issue relevant to the topic of climate change skepticism relates to the controversy surrounding the validity of the so-called "hockey stick" curve. Statistical climatologist Michael Mann of Pennsylvania State University and colleagues attempted to mathematically represent the average global temperature over the past millennium.[88] Direct temperature measurements only go back to 1860 so indirect or proxy measures must be used in temperature analysis. The proxy methods include tree rings, corals, ice cores, marine and lake sediments, boreholes, glacier length records among other methods. Mann's research was used by the IPCC to conclude in its Third Assessment Report that the increase in temperature in the 20$^{th}$ century is *likely* (meaning as having an estimated confidence level of 66-90 per cent, much less than the standard high confidence level of greater than 95 per cent usually accepted in statistical analysis) to have been the largest of any century during the past 1,000 years. The IPCC also concluded that it is *likely* that for the Northern Hemisphere, the 1990's was the warmest decade and that 1998 was the warmest year of the last millennium.[89]

Various criticisms have been made of the "hockey stick" curve, such as the graph doesn't show the Medieval Warm Period after 1000 or the "Little Ice Age" from 1550 to 1850. The reply made to this is that the Medieval Warm Period and Little Ice Age are European and North Atlantic phenomena and that the southern temperatures tended to average this out.[90] More searching criticisms have been made by Stephen McIntyre and Ross McKitrick, who in a number of technical papers challenged the validity of Mann's methodology, arguing that Mann's methodology produced results with spurious significance.[91] This has generated scientific debate.[92] The US House Energy and Commence Committee's Chair, Representative Joe Barton (Republican, Texas) commissioned a statistical analysis of Mann's work, resulting in the *Ad Hoc Committee Report on the 'Hockey Stick' Global Climate Reconstruction.*[93]

The frame of reference of the report was for qualified statisticians and mathematicians to assess the correctness of Mann (*et al*)'s work and not whether global climatic change is occurring, something which seems to have been missed by a number of climate change skeptics who have cited this report. The Ad Hoc Committee concluded that "Mann's assessments that the decade of the 1990's was the hottest decade of the millennium and that 1998 was the hottest year of the millennium cannot be supported by his analysis."[94]

The US National Academies National Research Council (NRC) also reviewed Mann's work and concluded that the hockey stick graph was questionable but that there was still independent evidence of global climate change and global warming.[95] In particular other supporting evidence went a considerable way to supporting some of Mann's conclusions. The NRC made these conclusions which we cite from the text:

- The instrumentally measured warming of about 0.6 C during the 20th century is also reflected in borehole temperature measurements, the retreat of glaciers, and other observational evidence, and can be simulated with climate models.
- Large-scale surface temperature reconstructions yield a generally consistent picture of temperature trends during the proceeding millennium, including relatively warm conditions centered around A.D. 1000 (identified by some as the "Medieval Warm Period") and a relatively cold period (or "Little Ice Age") centered around 1700. The

existence and extent of a Little Ice Age from roughly 1500 to 1850 is supported by a wide variety of evidence including ice cores, tree rings, borehole temperatures, glacier length records, and historical documents. Evidence for regional warmth during medieval times can be found in a diverse but more limited set of records including ice cores, tree rings, marine sediments, and historical sources from Europe and Asia, but the exact timing and duration of warm periods may have varied from region to region, and the magnitude and geographic extent of the warmth are uncertain.

- It can be said with a high level of confidence that global mean surface temperature was higher during the last few decades of the 20th century than during any comparable period during the preceding four centuries. This statement is justified by the consistency of the evidence based on a wide variety of geographically diverse proxies.
- Less confidence can be placed in large-scale surface temperature reconstructions for the period from A.D. 900 to 1600. Presently available proxy evidence indicates that temperatures at many, but not all, individual locations were higher during the past 25 years than during any period of comparable length since A.D. 900. The uncertainties associated with reconstructing hemispheric mean or global mean temperatures from these data increase substantially backward in time through this period and are not yet fully quantified.
- Very little confidence can be assigned to statements concerning the hemispheric mean or global mean surface temperature prior to about A.D. 900 because of sparse data coverage and because the uncertainties associated with proxy data and the methods used to analyze and combine them are larger than during the more recent time periods.[96]

The reports of both the Ad Hoc Committee and the National Academies Synthesis Report are clear that the evidence for anthropogenic impacts on the climate system is independent of the mathematical constructions of one research team. If the hockey stick is incorrect, then global climatic change is still supportable on other evidential bases.[97] For example, the Ad Hoc Committee has said about the current scientific consensus of the temperature record of the last 2000 years: "There is strong evidence from the instrumental temperature record that temperatures are rising since 1850 and that global warming is a fact. How accurate the reconstructions over the past millennium are is a matter of debate and we do not believe there is a consensus on this issue."[98] The Ad Hoc committee also pointed out that average global temperature issues are not the real focus of concern:

> It is the temperature increases at the poles that matter and average global or Northern Hemisphere increases do not address the issue. We note that according to experts at NASA's JPL, the average ocean height is increasing by approximately 1 millimeter per year, half of which is due to melting of polar ice and the other half due to thermal expansion. The latter fact implies that the oceans are absorbing tremendous amounts of heat, which is much more alarming because of the coupling of ocean circulation to the atmosphere.[99]

Thus the Ad Hoc Committee recognized that even if Mann's work has statistical and methodological defects, global climatic change is real and represents a major hazard. Indeed, even if the hockey stick graph critics are right, and the world may have been as warm or nearly as warm as it is today at some other points in the last 2,000 years, this does not mean that global climatic change is not a natural hazard. The argument shows on the contrary the sensitivity of the global climatic system - contrary to some reports[100]- and raises the possibility of abrupt climatic change.[101]

**Dangerous and Abrupt Climatic Change**

On Tuesday 23 May 2006, BBC News carried an item on its website by its environmental correspondent, Richard Black, entitled "Global Warming Risk 'Much Higher.'"[102] This article stated that various scientific research indicated that global average temperatures would increase to a greater extent than predicted by the IPCC, with the consensus position estimate of a rise of between 1.5 and 4.5 C, with a most likely figure being 3 C on a business as usual scenario over the 21st century, could be too low by up to 75 per cent. Black quoted the Australian Greenhouse Office saying that current temperature estimates were challenged by this research: "The evidence for a warming Earth is stronger and the impacts of climatic change are becoming observable in some cases."[103] Let us examine a small amount of this recent evidence.

The world's oceans have absorbed about 50 per cent of the carbon dioxide released into the atmosphere by human activities over the past 200 years.[104] This has resulted in acidification of the oceans as carbon dioxide is absorbed by the oceans faster than natural assimilation rates.[105] Carbon dioxide forms carbonic acid when dissolved in water and this acidification of the world's oceans is already affecting a wide variety of marine processes. Major effects are expected upon coral reefs as the more acidic waters start to dissolve calcium carbonate. Organisms find it harder to build their shells.[106] Plankton species may also be adversely affected. Turley (*et al*) point out that planktonic microalgae such as coccolithophores may be adversely affected. These organisms are important in the global carbon cycle, transporting calcium carbonate to marine sediments.[107] Coccolithophores are also thought to play a part in climate regulation through the production of dimethyl sulphide (DMS) which has a role in producing the cloud condensation nuclei.[108] A reduction in their numbers could lead to a decrease in dimethyl sulphide reaching the atmosphere from the oceans and through cloud changes result in further increases in global temperatures.[109]

Another important area of controversy in the field of "dangerous" climatic changes relates to alleged trends in hurricane destructiveness.[110] Some researchers have claimed that there is a trend to increasing destructiveness; 2004 and 2005 produced a record hurricane season in the United States, including Hurricane Katrina which devastated New Orleans. Brazil had its first cyclone; there were five in the Cook Islands in just five weeks and 10 in Japan within this two year period.[111] Some argue that the alleged dramatic rise in the power of the Atlantic tropical cyclones is correlated with an increase in the late summer/early autumn (fall) surface temperature over the North Atlantic and this is in turn due to anthropogenic caused global climate change.[112] Another group of researchers believe that this alleged trend is due to natural climatic variation known as the Atlantic Multidecadal Oscillation (AMO).[113] A report submitted to the World Meteorological Organization's Commission for Atmospheric Science in Cape Town in 2005 concluded that "[no] single high-impact cyclone event of 2004 and 2005 can be directly attributed to global warming, though there may be an impact on the group as a whole."[114] However more recent research papers indicate that there is a trend of an increasingly number of category 4 and 5 hurricanes since 1970 and that this trend is directly linked to the trend of sea-surface temperature (SST) rises, so that global warming is causing an increase in hurricane intensity.[115] One recent paper finds an 85 per cent chance that external forcing explains 67 per cent of sea-surface temperature increases in the Atlantic and Pacific tropical cyclogenesis regions.[116]

Professor Chris Rapley, director of the British Antarctic Survey said on September 19 2006 in a media release that sea levels are rising faster than predicted by the IPCC report of 2001 because of the melting of land-based ice such as that of Greenland and Antarctica.[117] Nicolls and Lowe point out that the IPCC Third Assessment Report[118] of a 9 to 88 cm range for global sea level rises "does not embrace the full range of uncertainties, including those associated with changes in the major ice sheets, particularly the maritime West Antarctic Ice Sheet (WAIS), which contains enough water to raise global sea levels by up to 6 m."[119] The issue of the stability of the West Antarctic Ice Sheet (WAIS) will be discussed shortly.

Most of the world's glaciers outside of the polar regions are retreating including the Rockies, the Sierras, Andes, the Alps and Tibetan Plateau. For example in the 1850s, according to the World Glacier Monitoring Service in Zurich, about 4,474 sq km of the Alps were glaciated; by the 1970s 2,903 sq km were glaciated and in 2000, this figure fell to 2,272 sq km.[120] This is an average loss of 2.9 per cent from 1850 to the 1970s and an average loss of 8.2 per cent per decade from the 1970s until 2000.[121] A research team led by glaciologist Lonnie Thompson of Ohio State University hypothesized that climate is forcing abrupt changes that could lead to the melting of most low latitude, high altitude glaciers in "the near future," meaning decades.[122] The evidence of this research team was based on the analysis of ice cores, aerial photographs, maps and plant remains.

There are a number of important consequences of this accelerated pace of low latitude, high altitude glacial melts. Glaciers function as freshwater reservoirs where winter snowfall is stored and released over the summer rather than being released in a mass during springtime when the snow melts. This will threaten the freshwater supplies of many regions on Earth. For example, the Tibetan Plateau ice field has100 times more ice than the Alps and supplies almost half of the drinking water to 40 per cent of the world's population. Seven river systems originate there: the Indus, the Ganges, the Brahmaputra, the Salween, the Mekong, the Yangtze and the Yellow River.

It is also known that the pressure variations may alter the strength of faults and earthquakes may occur from pressure variations due to precipitation if the Earth's crust is in a critical state.[123] Glacial melting may also trigger earthquakes by a process known as *isostatic rebound,* a rebound of the Earth's crust from the removal of pressure caused by ice. Climate change may cause extreme geological events, triggering seismic and volcanic activity.[124] Thus warming at the beginning of the interglacial period 10,000 years ago resulted in increased volcanic activity in Iceland.[125] Earthquakes could produce sea-floor landslides which could in turn accelerate the extreme release of methane gas from solid methane hydrate on the ocean floor.[126]

The Arctic average temperature has risen at almost twice the rate of the rest of the Earth over the last few decades.[127] Hassol and Correll summarize four reasons why the Arctic warms faster than lower latitudes:

> First, as arctic snow and ice melt, the darker land and ocean surfaces that are revealed absorb more of the sun's energy, increasing arctic warming. Second, in the Arctic, a greater fraction of the extra energy received at the surface due to increasing concentrations of greenhouse gases goes directly into warming the atmosphere, whereas in the tropics, a greater fraction goes into evaporation. Third, the depth of the atmospheric layer that has to warm in order to cause warming of near-surface air is much shallower in the Arctic than in the tropics, resulting in a larger arctic temperature increase. Fourth, as warming reduces the extent of sea ice, solar heat absorbed by the oceans in the summer is more easily transferred to the atmosphere in the winter, making the air temperature warmer than it would be otherwise.[128]

Declining sea ice is an indicator of climate change and impacts upon a number of other factors such as exchange of heat and moisture at the interface of the atmosphere and the ocean, humidity, cloudiness, surface reflectivity and sea currents.[129] Arctic annual average sea-ice extent has declined by approximately eight per cent since 1976 and the summer sea-ice extent has decreased even more rapidly, losing 15-20 per cent of ice coverage.[130] Some scientists believe that in the summertime the Arctic Ocean may be free of ice by the end of the 21$^{st}$ century.[131] Arctic winter ice has previously been stable but during the winter of 2005 and 2006 a retreat in the ice cover occurred. In the winter of 2006 in the Arctic, the region had some 300,000 sq kms less ice than 2005.[132] Satellite data indicates a trend of a rise of winter temperatures and a decrease in the length of the seasonal ice growth period.[133] Further, winter ice retreats are predicted and if the trend continues the Arctic Ocean will lose all of its ice by 2030.[134] Walt Meier of the US National Snow and Ice Data Center in Colorado believes that there is a "good chance" that a "tipping point" has already been reached in the Arctic.[135] Positive feedback mechanisms are operating, where increased global warming occurs due to changes in surface reflecting properties. White ice, which was a good reflector of solar radiation, is replaced by darker surfaces which absorb relatively more solar radiation. Seawater will absorb more solar radiation than ice.[136] Further, the replacement of Arctic tundra by darker trees and scrubs will also accelerate global warming.[137] Worse, the melting of permafrost in the Arctic tundra and boreal forests is already triggering the release of methane from layers of thawing peat.[138] In the past the release of methane from the melting of permafrost has been a highly significant mechanism for global warming. James Hansen of NASA's Goddard Institute for Space Studies believes that beyond a global warming of 1 C further than present global average temperatures, a critical threshold or tipping point will be reached where positive feedback mechanisms will accelerate warming.[139]

Warming patterns in Antarctica are also a matter of concern. Summertime Arctic ocean ice melts will not raise sea levels as the ice is already floating in the water; and likewise for floating ice shelves in Antarctica. Melting water from receding mountain glaciers will only raise sea levels slightly. This will not be so for some melt scenarios of ice masses on land in Antarctica and Greenland.[140] There has been an accelerating decay of ice sheets in Antarctica and Greenland,[141] with an increase in glacial seismicity,[142] indicating a more rapid response of ice sheets to global warming than previously expected by scientists.[143] Several large glaciers on both Antarctica and Greenland are exhibiting increased water flows into the ocean, due some scientists believe, to warmer subsurface waters melting the submarine bases of the glaciers.[144] As only around one half of the Earth's existing radiation imbalance is detectable in rising atmospheric temperatures, it is hypothesized that the rest of this heat must be stored in the sea and various research teams have detected ocean warming in the tropics and mid-latitudes.[145]

Further, 30 years of weather balloon data have indicated that the air over Antarctica is warming faster than in any other area of the world: warming 0.5 C to 0.7 C per decade over the last 30 years, compared to a worldwide temperature increase of 0.1 C per decade.[146] The winter air temperature of the middle troposphere, of altitudes of 10 kilometers has been increasing at a rate of 0.5 to 0.7 C per decade over the last 30 years.[147] This is larger than any tropospheric warming anywhere on Earth. Although research cannot at the present time assign an exact cause to this from the radiosonde observations, the temperature change is thought to be due to changes in greenhouse gas concentrations and cloud quantities.[148]

Velicona and Wahr[149] used gravitational survey data from the Gravity Recovery and Climate Experiment (GRACE) satellites, to measure, through estimates of the Earth's global gravitational field, changes in the mass distribution of the Antarctic ice. After a removal of various bias factors it was concluded that for the period 2002-2005 there was a significance decrease in the mass of Antarctic ice at a rate of 152 ± 80 cubic kilometers of ice per year, with most of the loss in ice mass coming from the West Antarctic Ice Sheet. This loss is equivalent to 0.4± 0.2 millimeters of global sea level rise per year.[150]

Jim Hansen, director of NASA's Goddard Institute for Space Studies, believes that on the business as usual scenario, there could be a very rapid melting of the Greenland ice cap with sea level rises of a couple of meters this century and several more in the 22nd century.[151] Gregory (*et al*)[152] have concluded that the Greenland ice-sheet is likely to be eliminated if the annual average temperature in Greenland increases by greater than 3 C, raising global average sea levels by about seven meters over 1,000 years. Other researchers have hypothesized that the melting of the Greenland ice cap is irreversible above 2.6 C warming.[153]

The total annual loss of the Greenland Ice Sheets, allowing for snowfall is 220 cubic kilometers, which is twice the loss of the previous decade. According to Rignot: "The glaciers are sending us a signal. Greenland is probably going to contribute more and faster to sea-level rise than predicted by current models."[154] Rignot and Kanagaratnam[155] use satellite-borne radars to make observations about the changes in the velocity structure of the Greenland Ice Sheet between 1996 and 2005. It was found that the three kilometer thick ice mass on Greenland in the last few years has doubled its speed towards the sea. The Kangerdlugssuaq Glacier in central east Greenland increased its speed between 2000 and 2005 from six kilometers per year to 13 kilometers per year. The Jakobshavn Isbrae Glacier in West Greenland almost doubled its speed between the same years. In the south, the Helheim Glacier accelerated by 60 per cent. Rignot and Kanagaratnam found that the loss of mass of Greenland ice had doubled from 1996 to 2005, being a loss of 224± 41 cubic kilometers per year, approximately enough water to supply 224 cities the size of Los Angeles for a year.[156] Meltwater is entering ice crevasses and acting as a lubricant for the glacier, allowing ice to slide into the sea. This mechanism has not been included in the usual models according to Rignot.[157]

Greenland freshwater melts, from a deglaciation of Greenland, may lead to a slowdown or collapse of the thermohaline circulation system.[158] The thermohaline system is an ocean temperature/salt-based system which transfers heat from the tropics to higher latitudes and makes Europe up to 8 C warmer than at other longitudes at its latitude. The thermohaline system has shutdown in the early-Holocene period due to the addition of freshwater[159] and there are more concerns that it could shutdown again within modern times, especially since in December 2005 a research team led by Harry Bryden of the National Oceanography Centre in Southhampton UK detected from their measurements, compared to earlier measurements, a weakening by 30 per cent of the cold deep return section of the thermohaline circuit. They concluded that the northward branch of warm waters had also slowed.[160] The slow down could be merely a short-term fluctuation or it could be the herald of an imminent human cataclysm that could bring on a mini-ice age in Western Europe. Although *standard* climatic models do not predict a slowdown in the thermohaline system until the end of the 21st century, some researchers have disagreed. Schlesinger (*et al*) used a computer model to stimulate the present day thermohaline circulation system and its response to the addition of freshwater to the North Atlantic Ocean.[161] They estimated the probability of a thermohaline system shutdown between 2006 and 2205 without a carbon tax on fossil fuels and on the business as usual scenario, without a climate policy. Without a climate policy it was found that there was a greater than 50 per cent chance of a shutdown in this system, but even with maximal policy intervention the probability was still 25 per cent. This research team concluded that the risk of a thermohaline system shutdown is "unacceptably large."[162]

There are many other studies of positive feedback mechanisms, mechanisms that could lead to "dangerous" rises in global average temperatures. In general, climate change itself may trigger additional rises in greenhouse gas emissions, which in turn amplifying global warming. For example Torn and Harte[163] have examined historical evidence, based on the study of Antarctic ice cores which contain a 360,000 year record of global temperatures and greenhouse gas levels. This evidence demonstrates that greenhouse gas emissions increase during warming period, indicating that positive feedback mechanisms are at work. Torn and Harte used a General Circulation Model (GCM) of climate with positive feedbacks and found that the warming the IPCC in its Third Assessment Report associated with an anthropogenic doubling of carbon dioxide, that is 1.5- 4.5 C, was amplified to 1.6 to 6 C. They concluded: "a symmetrical uncertainty in any component of a feedback, whether positive or negative, produces an asymmetrical distribution of expected temperatures skewed toward higher temperatures."[164] In other words, the uncertainties involved in the complex interplay between positive and negative feedback mechanisms is likely to still result in a higher probability that the future will be hotter than the IPCC predictions, with the upper projected warming for the end of the 21st century being 7.7 C.[165] Scheffer (*et al*) estimate that positive feedbacks between global warming and atmospheric carbon dioxide concentration, based on data from past climatic change, will produce over the 21st century warming of an additional 15-78 per cent relative to the IPCC Third Assessment Report predictions.[166]

Terrestrial ecosystem feedbacks contribute both positive and negative feedback mechanisms to the climate system which may respectively increase or reduce the extent of global climatic change.[167] For example, one hypothesis is that a negative feedback mechanism exists, whereas higher levels of carbon dioxide in the atmosphere will act as a fertilizer for plants thus increasing plant growth and in turn "sinking" carbon dioxide into plants and vegetation.[168] However, research indicates that higher temperatures will offset the carbon sink function because increased respiration will significantly exceed any CO2-enhanced photosynthesis effect. Plant growth will also be contingent upon the complex interplay of nutrient and water availability.[169] Carbon-cycle models also predict that the ocean sink will also weaken, but to a lesser degree than the land sink.[170]

It has been hypothesized by Lindzen (*et al*)[171] that water vapor could have a part to play as a negative feedback mechanism. The hypothesis is that although a warmer atmosphere in general will hold more water vapor and thus trap more heat, the upper atmosphere may dry out as it warms. More recent satellite data has indicated that the upper atmosphere is not drying out as it warms, but it is becoming wetter. Water vapor then is a positive feedback mechanism.[172] Increased water vapor in the atmosphere is reinforcing human-induced global warming.

Sometimes in the "game" of human-induced climate change, even when one "wins", one loses. Between 1950 and 1980 solar radiation entering the Earth's atmosphere and reaching the surface was declining at about two per cent per decade. "Clean air" laws in the West and a decrease in industry in the former Soviet Union reduced the concentration of aerosols in the atmosphere. These aerosols had directly blocked solar radiation and/or aided in cloud formation which had the same type of blocking effect. This is the so called "global dimming" or "solar dimming" hypothesis.[173] However, according to research by Professor Martin Wild of the Institute of Atmospheric and Climate Science in Zurich Switzerland: "The enhanced warming we have seen since the 1990s along with phenomena such as the widespread melting of glaciers could be due to this increased intensity of sunlight compounding the effects of greenhouse gases."[174] Until 1990 air pollution had shielded the atmosphere from around 50 per cent of the global warming that otherwise would have occurred.[175] "Global brightening will add to the impacts and could mean a six per cent effective increase in the amount of solar radiation reaching the Earth.[176] In addition to these considerations, a number of scientists, including Professor Sami Solanki of the Max Planck Institute Germany, have documented that there is at present a slump in the sun's magnetic activity, but when this activity intensifies, according to Professor Leif Svalgaard of Stanford University, "global warming will return with a vengeance" if business is as usual.[177]

Another problem which is sometimes not considered in present investigations of global climatic change, is the nitrogen problem. Humanity, through agricultural and industrial activity has released a "cascade" of reactive nitrogen. This is a problem not only because of nitrogen compounds' contribution to global climatic change but also because of ill-effects on biodiversity, acid rain, human health and eutrophication problems from nitrates entering water systems.[178] Nitrous oxide, molecule for molecule has over 300 times the global warming potential of carbon dioxide, and it also destroys stratospheric ozone.[179] Although nitrogen compounds (NOx) may initially stimulate plant growth through aiding photosynthesis, denitrification processes in the soil are likely to swamp this effect, with their being a net increase of NOx in the atmosphere.[180] Reflecting on the magnitude of the nitrogen problem in the context of the environmental crisis, Professor Brian Moss of the University of Liverpool UK's Botany Department has said:

> By and large the nitrogen problem is very much one of technology-driven societies, and the only salvation is a revolution to less consumptive lifestyles. This won't happen voluntarily but it may be forced by the combined effects of climate change, the end of the oil economy, rising populations, economic and environmental refugees and the loss of goods and services from the 70 per cent or so of natural ecosystems predicted to have disappeared by 2050.[181]

Whether this is so will be examined in this book

**The Impact and Harm of Global Climatic Change**

Available evidence indicates that the climate system is more sensitive than it is generally thought to be.[182] The Stern Review, *The Economics of Climate Change* - a report by the UK Government's chief economist Sir Nicholas Stern - concluded in summary of this evidence that "there is up to a one-in-five chance that the world would experience warming in excess of 3° C above [the] pre-industrial [level] even if the greenhouse gas concentrations were stabilized at today's level of 430 ppm CO2e [carbon dioxide equivalent]."[183]

Climate sensitivity ("S") is the equilibrium warming of the atmosphere arising from a doubling of carbon dioxide levels.[184] Myles Allen (*et al*)[185] have argued that there are fundamental theoretical reasons why it is not possible to objectively quantify estimates of risks, dependent on the shape of the upper tail of the climate sensitivity distribution. These estimates are dependent upon prior assumptions about the climate. First, climate sensitivity cannot be *directly* measured because the real climate system cannot be manipulated by having carbon dioxide levels doubled and then brought into equilibrium. Allen (*et al*) also claims that there are no observable quantities directly proportional to climate sensitivity for all the values consistent with existing observations.[186] On the other hand, various studies indicate that the observable properties of the climate system tend to be proportional to 1/S rather than S. If uncertainties follow a normal (or Gaussian) distribution curve, there will be no formal upper bound on S.[187] Thus; estimates that any particular greenhouse gas stabilization level will not result in dangerous global climatic change must be dependent upon prior "subjective" assumptions of climatologists. This, however, does not mean that these assumptions are unscientific or untestable, as these assumptions can be tested by a variety of indirect measures.[188] Further, debate on climate sensitivity assumes that there is a single unchanging value for S. However, global warming may itself change the sensitivity, as the climate system is likely to include non-linear change and threshold effects. Hence there cannot be a firm upper limit placed on climate sensitivity.[189]

Given this qualification, the rise in global mean temperature ΔT, relative to pre-industrial times, is predicted by the overwhelming majority of peer-reviewed scientific papers and reports, to be substantial and to impact upon most aspects of human life and society.[190] At present ΔT = 0.6 C and already, as this chapter details, environmental effects are being observed. For ΔT= 1 C some of the predicted effects include an increase in bushfires[191]; an increase in heatwaves and heat-related mortality; an increase in extreme rainfall patterns causing droughts, floods and landslides with the likelihood of a more intense El Nino; 18-60 million additional people at risk of hunger and 300-1600 additional million people suffering an increase in water stress.[192] At ΔT = 2 C agricultural yields fall in the developed world and 1-2.8 billion people experience an increase in water stress and an additional 220 million people are at risk of hunger.[193] At ΔT= 3 C few ecosystems are capable of adapting, losing between seven and 74 per cent of the extent of their biodiversity.[194] Up to 400 additional millions are at risk of hunger with 70 to 80 per cent of these people being in Africa.[195] At ΔT= 4 C a collapse of many agricultural systems will occur, including Australia, with up to 600 million additional people being at risk of hunger.[196]

The Stern Review states that for at ΔT= 5 C and above:

> The latest evidence suggests that the Earth's average temperature will rise by even more than 5 or 6° C if emissions continue to grow and positive feedbacks amplify the warming effect of greenhouse gases (e.g. release of carbon dioxide from soils or methane from permafrost). This level of global temperature rise would be equivalent to the amount of warming that occurred between the last [ice] age and today-and is likely to lead to major disruption and large-scale movement of population. Such "socially contingent" effects could be catastrophic, but are currently very hard to capture with current models as temperatures would be so far outside human experience.[197]

The Stern review notes that the current level of greenhouse gases in the atmosphere is about 430ppm (carbon dioxide equivalent) and could reach 550ppm CO2e by 2050 at current rates of increase, but this level could be reached as early as 2035. A 550ppm level will give a 77-99 per cent chance of an increase about 2

C. Business as usual, doing nothing scenarios, give a 50 per cent chance of a 5 C rise by the end of the 21st century.[198] According to the Stern Review the cost of inaction is 5-20 per cent of the world's gross domestic product a year, forever, so that the cost of climate change will be more than World War I, World War II or the Great Depression.[199] The cost of inaction, the Stern Review put at around one per cent of global GDP each year.[200]

Although the Stern Review has been criticized by a number of economists and by global warming skeptics,[201] its general conclusions are also reflected in a number of other recent reports. For example, the CSIRO Consultancy Report, *Climate Change in the Asia/Pacific Region*[202] recognizes that climate change models indicate that temperature increases in the Asia/Pacific region will be in the region of 0.5- 2 C by 2030 and by 1-7 C by 2070.[203] The region is likely to be affected by a rise in global sea level of about 3-16 cm by 2030 and by 7-50 cm by 2070 with regional sea level variability.[204] The Asia Pacific region is highly vulnerable to climate change; the report states: "A review of 186 different regional and national estimates of the potential impacts of future climate change to various sectors within the Asia/Pacific region confirms that there is little room for optimism."[205] Sixty two per cent of impact estimates indicated "clear adverse/consequences" and only 19 per cent clear benefits, with 20 per cent indicating that impacts could, depending upon circumstances "lean either way,"[206] Adverse impacts will be seen in respect to coastal communities, ecosystems and biodiversity, disease and heat-related mortality, water resources and agriculture and forestry. In general there will be a large negative net effect on regional and national economies and "even with growing regional prosperity, localized climate impacts, such as the collapse of a fishery or the inundation of core cropping land, could devastate local economies"[207] Finally, all of these impacts brought by climate change will have a significant impact upon the security of the Asia-Pacific. The report states:

> Existing challenges to human security in the Asia/Pacific region may be significantly exacerbated by the broad range of impacts that climate change may bring. Chronic food and water insecurity and epidemic

> disease may impede economic development in some nations, while degraded landscapes and inundation of populated areas by rising seas may ultimately displace millions of individuals forcing intra and inter-state migration. The implications of such challenges to human society are difficult to anticipate, but there is currently little awareness of the implications and regional management frameworks for addressing climate change-induced security and migration issues are lacking.[208]

The Stern Review also recognized that climate impacts may ignite violence as has already occurred in many parts of Africa, through exacerbating the competition for scarce resources such as water.[209] Dupont and Pearman in an important study of the relationship between national security and climate change for Australia, conclude that climate change is a bigger security risk to Australia than "terrorism" and could produce destabilizing civil conflict and unregulated population movements:

> The cumulative impact of rising temperatures, sea levels, and more mega droughts on agriculture, fresh-water and energy could threaten the security of states in Australia's neighborhood by reducing their carrying capacity below a minimum threshold, thereby undermining the legitimacy and response capabilities of their governments and jeopardizing the security of their citizens. Where climate change coincides with other transnational challenges to security, such as terrorism or pandemic diseases, or adds to pre-existing ethnic and social tensions, then the impact will be magnified. However, state collapse and destabilizing internal conflicts is a more likely outcome than interstate war. For a handful of small, low lying Pacific nations, climate change is the ultimate security threat, since rising sea-levels will eventually make their countries uninhabitable. Far from exaggerating the impact of climate change it is possible that scientists may have underestimated the threat.[210]

Strategic theorists Schwartz and Randall have stated that if the worse case scenarios of abrupt climate change occur environmentally damaged states, many of them armed with nuclear weapons, may wage war on wealthier or more resource- rich neighboring states.[211] Dupont and Pearman believe that such an apocalyptic scenario is "improbable, although not entirely out of the question".[212]

Abrupt climate change could be the spark that ignites a nuclear war.

**Human Health and Global Climatic Change**

A large technical literature exists on the human health consequences of global climatic change, so large in fact that it cannot be adequately summarized in a book length treatment, *let alone* in the section of one chapter of a book.[213] Indeed, approaching this matter from the holistic scientific paradigm of conservation medicine/ecological health, this issue can be seen to be vastly more complex than presently existing approaches to the human health ramifications of global climatic change reveal. As we have seen, the human health effects of global climatic change need to be understood within the context of the globalization of economic activity occurring across the Earth, which in itself has substantial impacts upon human health as well as the transmission of disease.[214] Closer interactions and interconnections between people through international travel and immigration has many benefits, but there may also be many costs as Ornstein and Ehrlich have observed:

> For decades now humanity has been setting itself up as the progressively ideal target for a worldwide epidemic. The combination of rapidly increasing number of malnourished people, who live in conditions of poor sanitation and with impure water, with evermore-rapid transportation systems has been making the human epidemiological environment, ever more precarious. We have created a giant crowded "monoculture" of human beings, millions of whom are especially vulnerable to disease and among whom carriers can move with unprecedented speed.[215]

For example, a global pandemic of avian influenza, if and when it occurs, perhaps involving new strains of H5N1,[216] will have major impacts upon mortality and morbidity, with dramatic short-term impacts upon modern economies in the developed world.[217] In one study of the macroeconomic effects of an influenza pandemic on the Australia economy by Australian Treasury economists Kennedy (*et al.*), [218] the effects of such a pandemic was simulated using the Treasury Macroeconomic (TRYM) computer model of the Australian economy. The number of deaths in the first quarter of the first year was assumed to be 40,000, resulting in a proportional 0.2 per cent decline in Australia's working age population. These deaths were seen to have little direct short-run economic impacts compared to factors such as work absenteeism and confidence effects on consumption. In the simulation household consumption falls by 6.3 per cent over the year, resulting in a recession. Considering factors such as a fall in business and dwelling investment results in a fall in GDP of over 5 per cent, a recession half the size of the Great Depression. McKibben and Sidorenko in their study of the global macroeconomic consequences of pandemic influenza, [219] using the Asia Pacific G-Cubed (APG-Cubed) computer model, found that even a mild influenza pandemic can result in 1.4 million deaths and a 0.8 per cent loss (approximately $US 330 billion) in a global economic output. The "ultra" pandemic could see 142.2 million people killed and a loss of global GDP of $US 4.4 trillion - with the real possibility of severe socio-economic dislocation occurring.

Emerging infectious diseases, being diseases that have recently expanded their geographic or host range, or caused by newly evolved pathogens (including a new strain of cholera, Ebola, Lyme disease, legionnaires disease, multi-drug resistant bacterial infections, to name but a few,[220] as well as some older diseases (such as malaria) may be assisted in their spread by global climatic change. Climate change may impact upon previously existing biodiversity, breaking down biological control systems which once limited the spread of vectors and pathogens. Habitat loss could bring humans in contact with once isolated pathogens, or alter predator/ prey ratios, allowing relatively more prey to carry pathogens. Climate change may lead to disease-carrying vectors shifting their geographical range.[221]

These examples illustrate the importance of taking a holistic, ecological view of the relationship between human health and global climatic change. As recognized in the Millennium Ecosystem Assessment Report, *Ecosystems and Human Well-Being*[222] the "causal link between environmental change and human health are complex because often they are indirect, displaced in space and time and dependent on a number of modifying forces".[223] Thus, for example, during the hot summer in Europe of 2006, farms in the Netherlands, Germany and Belgium were affected by bluetongue disease, an insect-borne viral infection of ruminants (e.g. cows, sheep, goats and deer), which originated in sub-Saharan Africa.[224] Although it is not known how the disease got to Europe, the warm European weather aided the spread of the disease as the temperature sped up the life cycle.

Global climatic change will impact upon human health both directly and indirectly.[225] We will discuss direct effects first. The direct health consequences are illnesses and conditions arising from an increase in extreme weather events, such as torrential rains and flooding, droughts, storms and thermal stress from heat waves. These events may mechanically harm human health, as in the case of direct deaths and injuries from torrential rains and floods, or the impacts could be more complex, such as ill health arising from a contamination of groundwater reserves. As an example, the August 2003 heat wave in France was the worse since 1873, when record-keeping began. From August 1 2003 to August 20 2003, almost 15,000 excess deaths occurred, primarily among elderly people.[226] A study of the projected health effects of climate change in the Asia/Pacific region by Preston (*et al*) [227] concluded that climate projections indicate that the Asia/Pacific region will become warmer and wetter in the next few decades, resulting in a significant increase in heat-related illnesses. Events such as the May 2002 heat wave in the Indian State of Andhra Pradesh, where temperatures reached 49 C and over 1,000 deaths occurred, will become more common.[228] A 2 C rise in the average temperature of the regions of northern Pakistan, India and western China by 2030, will have a "devastating impact" upon almost 60 per cent of the world's population Preston (*et al*) conclude.[229]

Dupont and Pearman have summarized literature relating to food security, water scarcity and human health as a product of global climatic change.[230] The latest climatological evidence indicates that tropical zones as well as temperate zones will be significantly affected - and tropical and sub-tropical regions have the bulk of the world's people. Much of the world's food is produced in this region by subsistence farmers.[231] According to the Consultative Group on International Agricultural Research, the adverse impacts on climate change upon agro-ecosystems is likely to result in a 20 per cent decrease in food productivity over the next few decades, at a time when Asia needs to increase food productivity to meet the needs of growing populations.[232] Further, although global climatic change is likely to produce an average increase in rainfall over the Earth, rainfall distribution patterns will change, with a pole-ward advance of mid-latitude desert regions and a possible short-term increase in crop yields, depending upon the extent of temperature rise. In the tropics major changes to the frequency and strength of El Niño weather events could lead to a greater frequency and intensity of destructive storms.[233]Less water for irrigation could result from an increased loss of water from evaporation caused by higher temperatures.[234] As Dupont and Pearman observe, in "a world where over two billion people already live in countries suffering moderate to high water stress, and half the world's population is without adequate sanitation or drinking water, relatively small shifts in rainfall patterns could push countries and whole regions into deficits, leading to a series of water crises with global implications".[235] The World Bank believes that most nations in the Asia/Pacific region will face a water crisis by 2025 unless action is taken now to avert the problem.[236]

Small-island states, especially small-island developing states (such as, in the Pacific, the Cook Islands, Fiji, Kiribati, the Marshall Islands, Tuvalu, etc.) are likely to be the nations most seriously affected by global climatic change according to the IPCC Working Group II, Fourth Assessment Report.[237] Low-lying small-island states are highly vulnerable to sea-level rise, flooding, storm damage, beach erosion and tropical cyclones.[238] Typically these low-lying islands are only a few meters above mean sea level and islands with higher-altitude lands still have vulnerable coastal areas. The rate of increase in air temperature in the Pacific and Caribbean has been around 0.1 C per decade, exceeding the global average, and the mean sea level rise has been approximately 2 mm per year.[239] Most small-island states have limited freshwater resources, especially atoll countries and limestone islands and are already facing water stress. Pollution of water from storm surges is a major threat to the health of people in these lands through a degradation of water quality and the spread of water-borne diseases.[240]

We mentioned earlier in this discussion that global climatic change will impact upon human health both directly and indirectly and we then briefly discussed direct impacts. Indirect health effects of climate change occur through ecosystem change or disruption. These effects alter the range, seasonality and intensity of water-borne pathogens and vector-borne diseases.[241] This indirect mechanism has already been mentioned in the opening discussion of this section, using the illustration of emerging infectious diseases. However, climate change is predicted to impact upon already existing diseases.

Malaria, for example, is responsible for the deaths of one million people in the world today and 300-500 million people get a less severe form of the disease.[242] In Mozambique, the entire population is at risk from malaria and one in four children die before the age of five from this disease.[243] The economic cost to sub-Saharan Africa is US $ 12 billion a year from this disease.[244] Unfortunately malaria is returning to regions where it was previously eliminated due to the evolution of drug-resistant parasites and a decrease in vector control activities.[245] These factors at present are the main forces driving malaria resurgence rather than climate change, but long- term climate change is likely to adversely impact upon the geographical distribution of malaria and its intensity of transmission in Africa in the future.[246] In this context it is interesting to note that although some computer models predict that warmer temperatures will result in malaria transmission in Europe and North America,[247] malaria was widespread in temperate regions until the second half of the 20th century and during the coldest period of the "Little Ice Age," from 1564 to 1730, malaria was a significant cause of death and illness in England.[248] The elimination of malaria was due to various public health and technological activities (e.g. elimination of mosquito habitat) rather than climate change.[249] Nevertheless, as the case of Africa shows, malaria could easily reemerge in Europe and North America with a lapse in public health preventative measures combined with higher temperatures.[250]

**Conclusion: State of the Argument**

This chapter has given a brief overview of the "crisis of civilization" - the multitude of converging environmental threats that confront humanity. Although global climatic change is generally regarded as the most significant of these threats, it is important to understand, as this chapter has emphasized, that problems such as "peak oil," and oil depletion biodiversity decline, the degradation of water resources, desertification, natural disasters and so on, all have complex interactions with the global climate. The first section of this chapter, written from the perspective of the conservation medicine or ecological health paradigm, attempted to briefly chart some of these problems, and their interaction with other environmental crises. There are often very strong synergistic interactions between these factors; for example ozone depletion and global warming may interact to produce greater than would otherwise be expected rapid climate change.[251] Thus, for example if some of the most pessimistic predictions regarding "peak oil" and oil depletion are correct, and if there is no comprehensive substitute for oil to fuel our globalized industrial society, the possibility of some form of economic collapse is real, although with limited data (due to the very limited amount of research done on this topic of collapse) it is impossible to give a probability estimate of this scenario. It remains one possibility, among others that may need to be dealt with by humanity.

The risks of climate change are greater than is generally thought, not only because of the existence of more general environmental and resource crisis, but also because of the possibility of abrupt and dangerous climatic change. Climate change skeptics are prone to dismiss such discussions as "pessimistic", "alarmist" and "unbalanced". Thus in the second section of this chapter we addressed the issue of climate change skepticism and showed that in the light of recent scientific research, the position of climate change skepticism is flawed. In many cases climate change skeptical arguments prove exactly the opposite of what their proponents hope to show: critical uncertainties in climate data and models may well indicate that the risks of climate change are much greater than is generally thought. The third and fourth sections of this chapter then set out to show that this is in fact the case. The risks of dangerous and abrupt climate change are higher than is generally thought to be the case on the "do nothing (or little), business as usual" scenario. This being so, it is concluded that addressing the issue of global climatic change is a matter of fundamental importance to humanity.

The fifth section has given a brief discussion of the human impacts of global climatic change. The current and projected future impacts of climate change on human health are generally negative, with exceptions such as lower cold-related mortality in some regions. Thermal stress, deaths and injuries from weather disasters and climate-induced changes to the distribution and incidence of food-borne and water-borne diseases are all important factors. The health impacts in developing regions of a reduction in the yields of agricultural crops and a degradation of water resources are also significant. Developing countries, less capable of technological adaptation will face a larger burden of disease. Thus, Robin Stott, is not exaggerating when stating that: "Climate change related to global warming is the world's most urgent public health problem".[252]

# Chapter 3

## Economics, Technology and Law: Confronting the Climate Crisis

> " *Technology will solve our problems*". This is an expression of faith about the future, and therefore based on a supposed track record of technology having solved more problems than it created in the recent past. Underlying this expression of faith is the implicit assumption that, from tomorrow onwards, technology will function primarily to solve existing problems and will cease to create new problems...[A]dvances in technology just increase our ability to do things, which may be either for the better or for the worse. All of our current problems are unintended negative consequences of our existing technology. The rapid advances in technology during the 20th century have been creating difficult new problems faster than they have been solving old problems: that's why we're in the situation in which we now find ourselves.[1]

**Introduction**

Chapter 2 examined the environmental crisis and the threat which global climatic change poses to human health and the sustainability of human civilization. The aim of that chapter was to survey scientific evidence indicating that the consequences of global climatic change are likely to be more severe than the consensus view recognizes, due to various positive feedback mechanisms. As well there are other environmental and research issues, such as peak oil, which will also impact upon human societies. The possibility of a pandemic of a mutated version of avian flu where there is no widely available vaccine - and which disrupts vital social functions as key personnel are killed off - is also a threat to human civilization.

In this chapter we will examine the responses made to global climatic change from an economic, technological and legal perspective. How can economics, technology and law aid humanity in confronting the climate crisis? We will argue that economics, technology and law - the three fields from which most people look for solutions to climate crisis- have much to offer, but individually or collectively these fields will not contribute a complete solution to the climate crisis. It will be argued that economic, technological and legal responses, which in general do not envisage a need for fundamental change in society and modes of living, have limits and although these responses have great merits there is a need to see such approaches within the framework of a wider, more comprehensive vision. To present our thesis, we will begin with a discussion of legal responses. It is necessarily the case that our discussion must be condensed because we have a vast quantity of material to consider in this book. It seems to us that it is more important to adopt a holistic perspective which attempts to see challenging environmental issues in something of a "big picture" and to comment on the global significance of these problems, than to focus upon narrow issues. As one of us (Smith) has said elsewhere: "In a nutshell, each discipline and field can make a contribution. But none of them is sufficient on its own to really carry the weight. It's got to be everybody working together at both the individual and international levels to deal with it. There's no one solution".[2]

**Global Climatic Change and the Limits of Law**

The law constitutes the rules of the "game" of society and it is therefore logical that the human attempt to deal with a threat will at some point find a legal expression. It is also logical that people will attempt to use legal mechanisms, such as litigation, to bring about social and legal change. In the case of climate change, litigation has been used with considerable success in both the United States and Australia, major greenhouse gas-emitting nations that have not ratified the *Kyoto Protocol.*[3] In *Climate Change Litigation: Analysing the Law, Scientific Evidence and Impacts on the Environment, Health and Property*[4] two of the present authors (Smith and Shearman) summarized existing case law up to June 2006. Since that time there have been two further climate change litigation cases decided in Australia by environmental courts, and more importantly, the US

Supreme Court by a 5-4 vote ruled in the case of *Massachusetts v. Environmental Protection Agency*[5] that the US Government's Environmental Protection Agency has a legal power under the US *Clean Air Act* to regulate tailpipe emissions of greenhouse gases. Another case, which at the time of writing is to be argued before the US District Court for the Northern District of California, involves the State of California suing six major car companies for the significant contribution that their motor vehicle emissions have made to global warming, harming the resources, infrastructure and environmental health of California.[6]

First, however, let us review the recent Australian decisions. In *Gray v The Minister for Planning,*[7] heard by Pain J before the Land and Environment Court of New South Wales, the applicant Peter Gray, a 26-year-old student, challenged decisions of the Director- General of the New South Wales Department of Planning, made under part 3A of the *Environmental Planning and Assessment Act 1979* (NSW) relating to a proposal to build the Anvil Hill Coal mine. The applicant sought a declaration that the environmental assessment prepared by the developer of the mine, Centennial Hunter Pty Ltd, adequately addressed the environmental assessment requirements, was void and without effect. The Anvil Hill site is a deposit of approximately 150 million tonnes of thermal coal and the proposed open cut mine would produce 10.5 million tonnes of coal per annum.

Section 75 H (1) of the *Environmental Planning and Assessment Act* requires that the proponent of a project to submit to the Director-General an Environmental Assessment for approval of the project. Centennial had lodged the required assessment with the Director-General, and the assessment of greenhouse gases was conducted in accordance with the World Business Council for Sustainable Development and World Resources Institute Greenhouse Gas Protocol 2004. The Environmental Assessment considered direct greenhouse gas emissions arising from sources owned or controlled by the company and greenhouse gas emissions from the generation of electricity purchased by the company and then consumed. However, the Environmental Assessment did not consider indirect emissions from sources not owned or controlled by the company, such as third parties burning coal.

Gray argued that the Environmental Assessment did not comply with the statutory environmental assessment requirements. The Court rejected this argument and also held that there was no legal test by which the Director-General's judgment about the adequacy of an environmental assessment can be assessed, because there were no objective criteria in the *Environmental Planning and Assessment Act* which permit the assessment of the Environmental Assessment requirements prepared by the Director-General. Nevertheless the Court accepted the applicant's argument that greenhouse gas emissions from the burning of coal should be considered in the Environmental Assessment due to the contribution to global climatic change. Even though it may not be currently possible to accurately measure the impact of the burning of this coal, it does not follow from this, that the environmental impact is therefore insignificant. Further, the Court found that the Director-General failed to consider principles of ecologically sustainable development, such as the precautionary and intergenerational equity principles, which was a failure to comply with a requirement at law.[8] Consequently, the Court held that "the view formed by the Director-General... that the environmental assessment lodged by Centennial Hunter Pty Ltd in respect of the Anvil Hill Project adequately addressed the Director-General's requirements is void and without effect".[9]

An editorial in the *Australian* newspaper said that *Gray v The Minister for Planning* was a "misguided judgment" and that it puts the New South Wales coal export industry at risk and "[even] more concerning is Justice Pain's apparent use of her judicial authority in what looks like an attempt to dictate government policy from the bench".[10] However, the New South Wales Parliament on November 23 2006 introduced amendments to the *Environmental Planning and Assessment Act* which will allow the Anvil Hill coal mine to be approved in its present form. The amendments include changes to section 75 J (1) (6) of the Act which removal the current requirement that "the environmental assessment requirements have been complied with". All that now needs to be done is that an application be made and in turn the Director-General give his/her report to the Minister for Planning. The Director-General's review of the Environmental Assessment, is thus of lesser importance and the importance of the decision in

*Gray v The Minister for Planning* is minimal.

The value of *Gray v The Minister for Planning* as a precedent is also minimized by the more recent decision in *Re Xstrata Coal Queensland*, heard before the Queensland Land and Resources Tribunal, with a decision delivered on February 15 2007.[11] The case concerned an application by the coal mining company Xstrata Coal Queensland under section 275 of the *Mineral Resources Act 1989* (Queensland) for the grant of an additional surface area for the development of a new open cut coal mining operation and for the grant of an environmental authority via a mining lease under the *Environmental Protection Act 1994*. The Queensland Conservation Council Inc and the Mackay Conservation Group Inc objected to the proposal with respect to adverse environmental impacts, prejudice to the public right and interest, any good reason shown to refuse, and ecologically sustainable development principles, arguing that the proposed mine would contravene these factors unless conditions were imposed on the grant to "...avoid, reduce or offset the emissions of greenhouse gases that are likely to result from the mining, transport and use of coal from the mine".[12] The Queensland Conservation Council particularized its objection to require the applicant to avoid, reduce or offset 100 per cent of the greenhouse gas emissions, which would render the mine unviable. At the start of the hearing the Queensland Conservation Council sought to amend the 100 per cent figure to only 10 percent, but the Court refused the amendment because the applicant had prepared its case on the 100 per cent figure and the late amendment would prejudice its case. The Mackay Conservation Group Inc did not attend the hearing and did not particularize the conditions it sought, wanting the imposition of unspecified conditions to avoid, reduce or offset the greenhouse gas emissions.

In the course of delivering judgment Koppenol P, the presiding member of the Queensland Land and Resources Tribunal, was critical of scientific evidence presented by the environmentalist groups. The Stern Review (to be discussed later in this chapter) was objected to because of critical papers published by Professor Robert Carter *et al* and Professor Sir Ian Byatt *et al*, papers which allegedly contested the view "that the scientific evidence for GHG-induced serious global warming and climate change was overwhelming was just an assertion and was wrong".[13]

Koppenol P also accepted criticism of the Fourth Assessment Report of the Intergovernmental Panel on Climate Change's *Summary for Policymakers*, saying that "a temperature increase of only about 0.45 C *over 55 years* seems a surprisingly low figure upon which to base the IPCC's concerns about its inducing many serious changes in the global climate system during the 21st century".[14] With all respect, such a conclusion is at variance with the consensus view of climate science.[15] Koppenol P stated that although the Carter-Byatt critique of the Stern Review was not mentioned in the hearing, only the Queensland Conservation Council objected to the Carter-Byatt critique being considered. It is not stated in the case whether this hearsay evidence was subjected to cross-examination by Carter-Byatt *et al* being present.[16]

It was concluded by the presiding member that ecologically sustainable development principles are "based upon an *assumption* concerning the cause and effect of global warming".[17] That assumption was a demonstrated causal link between the mine's greenhouse gas emissions and discernable harm. If the mine's greenhouse gas emissions were completely eliminated, there was no evidence that this would have the slightest effect on climate change.[18] Finally, apart "from having no demonstrated impact on global warming or climatic change, any such condition would have... the real potential to drive wealth and jobs overseas and to cause serious adverse economic and social impacts upon the State of Queensland". [19] The logical conclusion of this judgment is that as far as coal mining goes, it is business as usual in Queensland. On the basis of the reasoning in this judgment there seems to be no reason to be concerned about global climatic change, and no reason to regulate the expansion of new coal mines.[20]

*Re Xstrata Coal Queensland* can be contrasted with the decision handed down by the United States Supreme Court in *Massachusetts v Environmental Protection Agency.*[21] In this case twelve US states and several cities took action against the US Environmental Protection Agency (EPA) in an attempt to force that agency to regulate the emissions of four greenhouse gases, including carbon dioxide under section 202 (a) (1) of the US *Clean Air Act*. This Act requires that the EPA "Shall by regulation prescribe...standards applicable to the emission of any air pollutant from any class... of new motor vehicles...which in [ the EPA Administrator's] judgment cause[s], or contribute[s] to, air pollution...reasonably...anticipated to endanger public health or welfare". "Air pollutant" is defined by the Act to include "any air pollution agent...including any physical, chemical...substance...emitted into...the ambient air". The EPA denied the petition on three grounds. First, it argued that the *Clean Air Act* does not authorize it to issue mandatory regulations to address global climatic change. Second, even if it did, it would not be wise to do so because a causal link between an increase in anthropogenic greenhouse gases and a rise in global surface air temperatures was not unequivocally established. Third, the EPA argued that any such attempt at greenhouse gas emission regulation would conflict with the President's comprehensive approach to climate change based upon technological innovations and the use of non-regulatory programs to encourage the private sector to make voluntary reductions in greenhouse gas emissions.

Review was denied by the D.C. Circuit. Of the three judges, two agreed that the EPA had properly exercised its discretion in denying the petition. One of those judges was of the opinion that the petitioners had failed to establish standing (the right to bring a matter before a court) under Article III of the US Constitution, which requires the demonstration of particularized injury. The US Supreme Court held that the petitioners did have standing because at least one of the petitioners, such as the State of Massachusetts, had suffered a concrete and particularize injury that was either actual or imminent, and that the injury is fairly traceable to the defendant, and that a favorable decision will be likely to redress that injury. Massachusetts' affidavits were unchallenged and they claimed that global sea levels rose between 10 and 20 centimeters over the 20th century. A substantial part of the coastal land of Massachusetts has been lost and remediation costs were already hundreds of millions of dollars.[22]

The US Supreme Court held that greenhouse gases fit well within the *Clean Air Act's* definition of "air pollutant" under section 7602 (g) which includes "any air pollutant agent...including any physical, chemical,...substance...emitted into...the ambient air...." Thus the EPA has statutory authority to regulate emissions of greenhouse gases from new motor vehicles. The Court also rejected the EPA's argument that it would be unwise for it to act because of a "laundry list" of reasons based around the Executive Branch's own program for greenhouse gas regulation. Under the Act the EPA can only avoid promulgating regulations if it determines that greenhouse gases do not contribute to climate change or if it provides reasonable explanation as to why it will not exercise its discretion to decide the question of whether greenhouse gases contribute to climate change. The EPA did not show that the uncertainty in the field was so great as to preclude a decision. However, the US Supreme Court did not explicitly decide the question of whether global climatic change is occurring and if so, that the cause is anthropogenic greenhouse gas emissions. Only the issue of whether scientific uncertainty was a valid basis for the EPA to decline to regulate was addressed within the statutory framework of the *Clean Air Act.*

The decision in the *Massachusetts v Environmental Protection Agency* has been hailed by environmental groups as a watershed that could lead to environmental legal change in Australia as well as the US.[23] There is hope, for example, that the EPA will approve California and eleven other US state's programs, beginning in 2009 to limit tailpipe emissions. The motivation behind the State of California suing six of the world's largest auto makers over the harm caused by global warming[24] is to respond to auto makers who have blocked with their own legal actions, California's 2004 rules requiring care makers to cut tailpipe emissions from cars and trucks. This action by California is the first lawsuit seeking to hold car manufacturers liable for the damages caused by their vehicle's greenhouse gas emissions. On this case, Roda Verheyen, Co-Director of the Climate Justice Programme has said:

> This was a case waiting to happen. It is the most significant piece of climate change litigation that has ever been brought. It shows that those who suffer damage from climate change can seek compensation in the courts. More of these cases will happen until governments and companies make the deep cuts in greenhouse gas emissions that the planet needs. Car makers should realize that they cannot continue business-as-usual and realise high profits while destroying the climate by selling high-consumption engines.[25]

The South Australian government introduced the *Climate Change and Greenhouse Emissions Reduction Bill 2006* into Parliament in December 2006 and at the time of writing of this book, the Bill has passed through the House of Assembly with amendments. The Bill sets out the goals: (1) to reduce, by December 31 2050, greenhouse gas emissions within the State by at least 60 per cent to an amount that is equal to or less than 40 per cent of the 1990 levels as part of a national and international response to climate change; (2) to increase the use of renewable electricity so that it comprises 20 per cent of total electricity consumption within the state of South Australia by December 31 2014 and (3) to promote commitment to action within the State of South Australia to address climate change through the development of specific or interim targets, as appropriate, for various sectors of the state's economy and the development of policies and programs for the reduction of greenhouse gas emissions. South Australia, joins Alberta's *Climate Change and Emissions Act 2003* and California's *Climate Act 2006*, as being one of a small number of jurisdictions establishing greenhouse gas emission reduction targets in legislation.

At the time of writing Britain is undertaking public consultations on the *Climate Change Bill*, a draft of which was published on March 13 2007. The aim of the proposed law is to move Britain towards becoming a low-carbon economy, and to make Britain the first nation in the world with legally binding carbon reduction targets. According to former British Prime Minister Tony Blair, global warming is "the biggest long-term threat facing our world" and he wanted to leave office on a high note, leading a new global agreement on climate change.[26] The *Climate Change Bill* is presented as a model of the way nations should go. A carbon budget will be set for every five years, with a goal of cutting emissions by 60 per cent by 2050, relative to 1990 levels. If future governments miss the targets they could be taken to court.

The proposals have been welcomed by the Tories and Lib Dems but they wanted the budgets to be set annually. Shadow environment secretary Peter Ainsworth said in March 2007 that the five year approach rather than a system of rolling annual rate of change targets, "will enable responsibility for failure to be shunted from one government to another". [27] This criticism is supported by statistics which indicate that UK carbon dioxide emissions rose in 2006 and are at present higher than they have been since Labour came power.[28] UK carbon dioxide emissions were in 2006 2.7 per cent above 1997 levels and 1.2 per cent higher than in 2005.[29]

This situation well illustrates the necessary limits of the law in an area as complex as global climatic change. Through either community concern or political opportunism - however one wants to view the matter - the appropriate legal changes to the laws of society may be made. This could come about, as we saw earlier in this section, through the long and winding road of climate change litigation, or it may come more easily through forward-thinking governments making legislative changes. However once the laws are in place, it is another thing entirely for advanced capitalist societies in a globalized economic system to succeed in meeting such conditions of sustainability.[30] For example the South Australian government, as we have seen wants to reduce, by December 31 2050, greenhouse gas emissions within the state by at least 60 per cent to an amount that is equal to or less than 40 per cent of 1990 levels. However, at the same time, the present state government wants to double the population of the state primarily by immigration, under pressure from the powerful ethnic and business lobbies.[31] The lack of a systematic approach to sustainability, especially the need of producing an ecologically sustainable population policy, seems to ensure that future failure to reach greenhouse gas emission reduction will occur.

Legal responses then are ultimately dependent for their success upon broader social, economic and technological responses to the problem of global climatic change, so while recognizing that the achievement summarized in this section are significant and important, they are only part of a much more complex set of strategies which humanity must consider. The law has not been at the frontlines of the debate in promoting the importance of energy efficiency and renewable energy resources.[32]

### "Big Picture" Strategies for Saving the World

In this section we will examine some "big picture" approaches to dealing with the problem of global climatic change. All of the thinkers considered here support the clear and logical strategies of reducing energy use and embracing alternative energy sources. Below we will discuss whether renewable energy will be sufficient to fuel the consumer society as we know it. The major intellectual challenge today is not attempting to show the risks of global warming but to come up with realistic strategies for dealing with this immense problem.

Mayer Hillman has given us an indication of the immense social change to consumer behavior which is needed to avoid "knowingly handing over a dying planet to the next generation".[33] Hillman supports the idea of personal carbon allowances within the overall framework of the contraction and convergence paradigm.[34] We will discuss the idea of contraction and convergence shortly. According to Hillman a person's fair and equitable annual share of carbon dioxide emissions is about one tonne.[35] This is the level of carbon dioxide emissions such that all of the Earth's people have an equal share of emissions (i.e. per capita equity), whilst the planet's climate is not seriously destabilized. The UK at present has per capita average emissions of about 10 tonnes, which is about 2.5 times the present world average.[36] Hillman says that since it is impractical to hope to achieve this reduction immediately, individuals will need to attempt a year-by-year reduction. He believes that a system of trading allowances will be a more effective way of achieving greenhouse gas emission reductions than strategies aimed to encourage individuals to adopt conserver/green lifestyles.[37] Hillman proposes that personal carbon allowances[38] would enable individuals with carbon surpluses due to low greenhouse gas emission rates, to sell their surplus on the market so that the conventional "polluter pays" principle will be complemented by a "conserver gains" principle.[39]

The contraction and convergence principle was developed by Aubrey Meyer of the Global Commons Institute.[40] Although there are algorithms of some complexity associated with the principle,[41] in essence the principle proposes that a nation's annual emissions *contract* by some future date to a level much lower than today's emissions, these levels stabilizing greenhouse gas concentrations so as to prevent "dangerous and anthropogenic interference" with the global climate system. There will be *convergence* from the present day inequitable per capita greenhouse gas emissions to a situation of equal *per capita* emission rights. "Dangerous" levels of greenhouse gas emissions are determined by our best scientific inquiries, such as by bodies like the Intergovernmental Panel on Climate Change (IPCC), but the Global Commons Institute believes that greenhouse gas concentrations at a level higher than 450 ppm are not safe.[42] Countries would be allocated part of a carbon budget relative to the country's portion of global income. The carbon emissions budget for each country would be progressively reduced each year, at agreed rates, so that convergence to equality occurs at a agreed date. Developed countries facing contracting carbon allocations will be motivated to develop non-carbon based energy sources. There will also be a scheme where high per capita greenhouse gas emitting countries could purchase from low per capita emitting countries, unused credits - that is, carbon trading. Poorer, low-emitting nations could use such funds as a source of economic development.[43]

The Global Commons Institute is critical of the Kyoto Protocol mechanisms as ways of dealing with the problem of global climatic change.[44] Kyoto Protocol rules allow tradable emissions entitlements that permit the developed high greenhouse gas emitting nations to effectively escape making real and substantial greenhouse gas emission reductions. The Global Commons Institute believes that the Kyoto reductions, over the long term, will not result in the level of per capita reductions needed; the Global Commons Institute puts this figure at around 0.3 tonnes per capita by 2050.[45] To put this figure in perspective, the Oxford-based company Climate Care estimates that a global flight from London, to Toronto, Christchurch, Adelaide, then back to London, would emit six tonnes of CO2.[46] Ian Roberts, writing a review of Hillman and Facett's *How We Can Save the Planet* in the *British Medical Journal*, says that he read the book on a beach in Crete and on the way back to Britain was "burdened by the guilty secret that he had personally soiled the upper atmosphere with nearly two tonnes of carbon dioxide".[47] This gives us some indications of the necessary cut-backs.

The contraction and convergence paradigm is superior to the Kyoto Protocol insofar as it accepts the need for substantial contractions in greenhouse gas emissions as a matter of urgency. Nevertheless there are some major difficulties confronting this position. As seen with the problems facing the Kyoto Protocol (which does have enforcement mechanisms),[48] there is a real problem in getting nations to stick to greenhouse gas emission reductions when real economic pain results. For the contraction and convergence principle to work, there needs to be an international policing authority with "real teeth", that is, the power to punish non- complying nations.[49] Something like a world government would seem to be needed to enforce matters, with a world army ready and willing to use force, if necessary. A counter-revolution launched against the new world ecological order may result in further uncontrollable greenhouse gas emissions!

Another objection to the contraction and convergence position relates to the per capita equity assumption. The equity principle has been taken as the only way to get developing nations to co-operate. The developed world, the argument goes, has "got rich on the back of the two centuries of fossil-fuel-driven development".[50] So, what right "do Europeans (with a per-person average of eight tonnes of CO2 each year) have to lecture Indians (with just one tonne per person) about pollution"?[51] The assumption here is that the West has simply exploited the developing world *absolutely* and that the developing world has not benefited from "two centuries of fossil-fuel-driven development". But even granted the left's thesis of colonial oppression and imperialism,[52] it still does not follow that absolute exploitation has occurred and that the world of Asia and Africa has not benefited to any degree from the technological and industrial development of the West. Clearly it has, and medical advances and technological advances in agriculture have allowed millions of people in Africa and Asia to live who would never have lived. These benefits would need to be considered in any sort of historical accountability for greenhouse gas emissions.

The contraction and convergence position is concerned with per capita equity rather than nation-to-nation equity. However a blanket acceptance of global egalitarianism could have ethical consequences which are less ethical. Per capita greenhouse gas equity means that individuals in bitterly cold countries would have the same carbon quota as individuals in countries with a balmy climate. Or alternatively, individuals in countries facing environmental hardships would have to survive with the same carbon budget as individuals in countries not facing those hardships. Even in developed countries, it is far from clear that per capita equity is a universally correct moral principle of distribution. Some people, such as the disabled or the severely ill, may require a higher carbon budget just to survive than young fit people who can ride bicycles and grow vegetables in window boxes and home gardens.

Further to this, it is the environmental impact of nations which is the variable that should be considered in the contraction and convergence equation, not merely per capita emissions. Societies and nations are not sets of isolated individuals, but interacting systems.[53] As it is stated, the contraction and convergence position has a bias towards highly populated countries which have relative to the US, lower per capita greenhouse gas emissions, such as China and India.[54] However, population levels must also be considered in assessing the environmental impacts of nations. As Firor and Jacobsen have said: "continued population growth accelerates climate change, as each new person adds more heat trapping substances to the atmosphere".[55] According to the International Energy Agency, China with its 11 per cent annual growth rate will overtake the US as the biggest emitter of greenhouse gases in 2007. Previously the International Energy Agency thought that this take over would not occur before 2025.[56] Fatih Birol, the chief economist of the International Energy Agency, said that China's (and India's growth) will swamp emission cuts made by the developed countries by an "order of magnitude of difference".[57] China has lower per capita emissions than most Western nations and believes that it should be free to pursue economic growth as the Western nations have done for two hundred years. China assumes, quite falsely, that it has not been a beneficiary of Western development - falsely, because without Western industrialization it is unlikely that it would be at its present state of development.[58] The West, obviously has benefited the most from carbon-based industrial expansion, but all countries and peoples have received some benefits.

China then, under the contraction and convergence position could become the world's largest industrial producer of greenhouse gases, while still maintaining low per capita emissions. This could be done by ensuring that the majority of its population remain poor, while a ruling class of elites retain their power and top-of-the- food-chain lifestyles. No doubt part of the appeal of the contraction and convergence view is its implicit communism and egalitarianism, where the rich, in a form of ecological poetic justice are made equal with the poor, leveled down and humbled. However the history of Marxism shows that egalitarianism is never achieved, for there is always a class of ruling elites who preserve their power and privileges.[59]

Apart from these objections there are also major problems with Hillman's idea of personal carbon allowances. It goes without saying at this point in time that it makes economic and ecological sense for individuals to be as energy efficient as possible in the home and in transport through the three R's: reduce, recycle, reuse.[60] Nevertheless to reach the carbon allowance of one tonne will essentially involve people abandoning a consumer society and adopting conserver lifestyles based upon self-reliance and voluntary simplicity.[61] Thus air travel for holidays and the transportation of food across the globe are ultimately unsustainable.[62] However, so as well are many of the present consumer items of people in the developed world, such as fashion and cosmetics. Arguably, items such as dishwashers, microwaves and even kettles (which use two thirds as much energy as an oven or a hob)[63] fall on the list of items that may ultimately need to be culled from one's life to live within one's carbon limits. It is a problem of staggering difficulty to envisage how a policy of personal carbon allowances could be implemented in a way which effectively and substantially cuts greenhouse gas emissions so long as nations embrace the objective of continuous economic growth and a consumer society. Liberal democratic governments live in fear of electoral backlashes from voters (who are carbon producers) and would be most reluctant to implement legislation "with teeth", which challenged the life of the average consumer in such a radical way, even in the short term.

Australian scientist Ian Lowe in his book *A Big Fix*[64] also presents a "big picture" approach to the environmental crisis (but with an Australian focus) as does Lester Brown in *Plan B 2.0*[65] Both Lowe and Brown argue that our present civilization is ecologically unsustainable because of growing human populations, increasing consumption levels, lifestyle choices, technologies and economic pressures. Both Lowe and Brown agree, as do the present authors, that the transition to a sustainable society will involve "fundamental changes to our values and social institutions", [66] and both believe that with the will to change a sustainable society can be reached.

For Lowe the main obstacle "is the dominant mindset of decision makers who do not even recognize the problem, or see potentials solutions as threats to their short-term interests".[67] Thus, for example, Bjorn Lomborg's *The Skeptical Environmentalist*[68] was greeted with enthusiasm by economic and conservative commentators as the book defended the thesis that economic growth will save the world. As Lowe points out, Lomborg's methodology was to argue (often correctly) that some environmentalists exaggerate the extent of ecological problems such as species loss. On this basis Lomborg concludes that the problem in question is not as serious as environmentalists suppose, even though the lower estimates which Lomborg accepts, say in the case of species loss, are similar to those of the *Millennium Assessment Report*, which does see the matter as a serious problem.[69] The idea of "growth as good", Lowe rightly says "is a deep-seated myth of our society" and challenging it is a heresy.[70] Sustainability requires us making a transition from a growth economy/consumer society (where in "a morally deficient and spiritually bankrupt society, people are urged to find fulfillment in consumption"[71]) to a steady-state economy. Although he says many interesting and useful things about how a sustainable society may operate, the more pressing question of how to make the transition to such a society is not addressed. Indeed, even with respect to a less radical scenario such as a "policy reform" scenario where governments seek to build a sustainable future, beyond rhetoric, "the political will to implement such a strategy is nowhere in sight".[72]

Lester Brown's *Plan B 2.0* faces similar problems to Ian Lowe's *A Big Fix*. The book of course is full of interesting ideas for producing a sustainable society, such as redesigning cities,[73] reducing urban water use, advocating the use of renewable energy resources and raising energy productivity, among other agreeable ideas. However, Brown's overall mechanism to achieve these changes for the developing world, is to use aid to bridge the income gap for poor countries. But as Andrew Simms has said, this alone won't work:

> Brown conveys a sense that a few fiscal measures, combined with the goodwill of rich countries, will deliver. This is an approach that has been followed for the last three decades, and it has not worked. During the 1990's the share of benefits from global economic growth reaching those living on less than a dollar a day fell by 73 per cent, in spite of countless promises to end poverty. This is the problem with Plan B2.0.[74]

Brown says that the deal with the converging crises of civilization there needs to be something like the mobilization made for World War II. He points out that the US was rapidly able to re-tool to change from a consumer society to a war-time society. A rationing program was introduced for strategic goods such as fuel oil, tires and gasoline.[75] It is true that capitalist societies can mobilize resources within months, when getting into war mode, but it does not follow from this that the same type of mobilization could be done in the case of the environmental crisis, which is a set of interacting problems rather than one with a concrete identifiable foe who can be de-humanized, vilified and then killed. Reflecting on the gravity of humanity's plight Brown says, with anguish:

> It is hard to find the words to convey the gravity our situation and the momentous nature of the decision we are about to make. How can we convey the urgency of this moment in history? Will tomorrow be too late? Do enough of us care deeply enough to turn the tide now? Will someone somewhere one day erect a tombstone for our civilization? If so, how will it read? It cannot say we did not understand. It cannot say we did not have the resources. We do have the resources. It can only say we were too slow to respond to the forces undermining our civilization. Time ran out.[76]

George Monbiot in *Heat: How to Stop the Planet Burning*[77] has addressed this challenge. He argues that to hold global temperatures to below 2 C requires stabilizing greenhouse gas emission concentrations at or below the equivalent of 440 ppm of carbon dioxide, whilst the present day (2006) carbon dioxide equivalent is between 440 and 450 ppm CO2 e. This is necessary to prevent the possibility of runaway positive feedbacks, for if the planet's average temperature rose by 2 C then it is likely to then rise by 3 or 4 C. As argued in chapter 2 of this book, we believe that Monbiot is being over-optimistic in his fundamental premise. Further, the capacity of the biosphere to absorb carbon will reduce from the current value of four billion tonnes a year to about 2.7 billion tonnes a year by 2030 due to a diminishment in the carbon sink capacities of the biosphere.[78] Hence in 2030 the people of the world cannot emit more than 2.7 billion tonnes of carbon a year, a reduction of 60 per cent from 2006 emission levels. Thus, Monbiot argues, to achieve stabilization, the carbon emissions per person will have to be no greater than 0.33 tonnes per annum.The reduction needed for developed countries will be about 90 per cent of the present levels by 2030.[79] Monbiot in *Heat* attempts to show that this cut in emissions can be made by 2030, with the political will, whilst preserving industrial civilization and the luxuries of consumer society, along with biosphere survival.

Monbiot accepts the contraction and convergence principle but with modifications by Mayer Hillman and David Fleming.[80] Rather than a carbon account for all goods, people would have a carbon account for only two commodities, fuel and electricity, for otherwise the system is too complex. The carbon price would appear in other items such as food. The carbon rationing system will in effect create a currency (which Monbiot calls "icecaps") based upon an entitlement to pollute - with carbon units being traded on the market. Monbiot contrasts the proposal with the European Union's Emissions Trading Scheme, which gave free emission permits to the big European companies. In general, those producing the most carbon emissions were given the most permits: from the beginning of 2005 to May 2006 UK power firms made a windfall profit of approximately £ 1 billion without having any cause to reduce greenhouse gas emissions.[81] By contrast to this situation, the market created by carbon rationing would stimulate demand for low-carbon technologies such as renewable energy.

Monbiot's task is then to show, chapter by chapter, that a 90 per cent cut can be made in all of those sectors necessary to keep an industrial civilization in existence. This does not include presently structured military systems, as Monbiot believes that the armed forces of the world should be greatly reduced.[82] Obviously obtaining an efficient use of energy must be an important strategy in obtaining these substantial cuts, but there are some fundamental problems with energy efficiency in a market economy.

One such issue is the Khazzoom-Brookes postulate.[83] If energy efficiency improves, people and companies can produce more services with the same amount of energy. Money that would have been spent on energy can be redirected elsewhere. As well, previous energy intensive processes which become more efficient now become more profitable. Thus investors seeking to make a profit will invest in more energy-intensive processes than they would otherwise. The wonderous logic of the free market could therefore lead increased energy efficiency actually *increasing* overall energy use.[84] Monbiot points out that the economist Stanley Jevons in his book *The Coal Question-Can Britain Survive?* (1865), found that the efficiency gained by cutting the quantity of coal to produce a ton of coal by over two thirds "was followed in Scotland, by a tenfold increase in total consumption, between the years 1830-1863".[85] Since that time, apart from two periods where an energy crisis pushed up prices, there has been a rising total global energy consumption, that has washed away any environmental savings from energy efficiency.[86] Monbiot points out that the Asia-Pacific Partnership on Clean Development and Climate (AP6), involving the governments of Australia, the United States, China, India, Japan and South Korea, put as an alternative to the Kyoto Protocol, offers no binding emission reduction targets, but relies upon the development of new energy efficient technologies. In the light of the Khazzoom-Brookes postulate, it is unlikely to work.[87]

Can the 90 per cent reduction figure be reached? Monbiot in his discussion of housing, outlines the difficulties in obtaining energy efficiency in old houses.[88] Another problem is that while the energy efficiency of some household gadgets is improving, there are now more of them and some such as plasma TVs are highly energy inefficient relative to simpler TVs.[89] Monbiot, citing the Environmental Change Institute's figures, concludes that the maximum reduction in energy use in housing by 2030 will be only around 30 per cent. Thus he believes that the savings will primarily need to be at the source of energy production, using low carbon fuel and electricity. We will discuss this matter further below. However on page 203 of his book, Monbiot concludes that "we could cut our carbon emissions by around 90 per cent in all but one [aviation]" and includes the home among this, which seems to be an inconsistency.[90]

The major difficulty with Monbiot's book though is the question of political will. He says: "of course, the more energy the carbon ration allows us to use, the more politically acceptable the scheme becomes. If a 90 per cent carbon cut means a 90 per cent energy cut, it is hard to see how any democratic government could make it happen".[91] That, however, is precisely the problem which needs to be solved for his rationing scheme, and indeed for the contraction and convergence position to get off the ground. How can democratic societies find the political will to make the hard decisions, before the dire consequences of climate change and ecological degradation cause further harm to human civilization?

**Will Economics Save the World?**

In a fascinating book, *Economics as Religion: From Samuelson to Chicago and Beyond*,[92] Robert H. Nelson argues that the foundations of economics are not purely based upon rationally justified presuppositions, but are often unexamined propositions, resembling faith commitments. Economists, Nelson argues, are often more like theologians than scientists, with their basic role being to "serve as the priesthood of a modern secular religion of economic progress that serves many of the same functions in contemporary society as early Christian and other religions did in their time".[93] This is essentially the same conclusion which one of the present authors (Smith) reached in *The Bankruptcy of Economics*.[94] Mainstream economists - and their thought is generally reflected in the economics writings and analysis of most of the influential newspapers in the Western world - have a "faith" that free trade and market-driven economics will resolve the environmental crisis, if it is granted that there is an environmental crisis at all.[95]

Thus, for example Ronald Coase argued in a paper "The Problem of Social Cost", an idea which in part earnt him the 1991 Nobel prize in economics.[96] Transaction costs are those costs involved in economic transactions in discovering the identity of dealers, information about terms and contracts and a vast array of administration and negotiation costs.[97] Coase argued that in a world of no transaction costs it doesn't matter how legal and economic institutions are structured to deal with situations of externalities, spillover effects or market failures such as pollution. Policy decisions should be made on the basis of considerations of optimal resource allocation, rather than on legal or moral grounds of entitlement. Thus, Coase says that people living downwind from a polluting factory are also responsible for any harm they experience because they would not be harmed if they did not choose to live there in the first place! [98] The right to pollute ends up, ultimately with the party that is able to optimally allocate resource.[99] Needless to say, there is an extensive literature critical of these ideas, pointing out that the Coase theory depends for its validity upon assuming true many other controversial propositions such as perfect knowledge of utility functions, a costless court system and so on.[100]

An extensive literature also exists outlining the limits of a free market approach to environmental protection, arguing among other things that there is no "invisible hand" operating in markets ensuring that ecologically sustainable practices will arise merely because environmental assets are in private hands.[101] As Paul Hawken, Amory Lovins and L. Hunter Lovins put it in their book *Natural Capitalism*:

> Capitalism, as practiced, is a financially profitable, nonsustainable aberration in human development. What might be called "industrial capitalism" does not fully conform to its own accounting principles. It liquidates its capital and calls it income. It neglects to assign any value to the largest stocks of capital it employs - the natural resources and living systems, as well as the social and cultural systems that are the basis of human capital.[102]

It is not the aim of this section to begin a treatise on the limitation of environmental economics, having presented that elsewhere.[103] Our concern here is with the limits of an economic approach to the problem of global climate change and we have mentioned that other, broad philosophical limits by way of scene setting and background. To turn now to our topic of analysis, an appropriate starting point is with the UK Stern Review, *The Economics of Climate Change*.[104]

The Stern Review begins with an acceptance of the position argued for in chapter 2, that the existing scientific evidence indicates that global climatic change is a major threat to civilization.[105] However, the Review concludes that remedial action to deal with global climatic change would only cost nations one per cent of GDP per annum. The Stern Review concludes that if humanity does not act now then the negative impact of climate change on GDP would have a quantum of impact that increases progressively to about five per cent of GDP by 2200, or even 20 per cent of GDP by that year if the ill-effects of raised sea levels, droughts, floods and cyclones are considered.

The economic basis of the Stern Review has been subjected to criticism by a number of economists including William Nordhaus[106] and the research team of Byatt (*et al.*).[107] Central to the criticisms of these researchers has been the focus upon the assumption of near-zero social discount rate. Discounting is a method by which future monetary values are weighted by a value of less than 1 to determine the dollar value today of future costs and benefits. The social discount rate is a parameter calculated in per cent per year that measures the utility of future generations relative to the present generation.[108] Use of a zero social discount rate entails there is an equal treatment of future generations with the present generation. On the other hand use of a positive social discount rate entails that a "discounting" of the welfare or utility of future generations compared to present generations occurs. Many philosophers and economists have been skeptical about the use of a positive social discount rate (and more radically about the use of conventional cost benefit analysis) because it is argued that these economic approaches offer inadequate protection of the rights of future generations.[109] It is argued that while a case can be made for the use of a discount rate to give an accurate present valuation of the value of money in the future,[110] the discounting of commodities should be distinguished from the discounting of well-being because many such values do not decline over time or exhibit "time preference".[111] Further, with respect to climate change, conventional discounting and cost benefit analysis approaches result in an undervaluation of losses and an overvaluing of the economic gains of climate change, giving optimality to unsustainable policies.[112] Behind the high social discount rate are optimistic assumptions about economic growth without a consideration of negative environmental impacts and the negative effects of global climatic change on ecological systems. It is assumed that smooth marginal changes are occurring in carbon dioxide concentrations and on this basis orthodox economists conclude that climate change impacts will also involve smooth marginal changes, ignoring the prospects of nonlinear changes.[113]

Although the Stern Review, in the light of the work of the philosophers and economists cited in the above paragraph, is correct in treating the welfare of future generations on par with the present generation, [114] there are some counter-intuitive consequences of Stern's analysis, *at least from an orthodox economic position.* Nordhaus feeds into Stern's economic model "a winkle in the climate system" which will cause "damages equal to 0.01 per cent of output starting in 2200 and continuing at that rate thereafter".[115] The economic investment today needed to remove the "winkle" to avoid this damage starting after two centuries, is according to Nordhaus' calculations, 15 per cent of today's consumption, which is approximately $7 trillion. This result occurs, "because the value of the future consumption stream is so high with near-zero discounting that we would trade off a large fraction of today's income to increase a far-future income stream by a tiny fraction".[116] Thus, if Stern is right in adopting near-zero discount rates, then the claim that sufficient remedial action to prevent the ill-effects of global climatic change will only cost one per cent of GDP per annum must be rejected. Indeed, the Stern Review says:

> Climate change will reduce welfare even more if non-market impacts are included, if the climatic response to rising GHG emissions takes account of feedbacks, and if regional costs are weighted using value judgments consistent with those for risk and time. Putting these factors together would probably increase the cost of climate change to the equivalent of a 20% cut in per-capita consumption, now and forever.[117]

The Stern Review assume that by acting now the 20 per cent of GDP figure can be avoided by merely expending one per cent of GDP per annum. However, at other places in the Review it is recognized that long term stabilization will require a reduction to under 20 per cent of present emissions or more in the light of evidence of a weakening of the ocean "sink" absorption capacity.[118] To keep the concentration of $CO2$ in the atmosphere at 550 ppm requires a rate of reduction of one per cent, but to keep the concentration at 450 ppm would require a rate of reduction of seven per cent.[119] A mere 400 ppm, according to Stern's table 1.111 is associated with a 4.9 C temperature rise, without accounting for positive feedback effects. However the Stern Review says in effect that emission reductions to such levels will have an unacceptably high economic cost.[120]

Finally, although we are critical of some of Bjorn Lomborg's work, his criticism of one aspect of the Stern Review seems to the present authors to be insightful and to indicate a major limitation of the Review.[121] Even if most of the world's nations were able to follow Stern's multi-trillion dollar century long proposal, it is naïve to suppose that there would not be some major greenhouse gas emitting nations that would not seek to avoid their obligations. For example, in 2002 China decided to cut sulfur dioxide emissions by 10 per cent, but these emissions in 2006 are now 27 per cent higher.[122]

The Stern Review supports an emissions trading scheme to allow economies to take a smoother adjustment path to future carbon-restricted economies. A "cap and trade" emissions scheme caps emissions at same level in each period; allows permits to be issued to emit greenhouse gases for that period; supplies a penalty for non-compliance and finally, allows participants to trade these permits among themselves.[123] This will make it easier for some companies to reach emissions reduction goals by in the short-term buying permits from companies who have more easily made the transition to greenhouse gas emission targets. Emissions trading schemes allow the competitive market to set prices and achieve the lowest cost way to reach a particular emissions cap, without relying upon government "command and control" regulation, investigation and supervision. There are a large number of such "carbon offset markets" operating across the world at both the international, and national level, including Kyoto markets, the European Union Emissions Trading Scheme, the Chicago Climate Exchange (CCX) and in Australia, the New South Wales Greenhouse Gas Abatement Scheme (GGAS). New South Wales, for example, is the only Australian state to introduce a mandatory emissions reduction scheme, where all electricity retailers selling into New South Wales must reach annual greenhouse gas reduction targets or else buy New South Wales Greenhouse Abatement Certificates.[124] The present authors will not consider the matter of carbon trading further because the concerns of this chapter is how the *initial* reductions by firms and society in general can be made to the levels suggested by Monbiot (as discussed above) of a reduction of about 90 per cent of present levels by 2030 by developed countries. Clearly emissions trading will be only a small part of any such radical reduction.

Some economists advocate global carbon taxes as a general economic mechanism for addressing the problem of global climatic change.[125] Others support a mini-carbon tax as part of a emissions trading scheme - the Australian government is floating the idea of such a scheme at the present time (October 2007) as part of its proposed global emissions trading scheme through the Asia-Pacific Partnership for Clean Development and Climate (AP6).[126] Environmental taxes are an attempt to incorporate the real health and environmental costs of products that, it is argued, companies have escaped paying. The economists Kahn and Franceschi prefer a tax-based system for the global reductions of greenhouse gas emissions to a cap and trade system, because, among other things, taxes offer greater incentives for technological innovations in the creation of technologies for emissions reductions.[127] The economists Montgomery and Smith take an opposing position and argue that a carbon tax will not promote research and development because such a mechanism will not send an effective signal to stimulate the necessary funding.[128] Along with this, carbon taxes disproportionately affect the lower income groups.[129] It is yet to be shown how carbon taxes *alone* could address the challenge of global climatic change, but of course, the economic supporters of these taxes typically and ultimately hope for economic policy to stimulate a technological fix.[130] Let us turn now, to consider the technological answer to the problem of global climatic change.

**Will Technology Save the World?**

The problem facing the technologists is an extraordinarily difficult one: to reduce global emissions without dismantling globalization and the global economy and permitting the global economy to continue to grow, with China, India and the rest of the developing world ultimately adopting Western lifestyles - but of course, within a "greener" economy. Robert Socolow and Stephen Pacala writing in *Scientific American* say that this task is "daunting":

> Over the past 30 years, as the gross world product of goods and services grew at close to 3 per cent per year on average, carbon emissions rose half as fast. Thus, the ratio of emissions to dollars of gross world product, known as the carbon intensity of the global economy, fell about 1.5 per cent a year. For global emissions to be the same in 2056 as today, the carbon intensity will need to fall not half as fast but fully as fast as the global economy grows.[131]

Socolow and Pacala say that this task would be "out of reach" were it not that today's "notoriously inefficient energy system" can be replaced "if the world gives unprecedented attention to energy efficiency" over the next 50 years.[132] Much can be done at present to achieve greater energy efficiency with existing technology through conservation strategies, and many popular books and TV shows have focused on this strategy as "something to do to save the planet".[133] However, improvements in energy efficiency in isolation from other factors will not necessarily result in energy savings that reduces greenhouse gas emissions if "take back" occurs - energy efficiency merely leads to an increase use of the technology. There is also the issue of the Khazzoom-Brookes postulate discussed above. Estimates for the contribution that energy efficiency improvements could make for limiting greenhouse gas emissions vary greatly because of the uncertainty about the implementation of various technological innovations[134] - but one frequently cited IPCC paper puts the potential for efficiency improvements for the 2000-2020 period at 25 per cent of emissions.[135] What needs to be considered though is that there are diminishing returns from savings from efficiency over time because the best opportunities will be ceased first, leaving more difficult matters for the future.

Technological fixes to the increasing levels of greenhouse gas emissions by industrialized nations featured prominently in the Intergovernmental Panel on Climate Change reports.[136] Mitigation technologies focused on both "alternative" and low carbon energy sources, the production of new "clean coal" technologies, the reduction of methane and nitrous oxide emissions in agriculture, geosequestration (CO2 capture and underground burial) and many conservation strategies.

This "high tech-fix" approach[137] is the basis for the "Asia-Pacific Partnership for Clean Development and Climate", (APPCDC) signed by Australia, China, Japan, India, South Korea and the United States. The APPCDC relies upon the development of new technologies to limit greenhouse gas emissions rather than adhere to emission-reduction targets as in the Kyoto Protocol. The member countries use about 45 per cent of the world's energy and emit over half of the world's CO2 per annum.[138] Carbon capture and storage methods, especially through geosequestration, feature among the technological strategies.

Geosequestration involves capturing CO2 and burying underground in deep saline aquifers, depleted oil and gas reservoirs or deep unmineable coal seams. Instead of conventional smoke stacks, coal-fired power plants will be equipped with gas scrubbers in absorption towers. China intends to build 560 new coal-fired plants and India 213 in the near future, all without this technology, doubling their present emissions of CO2 within 20 years.[139] Thus at present one of the main problems associated with geosequestration, admitted by everybody, is how to achieve economic viability in this technology by reducing the carbon capture price per tonne from the existing figure of about US $40 per tonne, to a projected economically viable US $10 per tonne of CO2 removed.[140] Beyond this, all existing power plants need to be retrofitted to trap CO2 emissions efficiently.[141] Some industry leaders believe that this technology may require 10-20 years at least before commercial viability arises.[142] Other technologies for burning coal more efficiently can be implemented much sooner and some are being implemented at present, although "clean coal" technology is presently far from clean with only 25-30 per cent of carbon emissions captured.[143]

Skeptics of the APPCDC abound. We will mention some Australian critics. Don Henry of the Australian Conservation Foundation said: "We have voluntary federal measures in place in Australia and greenhouse pollution from the energy sector has risen by around 30 per cent over the last 10 years".[144] Greens Senator Bob Brown said that the pact was designed to protect the coal industries of the signing nations.[145] Former head of the Australian Greenhouse Office, Gwen Andrews, has also said: "One has to wonder whether it [i.e. APPCDC] is really an effort to tackle the issue or just an attempt to deflect criticism about inaction by taking action in a way that doesn't commit them to any restraints on emissions".[146] Andrews did not receive a request for briefing from the Australian Prime Minister John Howard even once during her four years in the job as chief executive officer of the Australian Greenhouse Office, and this was a time when the Australian government was deliberating about the ratification of the Kyoto Protocol.[147] Prime Minister Howard and Industry Minister Ian McFarlane, according to investigations of *The Age* (Melbourne) newspaper, secretly met with senior industry figures and adopted most of their policy suggestions, including non-ratification for the Kyoto Protocol. The fossil fuel and energy lobby groups have been big donors to the Australian Coalition Government.[148] Gwen Andrews said to *The Age*:

> In my view, large energy, mining and resource interests in Australia had a disproportionate effect on government policy making with regard to energy and climate change. With few exceptions, they had little commitment to the concept of sustainable development. The arguments they used to influence the Government framed the debate in a way that pitted environmental interests against economic interest. Their definition of economic interest was constrained by their businesses and shareholders. There was a consistent underlying theme to their arguments that if Australia implemented any policy measures that raised the cost of doing business for them, they would have to consider the wisdom of investing further in Australia....Government ministers seemed to be heavily influenced by this line of argument.[149]

Apart from these political considerations, even if all the recommendations of the Asia-Pacific Partnership for Clean Development and Climate are voluntarily adopted by businesses, there will be a doubling of carbon emissions by 2050 under the pact.[150]

At this point in the debate it is usually granted that reductions in greenhouse gas emissions cannot be made through improvements in energy efficiency alone. As Kammen puts it, because "economic growth continues to boost the demand for energy - more coal for powering new factories, more oil for fueling new cars, more natural gas for heating new homes - carbon emissions will keep climbing despite the introduction of more energy-efficient vehicles, buildings and appliances".[151] Consequently it is frequently argued that the world must quickly make a transition from carbon-generating energy systems to so-called renewable energy sources generating little or no carbon as waste. Can renewable energy technologies fuel the globalized consumer society?

**Renewable Energy Cannot Sustain Consumer Society**

Ted Trainer's *Renewable Energy Cannot Sustain Consumer Society*[152] is the first book to present an argument about the limits of renewable energy. There have been many books on the limits of *specific types* of renewable energy sources, such as solar energy[153] but Trainer has performed the important intellectual service of summarizing the limits of each source. The situation in the renewable energy literature appears to us to be that advocates of one source keenly outline the limits of other sources whilst exaggerating the scope of their own source and minimizing its problems. The German Advisory Council on Global Change (WBGU) in its report, *World in Transition: Towards Sustainable Energy Systems*[154] concludes that the world can make a transition to sustainable, renewable energy systems, without endangering economic growth or the global economy if only enough finance is forthcoming from conventional sources (e.g. the World Bank) and "enhanced incentives for private-sector investors".[155] However, it is worthwhile contrasting this relatively optimistic view with the view of two climate change skeptics and generally technological optimists S. Fred

Singer and D.T. Avery who nevertheless say on the issue of supplying energy for the future:

> The biggest problem is that the world's current 12 trillion watt-hours of energy used per year (85 per cent of it fossil-fueled) will need to be expanded to 22-42 trillion over the next fifty years in order to accommodate the world's growing population and provide economic growth for developing countries.Energy experts say that even nuclear power will not be enough, due to a shortage of uranium ore. We will need safe nuclear breeder reactors and, ultimately, fusion power. *That means developing very expensive energy technology we don't yet have.*[156]

The research team of Hoffert (*et al.*) conclude that it would be a Herculean task to replace fossil fuel supplies by renewable energy sources in the short-term.[157] They state: "All of these approaches currently have severe deficiencies that limit their ability to stabilize global climate…[and] that a broad range of intensive research and development is urgently needed to produce technological options that can allow both climate stabilization and economic development".[158]

Biomass, solar thermal and photovoltaic, wind, hydro-power, geothermal, ocean thermal and tidal are, excluding firewood and hydroelectricity, less than one per cent of the world's power, considered collectively.[159] There are well recognized limitations to these energy sources.

Geothermal energy is being harnessed in some parts of the world such as Iceland. The Icelandic government intends to drill 3-8 kilometers through the Earth's crust to reach hot basalt that will be used to create steam to drive turbines to generate electricity for up to 1.5 million homes in Europe.[160] In Australia, the generation of electricity from "hot rocks" in the Cooper Basin has been proposed by Brisbane based company Geodynamics Limited, which foresees the geothermal sector generating 10 per cent of Australia's electricity by 2030.[161] Geothermal energy will also soon be used to produce electricity in Western Australia.[162] A panel of experts from Massachusetts Institute of Technology, chaired by Jefferson Tester, has concluded that hydrothermal technology -pouring water onto hot underground rocks and using the steam generated to turn turbines-is the most promising environmentally sustainable form of renewable energy for the supply of electricity.[163] There are some limits at present due to shortcomings in drilling technology, but the technology is developing that could permit drilling almost anywhere on the surface. The MIT panel of experts optimistically proclaim that there is enough energy from hot rocks less than 10 kilometers beneath the surface of the United States to supply the US with electricity at its 2007 rate of consumption for the next two millennia.[164] Trainer points out that it will take considerable energy to drill the holes, fracture the rock and force the water from one hole to another. Water will emerge at around 270 C, which will have a low generating efficiency of about 15-20 per cent.[165]

Biomass energy, such as biodiesel (made from oilseed, canola or fat from sheep or cattle), ethanol (using sugars extracted from wheat and corn) and cellulosic ethanol (still experimental, made from agricultural waste such as forest products) has an important role to play in humanity's energy future.[166] However even if the US turned all its corn and soybeans into biofuel this would only account for five per cent of US fuel needs, according to a study done by Hill (*et al.*).[167] There is insufficient land available to produce the fuel crops and already there are fears that the extensive use of biofuels could starve the poor.[168]

The hydrogen economy has been embraced by many as the ultimate answer to humanity's dual climate and energy problems.[169] The vision is to produce hydrogen without the use of carbon polluting means, say by algae producing hydrogen on "hydrogen farms" in desert regions. Such technology is still experimental and has a very low efficiency. Hydrogen could also be produced by electrolysis of water using nuclear power, hydroelectricity or wind generated electricity. Stephen Lincoln details the difficulties facing this approach:

> ...the magnitude of the change in the present energy supply structure required to achieve this is colossal. It is estimated that 230,000 tonnes of hydrogen would be required daily to replace the oil used in surface transportation in the United States in 2003. This is an optimistically low estimate as it assumes a presently unachievable sixty-four percent energy conversion into vehicular motion. If all of the hydrogen was generated through electrolysis of water it would require either the building of 500

> natural gas fuelled power plants generating 800 mega-watts each, 500 coal fuelled power plants generating 800 megawatts each, 200 Hoover Dams generating 2,000 megawatts each, 400 nuclear power plants generating 1,000 megawatts each, or a mix thereof, costing US$400 billion, or one twentieth of the 2003 gross domestic product of the United States. If the fossil fuelled power plant options were adopted this would increase the amount of carbon dioxide that would have to be sequestered if the advantages of using hydrogen as a fuel with respect to global warming were to be realized. The magnitude of the energy infra-structure changes required globally to switch to hydrogen for transportation appears even more daunting using current hydrogen production technologies.[170]

Hydrogen fuel expert Joseph J. Romm says in his book *The Hype About Hydrogen* that the difficulties facing a hydrogen economy are enormous: " *Neither government policy nor business investment should be based on the belief that hydrogen cars will have meaningful commercial success in the near-or medium term"*.[171]

The limitations of wind, tidal, solar thermal electricity and photovoltaic solar electricity have been outlined clearly by Trainer in *Renewable Energy Cannot Sustain Consumer Society*.[172] We would add that there are some high technology proposals to attempt to escape the well recognized limitations of these particular forms of renewable energy. For example, wind power has a problem of variability or intermittency, but it is thought that this problem could be overcome by use of technologies using high-attitude winds - even the jet stream. Needless to say, these technologies are experimental.[173] Likewise for the use of space color power, which may deliver electricity to markets in the second half of this century.[174] Geoengineering programs such as filling the skies with sulfur have been proposed as a desperate "Plan B" strategy for dealing with climate change, but these proposals also face an array of scientific and politico-ethical objections.[175] Unexpected, and expected effects may prove worse than global climate change. This, therefore leaves us with the nuclear question.

If renewable energy resources cannot be a substitute for fossil fuels, sufficient to fuel a globalized consumer society with exponential economic growth[176] then can nuclear energy do this? Nuclear energy is a controversial matter and high profile environmentalists are divided on this issue. James Lovelock in *Revenge of Gaia*[177]is enthusiastic about nuclear energy as says with respect to nuclear waste generated by fission that it will not poison the biosphere: "the natural world would welcome nuclear waste as the perfect guardian against greedy developers".[178] Heavily contaminated places have a rich diversity of wildlife he observes.[179] On the other hand, the Australian environmentalist and anti-nuclear campaigner Helen Caldicott, in her book *Nuclear Power is Not the Answer to Global Warming or Anything Else*[180] is critical of all aspects of nuclear energy. She argues in detail that there are major storage, security and health issues associated with uranium mining, processing and nuclear energy, and there is considerable literature supporting her claims.[181] But there is a lack of objective, independent comprehensive scientific reports assessing the merits or otherwise of nuclear power, making any concise overview of this topic, virtually impossible.[182] Nevertheless many countries see nuclear energy as a way of meeting growing energy needs as well as dealing with the climate change problem, so here we will side step some of the debates about storage, security, and health and consider whether nuclear energy can fuel the growing, globalized consumer society.[183]

Nuclear fusion on Earth (by contrast to nuclear fusion in the sun) involves the merging or fusion of the hydrogen isotopes of deuterium and tritium to form helium to release energy. The deuterium reserves in sea water are vast and could last humanity several million years if the world's current electricity demand was completely met by nuclear fusion. At present the Tokomak reactor at the Culham Science Centre has sustained a fusion reaction for two seconds generating 16 megawatts of energy.[184] Optimists believe that fusion power may be available at the earliest in the second half of the 21st century and pessimists believe much later or never.[185]

The major difficulty with fission power is a lack of high grade uranium to meet long-term energy demands on a scale sufficient to address the climate change problem. A nuclear era would be a short one indeed. Current estimates of proven and ultimately recoverable global uranium resources are 3.4 and 17 million metric tons respectively.[186] The present energy use of the world is 12 TW of energy per annum and increasing. At this rate there is five to twenty five years of global energy supply at present consumption rates. Even if all ultimately recoverable reserves of high-grade uranium could be mined and used to produce energy, the peak CO2 concentration would only be delayed by 2.5 decades. [187] However, the construction of nuclear stations (each of which costs up to $US 2 billion and would take about a decade) and their decommissioning require considerable energy. According to Professor Stuart White, Director of the Institute for Sustainable Futures at the University of Technology in Sydney, the average energy payback time of a reactor using high-grade ore is seven to ten years from commencement of operation.[188] So although the operational life of a nuclear power station may be 40 years, the energy used in its construction and decommissioning erode its net carbon amelioration. If lower grade uranium ores are used as a fuel source, then the carbon amelioration value is even lower. This reduces the climate change amelioration value of the total global recoverable[189] supply of high grade uranium to little over two decades.[190] The reprocessing of nuclear fuel (which aims to reduce the volume of spent nuclear fuel that has to be disposed of safely by recycling it for use in new types of nuclear reactors) may extend the energetic usefulness of mined uranium, but at a very high cost. This is because it involves the separation of components that can readily be used to build nuclear weapons. This practice still occurs in France, and Japan is also trying to develop it. There are fears that the US may soon also try to revive this practice.[191]

Three scenarios are presented below in order to tease out these issues, and in particular to highlight the sustainability problems that would be posed if nuclear energy were to be rapidly expanded. These scenarios are (a) an "all-out" effort to increase nuclear fuel in order to abate climate change; (b) a continuation or a modest expansion of the nuclear industry, and (c) a reduction in the use of nuclear fuel.

First, let us consider an aggressive global expansion in the use of nuclear energy. This strategy is likely to have the most rapid effect of the three proposed scenarios in slowing the acceleration of climate change, at least temporarily. This dampening effect could, at best, persist over one or two generations and perhaps slow the atmospheric accumulation of CO2 by about 50 ppm. But this strategy is likely to *reduce* the likelihood of civilization sustainability for several reasons.

These include an increased risk of the leakage of fissile material to the control of rogue states and non-state actors (thus increasing the risk of nuclear conflict and terrorism including by the use of "dirty" bombs. Vigorous expansion of nuclear energy would involve the construction of thousands of new power stations. The number of nuclear power stations could be trebled in about 20 years by commencing a reactor every week. But this scenario could easily create a nightmare. The risks from such a large and rapid expansion of nuclear energy use are likely to increase as a non-linear function of the number of nuclear reactors. Inevitably, in this scenario many reactors would be designed and commissioned rapidly. Some would be faultily constructed. Many will be built in countries with a poor safety culture, some with a high risk of political insecurity. Even stable nations can become unstable and so offer no guarantee of ongoing proper management. Reprocessing attempts would also probably increase, heightening the chance of weapons proliferation.[192]

After twenty years, there would still be insufficient reactors to utilize all recoverable reserves of uranium (even by starting construction of a reactor every week), nor would there be sufficient reactors to meet the then predicted demand for global electrical energy. So (under this scenario) the strategy would have to continue for another generation. In the overall process the world would generate an enormous additional quantity of nuclear waste, much of it in potentially unstable countries. And we would substantially deplete supplies of high-grade uranium, analogous to how we are currently depleting our stocks of oil and gas. This strategy would also be extremely expensive.

An intense scaling up in the use of nuclear fuel is also likely to create major opportunity costs. These could include misleading the public with the idea that the risk of climate change is finally being seriously tackled, and at the same time diverting large amounts of money and resources away from safer and more prudent energy options. Aggressive expansion would be a desperate exercise and would not "solve" climate change. The accumulation of CO2 would be temporarily slowed, but CO2 levels would still continue to increase in this period, though at a slower rate than in recent years. As the supplies of high-grade uranium become scarcer towards the end of this scenario, the by then enormous investment in the extraction, processing and reaction of uranium would start to fall in value, eventually approaching zero, in the same way that an oil rig is worthless if there is no oil. By about 2050, the world would have to replace the then aging reactors with a new energy technology, as uranium supplies run down. Because (under this scenario) renewable and other greenhouse friendly energy promising technologies are unlikely to have been substantially developed in the same period, the world would then probably have to initially return mostly to coal. The rate of greenhouse gas accumulation would again start to accelerate.[193] Climate change will not have been solved.

The International Atomic Energy Agency (IAEA) also doubts that nuclear power can grow fast enough to combat global warming. In 2005, IAEA director general Dr Mohamed El Baradei forecast that even if the global economy grew strongly, nuclear power would only grow by about 70 per cent over the next 25 years and its share of world energy would proportionately fall because of the more rapid expansion of other electricity sources. El Baradei also concluded that for the rural poor in developing countries, off-grid, small-scale, localized renewables are the best power solution, but that nuclear suited the needs of big, expanding cities, and countries with large centralized power generation.[194]

A modest global expansion in the use of nuclear energy, in contrast to scenario A, is far more difficult to dismiss. The proposition is that a modest expansion of nuclear power, globally, can make a modest contribution to the amelioration of climate change, and perhaps even to enhancing sustainability. The main argument which nuclear opponents rely on to oppose this, in addition to the issues of waste and nuclear proliferation discussed above, is the high comparative cost of nuclear energy. While it is almost certainly true that nuclear power stations are not cost competitive with coal (especially if the principles of ecological economics are used, and if all subsidies are included to calculate the real cost), our world already squanders vast sums on ventures which are far less economically competitive and productive than are nuclear power stations. Because the real costs of nuclear power are unclear, hidden and contested, and because there are other perceived advantages (mainly to do with self-interest of the sectors in favor of nuclear energy, or undeclared military and strategic interests), supporters of nuclear power don't find these economic arguments against nuclear power compelling.

Of the traditional reasons to oppose nuclear power, the disposal of nuclear waste and the possible diversion of fissile materials for military and terrorist purposes should evoke the most concern. However, both of these evil genies are already out of the bottle. Large quantities of fissile material have already gone missing[195] and large amounts of nuclear waste already require storage. The world is already awash with nuclear weapons. It is true that expanding the scale of nuclear energy is likely to worsen these problems, but the abolition of additional nuclear energy generation will not eliminate these problems. It can also be argued that the establishment of international norms and rules which reduce the risk of the misuse of fissile material is a more important goal than trying to abolish the use of nuclear fuel.

Consider now a contraction in the global use of nuclear energy. The world's existing stock of nuclear reactors will last for several decades. This investment relies on the future extraction and use of fissile material for fuel. Failing to use this investment would be expensive financially and energetically. Even though the use of high-grade uranium ore has a high energy cost, it does slow down greenhouse gas accumulation. Closing down all existing nuclear stations will not solve the existing problem of nuclear waste and weapons. Furthermore, ceasing to use nuclear fuel would aggravate climate change, unless this could be replaced by technologies which use renewable energy sources. This is unlikely in the short term. It is more rational to first replace carbon-intensive methods of generating energy such as coal. It is acknowledged that this scenario is unlikely in the near future. However, in the long run, this strategy appears highly desirable on the grounds of safety and to stop expanding the stock of nuclear waste.

Optimists may argue that new technologies (e.g. more efficient reactors, ways to efficiently extract uranium from seawater or from lower-grade sources, reactors which safely breed sufficient quantities of additional nuclear fuel, ways to permanently prevent fissile material from being used for violent or threatening purposes, and ways to safely deal with nuclear waste materials), will render invalid most of the preceding arguments against an expansion of nuclear energy. This would be unsound. Nuclear energy is a more mature technology than many alternative methods. A treasure has been spent on techniques such as breeder technologies and nuclear fusion, with little success.[196]

If optimistic assumptions are used to justify an expansion on nuclear energy, then they can be used equally to justify the development of other forms of technology. Some, such as those which use renewable energy (wind, solar) or geothermal energy appear to have far fewer inherent risks than entailed by the increased use of nuclear energy, even given their limitations with respect to fuelling exponential economic growth. Furthermore, pessimistic scenarios also exist. For example, new technologies, or the breakdown of social constraints, could plausibly worsen many of the problems enumerated above concerning nuclear weapons and the disposal of nuclear waste. In summary, optimistic assumptions are not used in this policy to justify an expansion of the nuclear industry, nor, indeed of any particular pathway.

Evidence so far indicates that uranium is a limited resource and it is therefore important not to increase reliance on it. Michael Meacher, UK Environmental Minister from 1997 to 2003, was at the center of the assessment of UK energy needs and the examination of the UK need to replace existing reactors promptly. He has said: "The imminent uranium shortage has been admitted by the World Nuclear Association, which provided a chart of the unfolding crisis on its website last year. But while the nuclear industry is comfortable with debating the safety of nuclear reactors it will not discuss the uranium supply shortfall".[197]

**Conclusion**

This chapter has examined legal, economic and technological responses to the climate change challenge, and also to the problem of fuelling the global consumer society. There are limits to the capacity of the legal and economic systems to generate solutions to the environmental crisis of civilization. Consequently, many people who have examined these problems in any depth, typically hope that a technological fix may be possible to save the day. The chapter has given a skeptical response to the hope that a technological fix alone, in the short-term, will meet the climate change challenge as described by George Monbiot, and summarized in this chapter. Renewable energy cannot sustain the globalized consumer society which sees a future world of nine billion people having the present per capita resource consumption of the developed world.

Energy expert Jeremy Leggett in *The Empty Tank*[198] is less pessimistic than us about the prospect of avoiding disaster on the basis of a technological fix and sees humanity achieving a sustainable future. Nevertheless he believes that there will be a shortfall between the current expectation of oil supply and its actual availability, which no combination of renewable energy resources can meet and which may lead to an economic collapse and "global financial catastrophe".[199] There is general agreement amongst environmentalists that a transition must begin towards the creation of sustainable, conserver rather than consumer societies.[200] We have seen that the *supply side* technology fixes to the climate change and energy resource depletion problems are severely limited. Attention thus should be directed to the *demand side* and lifestyle changes. The next chapter, and the rest of the book, will explore this approach.

# Chapter 4

## Conservation Psychology and Human Nature: Strategies for Change

> The mass of mankind is ruled not by its intermittent moral sensations, still less by self-interest, but by the needs of the moment. It seems fated to wreck the balance of life on Earth - and thereby to be the agent of its own destruction. What could be more hopeless than placing the Earth in the charge of this exceptionally destructive species? It is not of becoming the planet's wise stewards that Earth-lovers dream, but of a time when humans have ceased to matter.
>
> - John Gray[1]

### The Problem of Social Responsibility

Can the gloomy fate facing humanity, anticipated by the philosopher John Gray be avoided? Ted Trainer, a social ecologist writer at the University of New South Wales, Australia, has given a precise statement of the problem which we need to address in this chapter and the rest of this book in his paper "Social Responsibility: The Most Important, and Neglected, Problem of All".[2] There are a alarming number of social and ecological problems including the depletion of resources, the degradation and destruction of ecosystems, wars, social inequality and the breakdown of social cohesion in most societies - resulting in what we have called in chapter 2 a "crisis of civilization". As we have also seen in the literature summarized in this book, modern consumer-capitalist societies are based upon the ideology of continuous economic growth, but this goal of rising living standards and spiraling GDP increases is incompatible with the achievement of ecologically sustainable societies. As Trainer observes:

> There are good reasons for thinking that we cannot now solve these problems and that Western civilization will quickly decline into a new dark age. Books are being written on the firm belief that chaotic breakdown and the die-off of billions of people is likely, and some believe inevitable. Unlike previous societies ours is extremely vulnerable and fragile because of its complete dependence on vast quantities of energy, and on its house-of-cards financial system. Either of these factors could trigger sudden collapse.[3]

Although there has been much reflection upon the nature of the global crisis, there has been much less work done on *why* this perilous situation has been allowed to happen and why it continues with seemingly little done to avert the crisis. Trainer's answer is that all of humanity's problems have dragged on, unsolved, because most people have little social responsibility and are not fundamentally interested in solving these problems at a deeply personal level of commitment. Thus, children in Africa who die or are ill from a wide range of diseases arising from polluted drinking water, could be saved by people in the West making only minor sacrifices to their consumer lifestyle. Yet, these sacrifices are seldom made, at least by a necessary number of people to deal with the problem. Many people in the West seem more concerned about the purchases of cosmetics and other non-essential items, than helping the millions of people in developing countries that lack even the most elemental medical care.[4] But the issue is not one of ignoring the plight of people in far-off lands: an increasing number of people in our developed societies are homeless and living in poverty.[5] Further, even from a "selfish" or individualistic perspective, a rational economic agent, faced with a crisis such as that of global climatic change, should be concerned with the future ramifications of such a problem. It may affect them. Normal, rational individuals should be expected to respond to the growing evidence of a crisis of civilization with alarm and a call for immediate action. In past ages, the sight of an approaching enemy army or navy would have resulted in the sounding of an alarm and an immediate response. Yet today, with the coming threat of a collapse of civilization itself, the community response is but a whisper, if that.

The perplexities of consumer society deepens when evidence is considered that indicates that levels of individual happiness and contentment among people in Western societies, beyond a certain point of material acquisition, does not increase in proportion to increases in disposable income and affluence.[6] There may even be psychological anxiety, associated with maintaining high levels of affluence.[7] For this reason, a movement of downscaling (voluntarily giving up a consumer lifestyle or aspects of it) and voluntary simplicity has occurred[8]- although to date it is very much part of a counter-culture to mainstream consumer-capitalist society.[9]

It is clear upon reflection on this question, that there are a number of dimensions to this problem. There are many aspects of our culture which prevent an individual from responding to the crisis of the times. For one thing, the problems may seem too immense and difficult to be tackled and individual effort may be seen to be lost in the process. The rest of this book will focus upon various aspects of the problem of responsibility, such as the failure of the intellectuals, universities and philosophy to deal with these challenges and the hopes that religion may hold an alternative to consumerism. Here we will consider the psychological dimensions of this problem and examine what the discipline of psychology can contribute towards the building of an ecologically sustainable society.

**Conservation Psychology and the Philosophy of Sustainable Consumption**

David W. Kidner published a paper in *Environmental Ethics* in 1994 entitled "Why Psychology is Mute About the Environmental Crisis".[10] At the time he noted that although there was a rapidly expanding literature dealing with environmental psychology, the impact of environmental variables on human behavior, there was little material on how psychology as a discipline could aid in the prevention of environmental destruction.[11] Before Kidner's paper, B.F. Skinner (1904-1990), one of the most influential psychologists of the 20th century, presented a paper to the American Psychological Association in August 1982 entitled "Why We Are Not Acting to Save the World".[12] Skinner also made the same point.

More recently psychologists have attempted to address this deficiency and a new field of psychology "conservation psychology", modeled upon conservation biology and conservation medicine,[13] has been formed. Carol D. Saunders of Brookfield Zoo Illinois' Communication Research and Conservation Psychology Department,[14] has written a foundational article on the topic and has defined conservation psychology as " *the scientific study of the reciprocal relationships between humans and the rest of nature, with a particular focus on how to encourage conservation of the natural world".*[15] The discipline uses psychological principles, theories and methods to address B.F. Skinner's question and other important human aspects of conservation, such as saving biodiversity.[16] This field has already generated an interesting body of literature [17] and we will discuss some relevant research from this field in the present chapter. Before doing so, some other introductory points will be made.

Conservation psychology differs from the discipline of *ecopsychology*, a theoretical perspective which emerged in the early 1990s.[18] Ecopsychology is essentially a philosophical psychological movement, inspired by various religious traditions such as Buddhism and environmental philosophical writings critical of the discontinuity between humans and nature that characterizes the modern world.[19] The psychological traditions drawn upon by ecopsychology include the work of Carl Jung (1875-1961), Gestalt psychology and various transpersonal psychologies.[20] While it would make an interesting discussion to examine the origins and philosophical significance of this movement, little has been written on the specific problem of global climatic change from this perspective, for the concerns of the movement are about more general aspects of the human condition and the environmental crisis, rather than specific empirically grounded problems.[21]

Another important field of study, overlapping with conservation psychology is that of sustainable consumption studies.[22] Sustainable consumption studies begins with a recognition that consumption patterns characterizing modern Western societies are ecologically unsustainable and that the planet cannot support all the people living at the present Western level of consumption. Further, these consumption levels are unjust and immoral as consumption patterns are in Western societies often based on non-essential and trivial items (vanity and status goods), whilst the poorest people of the world lack access to the bare necessities of life, such as basic healthcare.[23] The idea of sustainable consumption featured in chapter 4 of *Agenda* 21, the main policy document of the Rio Earth Summit in 1972, where it was recognized that one of the major causes of the present environmental crisis was unsustainable patterns of consumption, particularly in Western countries.[24] More recently, the UK government has been one of the first governments to launch a national strategy on sustainable consumption and production, and a number of publications have been produced by the UK Department of Trade and Industry[25] and the Department for Environment, Food and Rural Affairs.[26] Tony Blair, British Prime Minister in 2006, said in this context that lifestyle changes are an important aspect of dealing with global climatic change: "Making the shift to a sustainable lifestyle is one of the most important challenges for the 21st century. The reality of climate change brings home to us the consequence of not facing up to these challenges".[27]

The UK government's approach to sustainable consumption will be examined in the next section. For the remainder of this section we will look in more detail at some of the key publications in the field of conservation psychology and examine the merits of these approaches in addressing the problem of social responsibility, especially within the context of dealing with global climatic change.

Unlike other disciplines such as mathematics and physics, psychology lacks a unified theoretical structure and consists of a plurality of paradigms and perspectives.[28] Even for an area such as behavior change theories and models, there are a multiplicity of perspectives such as learning theories, relapse prevention models, social support models, health belief models, reasoned action and planned behavior approaches, ecological approaches, transtheoretical models, and social learning/social cognitive theory approaches, to name a few.[29] Some of these models have a specific area of application and do not readily aid us in areas outside of the standard usage area. For example, the health belief model[30] is primarily concerned with a person's health-related behavior and a person's perception of the severity of a potential illness, the susceptibility to an illness, benefits of taking preventive action, and the barriers to action. This does not readily generate insights directly relevant to human behavior bearing upon environmental and climate change issues, and of course, the model was not originally devised to do so.

Another model, the transtheoretical model or "stages of change approach"[31] has been said to have relevance to questions of changes in human behavior related to conservation issues.[32] The transtheoretical model has been applied to various health areas such as smoking cessation and alcohol abuse.[33] Behavior change is seen as a gradual process involving five stages. These stages are (1) the precontemplation stage: (2) the contemplation stage; (3) the preparation stage; (4) the action stage and (5) the maintenance and relapse prevention stage. These stages are exactly as their names suggest. In the precontemplation stage people do not consider changing and may be uninformed or underinformed about the risk behaviors. During the contemplation stage, people balance up the pros and cons of change and have an ambivalent attitude about changing their behavior. They procrastinate and may stay in this stage for a considerable period of time. In the preparation stage people prepare to make change in the immediate future and normally have formulated a plan of action. In the action stage people attempt to make changes to their lifestyles which if maintained are sufficient to reduce the risks posed by the initial behavior. Thus, for example, in relation to smoking reduction, the person has abstained from smoking cigarettes. Maintenance is then necessary because a relapse into the risk behavior is always possible. Over time successful maintenance has a positive reinforcing effect and people become less tempted to relapse. Ultimately the risk behavior can be eliminated. There are also processes of change, activities which people use as they go through the stages. These include consciousness raising (becoming aware); emotional arousal (reacting to information on an emotional level); social appraisal (reevaluating past behavior); social liberation (recognizing the benefits of behavioral change) and self reappraisal and reevaluation.

An example of the transtheoretical model - which can be applied to both individuals and populations[34]- in an environmental context, would be as follows.[35] In the pre-contemplation stage the person does not know or consider adopting ecologically sustainable behavior, such as sustainable consumption patterns. The contemplation stage sees the person thinking about these issues and considering adopting such behaviors. Environmentally sustainable behavior is adopted in the action stage. In the maintenance stage environmentally sustainable becomes a matter of custom, and the person may encourage others to change their behaviour.

The Prochaska-DiClemente transtheoretical model provides an excellent general schema for understanding all types of changes in human behavior, but the model is a very general one by nature of being "transtheoretical". Alone it does not aid us with strategies in dealing with the problem of social responsibility. It is a skeleton requiring covering by the flesh of more specific psychological theorization.

Social Psychologists Susan Clayton and Amara Brook, who have published a paper entitled "Can Psychology Help Save the World? A Model for Conservation Psychology", [36] argue that social psychology can be a great aid in addressing environmental problems. Their social psychological model of conservation behavior sees human behavior as a function of contexts, past experiences and knowledge as well as motivations such as control and belonging.[37] A person's past experiences, knowledge and motivations may change a person's interpretation of context, thus altering behavior. Thus the model directs us towards examining situational context in understanding a person's behavior, rather than narrowly and exclusively focusing upon the person's motivations, beliefs and dispositions on behavior. The situational context includes the social and physical environment, especially the behavior of and interactions with, other people.[38] An individual's interpretation of the situational context is also influenced by past experience and knowledge, as well as their basic motives. These factors all go together to form the basis of a person's identity, which is in turn a substantial determinant of behavior.[39] If people adopt an environmentalist self-identity, rather than a self-identity based upon consumer values such as acquisition, affluence and self-esteem, they are more likely to engage in environmentally sustainable behavior such as the adoption of sustainable consumption patterns.[40]

Clayton and Brook illustrate the applicability of their social psychological model to conservation issues by considering some concrete examples. Their first example is most relevant to our concerns: the apparent conflict between people's stated beliefs of attributing a high value to the natural environment, and their environmentally destructive behavior.[41] A conflict often occurs between farmers, such as ranchers, who may have endangered species on their land, and the environmentalists who wish to conserve them. Clayton and Brook direct us toward understanding the context of this conflict, such as the financial situation of the farmers and that farmers have often been characterized by environmentalists as environmentally destructive. Preexisting beliefs also need to be understood; such as whether the endangered species from the farmer's perspective is perceived to be a threat to his/her livelihood. Personal motives, especially affiliation motivations need to be understood, as farmers may feel marginalized and distrusting of those seeking to protect the endangered species. A solution, is to attempt to mediate these conflicts and try to construct some wider framework where other farmers and conservationists can feel a part of, rather than opposed to. The aim is also to seek common goals whilst preserving sub-group identities.[42]

We agree that Clayton and Brook's social psychology model is a valuable contribution to conservation psychology - applied to specific conservation problems. It is more difficult to see its applicability in fine detail to the problem of social responsibility discussed here, with reference to the issue of global climatic change. The problem is, of course, many levels of difficulty greater than the conflict between say farmers and conservationists, because necessarily we need to examine the situational context of the entire operation of modern society. The model suggests that the situation context- that of the globalized consumer-capitalistic society - needs to be changed in ecologically sustainable ways so that appropriate change in individual behavior occurs. Again, we agree, but it seems that we are now in a vicious cycle because the *initial* problem which we wished to address was how to change modern society in sustainable ways. We had hoped that psychology would shed an answer to this, not lead us back to the fundamental social problem of responsibility.

An important contribution to the psychological response to the environmental crisis has been made by Deborah Winter and Susan Koger in their book, *The Psychology of Environmental Problems*.[43] This book applies existing psychological theories and research to the domain of existing environmental problems. The authors both recognize that present day industrial and consumption patterns in the developed world are unsustainable. Environmental problems have a psychological basis because these problems have resulted because of the thoughts, beliefs, values and actions of human agents. Consequently, to solve environmental problems humanity cannot rely solely upon scientific/technological fixes or legal intervention. Winter and Koger examine six major psychological approaches - social psychology, behavioral psychology, physiological and health psychology, cognitive psychology and the holistic approaches of gestalt and ecopsychology. Reasonably enough, they see merits and limits in all of these approaches. It is problematic to attempt to choose a "best theory" because human behavior is complex and multidimensional. Hallin showed with research on household related behavior that there are many factors leading to environmentally different behaviors such as past experiences, altruism, convenience, rewards and role models.[44]

Winter and Koger believe that an attempt to work out which psychological model (s) is best takes the focus away from finding solutions to environmental problems. Further, as people are already holistic beings, it may not matter if the initial focus is upon physiology, thinking, behavior or information, because in a mutually interacting system, a change in one element will have some impact upon the others.[45] Consequently, they put forward six operating principles: (1) visualize an ecologically sustainable world; (2) work with big ideas and small steps; (3) think circle instead of line; (4) consider ways in which less is more; (5) practice conscious consumption and (6) act on the personal and political levels, in particular, community participation. These measures are fine as rules of thumb for environmental activism and embody a commonsense approach to action for environmental change. Clearly if one wishes to change human behavior for an environmental purpose one needs an idea of what the end goal is, an ecologically healthy world. This is a "big idea" that has to be achieved by small steps because "Rome was not built in a day". Considering ways in which less is more is practicing sustainable consumption, which to be effective needs to be conscious and practical. However, Winter and Koger note that "in order for psychology to make a viable contribution to building a sustainable world, it must also be practiced in political contexts to change the structural (economic, legal and political) dimensions of environmental decline".[46] Thus in conclusion, it seems that the problem of social responsibility with respect to the environmental crisis can only effectively be addressed by psychology if the larger politico-social issues of the sustainability of modern society are addressed. This however puts us in the vicious circle, noted earlier.

**Sustainable Consumption: Escaping the Vicious Circle?**

Important and substantial work has been done by the UK Department of the Environment, Food and Rural Affairs (Defra), on sustainable consumption and sustainable development.[47] This body of work is part of the UK government sustainable development strategy.[48] It would take a lengthy book to summarize and discuss all of this research, so what will be most useful for this chapter will be to summarize the most fundamental aspects of the UK government approach, followed by a consideration of the psychological and sociological research upon which it is based.

The UK government sought an alternative to classical command and control regulation to influence human behavior such as using economic instruments such as taxes and trading schemes, because regulation alone is insufficient to achieve sustainable development. The very broad theoretical model sketched in a number of Defra publications can be called the "Defra diamond", based upon the concepts of *Enable*, *Encourage* and *Engage*. Figure 4.1 reproduces the standard "Defra Diamond".[49]

**Figure 4.1**

**The new approach – a diagrammatic representation**

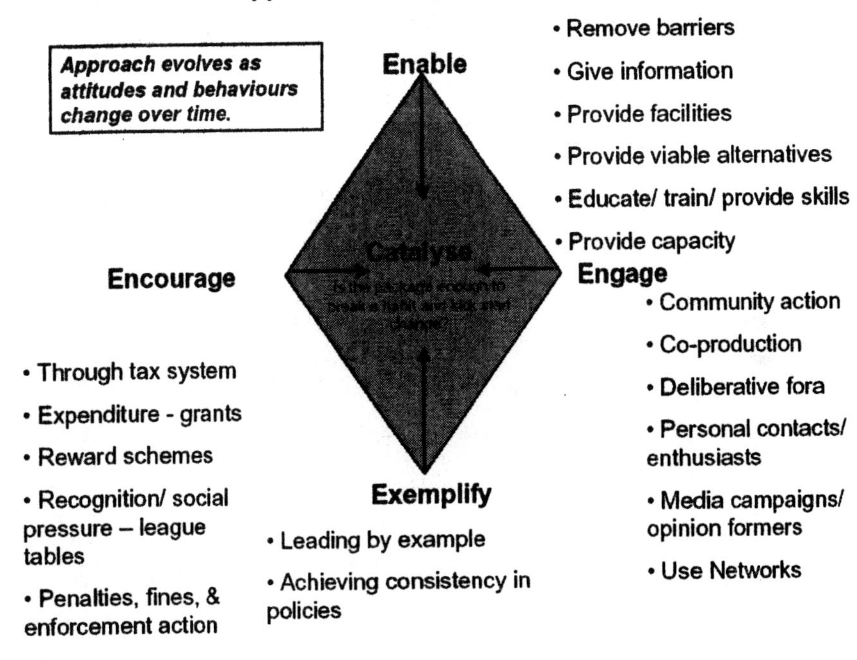


This approach is logical and based upon common sense psychology. First is the *Enable* step, enabling people to make responsible choices by providing them with education, skills and information. The *Encourage* step involves governments aiding, or if necessary enforcing, behavior change by various measures. The measures listed in Figure 4.1 are not the only possible measures that a government could use. For example, indirect measures such as increasing community cohesion, race equality and intergenerational justice, revitalizing neighborhoods and supporting young people's development all have a part to play. These measures strengthen the social capital of communities and essentially revitalize communities. The UK government's *Community Action 2020 - Together We Can*

program, is an attempt to tackle the environmental crisis as part of a process of revitalizing local communities.[50] This action plan can be summarized by another "Defra diamond", given in Figure 4.2.[51]

**Figure 4.2**

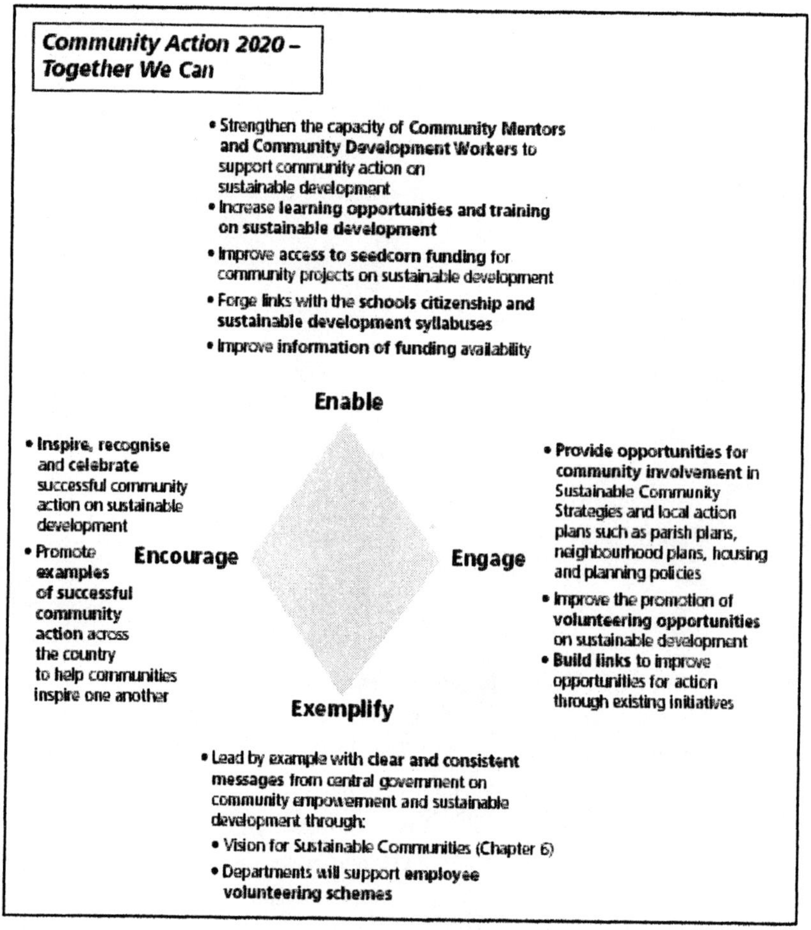

***Community Action 2020 – Together We Can***

- Strengthen the capacity of **Community Mentors and Community Development Workers** to support community action on sustainable development
- Increase **learning opportunities and training on sustainable development**
- Improve **access to seedcorn funding** for community projects on sustainable development
- Forge links with the **schools citizenship and sustainable development syllabuses**
- Improve **information of funding** availability

**Enable**

**Encourage**

- **Inspire, recognise and celebrate** successful community action on sustainable development
- Promote **examples of successful community action** across the country to help communities inspire one another

**Engage**

- **Provide opportunities for community involvement** in Sustainable Community Strategies and local action plans such as parish plans, neighbourhood plans, housing and planning policies
- Improve the promotion of **volunteering opportunities** on sustainable development
- **Build links** to improve opportunities for action through existing initiatives

**Exemplify**

- Lead by example with **clear and consistent messages** from central government on community empowerment and sustainable development through:
  - Vision for Sustainable Communities (Chapter 6)
  - Departments will support **employee volunteering schemes**

People are encouraged to take responsibility for their actions, to *engage* and to become involved in jointly developing policies with the government, a process known as co-production.[52] Deliberate forums involving face-to-face contact are one suggested avenue of communication, however, Tim Jackson's research[53] has revealed that information provision alone is insufficient to change people's behaviour, so that targeted communications must be part of a bigger process of coordinated interventions including regulation and the supply of appropriate goods, services and infrastructure.[54] People are often locked into unsustainable consumption patterns, and according to the report of the Sustainable Consumption Roundtable, *I Will If You Will*[55] (a roundtable funded by Defra and the UK Department of Trade and Industry), the "evidence based on consumer behaviour suggests that, often, we will need to have our unconscious routines shaken up before we can see the value in forming new ones".[56] The best method for this "is to drop new tangible solutions into people's daily lives, catalysts that will send ripples, get them talking, sweep them up into a new set of social norms, and open up the possibility of wider changes in outlook and behavior".[57] In particular, although the consumer economy is an inescapable part of capitalism, there is a pressing need to keep the economy going by decoupling economic consumption from material consumption: "if people purchased high-value services instead of resource-intensive artefacts, if consumer commodities became value heavy and materially light, then we could preserve economic stability and still meet environmental and social targets".[58] This, of course, is not a particularly easy thing to achieve and will require major economic and technological changes.[59]

The nature of radical social change, necessary to build a sustainable society is discussed in the Defra Behaviour Change Practical Guide, *Triggering Widespread Adoption of Sustainable Behaviour*.[60] Various models of change in large, open complex systems, such as societies, indicates that there is inherent uncertainty about change, with changes often rapidly occurring once some "critical mass" is reached.[61] Pro-environmental change is likely to take place among groups in society that are network-based because networks facilitate behaviour change more readily than an individual-to-individual approach. For governments "policy should be focusing on groups whose network properties best lend themselves to the diffusion of change, in order to boost the probability of success".[62] Thus because of factors of inherent uncertainty, networks, and the need for broad targeting, policies to promote pro-environmental behaviour cannot assume a simple linear connection between intervention and outcomes:

> Instead, the radical idea presents itself that *policy should be attempting perpetually 'seed' or catalyse* change, through a wide variety of mechanisms, in a wide variety of places. A range of fundamental features of the social system mean that a model of policy intervention predicated on the steady refinement of interventions towards a set of policies that 'work' may be ill-founded. Rather, given the complexities of 'behaviour change', a model of ceaseless innovation, within broad parameters of focus and in a network setting, offers a potentially valuable conceptualisation of how to move forward.[63]

A number of UK government sponsored publications have given surveys of the relevance of psychological theories to issues of promoting pro-environmental behavior change and have generally concluded that there are merits and limits to all such approaches.[64] One of the major reviews of evidence on consumer behavior change was done by Professor Tim Jackson of the University of Surrey, *Motivating Sustainable Consumption.*[65] Jackson sees consumer behaviour as the key to controlling the impacts of society on the environment, thus making consumption the "vanguard of history".[66] Consumption, apart from satisfying physical needs of food, shelter and safety also has an important role in identity formation and identification and the establishment of social meaning.[67] The rational choice model of consumer behaviour - which sees consumers as rational utility maximizers (or optimizers) calculating the costs and benefits of various actions - fails for many reasons, including that social factors shape individual preferences. Indeed, our concepts of self may be socially constructed, so that individual change itself is insufficient and infeasible.[68] People, Jackson argues, are often locked into unsustainable consumption patterns for a variety of social and institutional reasons.

The social construction of human behavior and social learning theory indicates, Jackson believes, that governments have a key role in supplying leadership on sustainable consumption issues. However, available psychological evidence indicates, Jackson says, that information intensive interventions as well as taxes and incentive schemes, have had limited success in triggering pro-environmental sustainable consumption behaviors.[69] The sheer complexity of human behaviour, the multitudes of interactions between individuals, groups and institutions, makes change of human behavior enormously difficult.[70] As Jackson puts it in summary, examining "consumer behaviour through a social and psychological lens reveals a complex and outwardly hostile landscape that appears to defy conventional policy intervention".[71] According to social-psychological theory

> Consumer behaviours and motivations are complex and deeply entrenched in conventions and institutions. Social norms and expectations appear to follow their own evolutionary logics, immune to individual control. Social learning is powerful but not particularly malleable. Persuasion is confounded by the information density of modern society. The rhetoric of consumer sovereignty is inaccurate and unhelpful here because it regards choice as entirely individualistic and because it fails to unravel the social and psychological influences on people's behaviour. But short of mandating particular behaviours and prohibiting others - an avenue that Government has been reluctant to pursue - it is difficult at first to see what progress can be made in this intractable terrain.[72]

Furthermore "the forms of governance familiar to the individualistic/entrepreneurial society are never, by themselves, going to be sufficient to achieve the kind of behavioural change demanded by sustainable development".[73] Pro-environmental behavior change will require a combination of government laws, regulations and interventions, education programs, small group and community management, as well as moral, religious and ethical appeals.[74] Integrating these strategies into a coherent framework remains an outstanding creative problem. Nevertheless, Jackson believes that the social-psychological evidence "offers a far more creative vista for policy innovation than has hitherto been recognized".[75] As he notes:

> Government policies send important signals to consumers about institutional goals and national priorities. They indicate in sometimes subtle but very powerful ways the kinds of behaviours that are rewarded in society, the kinds of attitudes that are valued, the goals and aspirations that are regarded as appropriate, what success means and the worldview under which consumers are expected to act. Policy signals have a major influence on social norms, ethical codes and cultural expectations.[76]

Importantly, governments can "lead by example" in environmental management initiatives and sustainable procurement programs.

These conclusions by Professor Jackson are generally agreed with by the academic authors of another Defra report *Promoting Pro-Environmental Behaviour*, undertaken by Andrew Darnton (*et al.*) of the Centre for Sustainable Development, University of Westminster, London.[77] Individuals, it is concluded, have the capacity to change, but usually changes occur through key individuals or "change champions" and "Engaging and nurturing key individuals may be more effective in bringing about system-wide change than targeting the behaviour of all individuals.[78] However social change is a complex process with interactive loops, such as feedback loops, double loop learning, non-linear effects and cycles of action and reflection.[79] Due to the complex interactions between elements of the social system, Darnton (*et al.*) adopt a systems theory approach, viewing organizations and networks as complex adaptive systems. Change processes need to occur horizontally, within organizations and vertically, between organizations.[80] Thus, the "complex inter-relations between these multiple factors and organisational levels support the argument for a 'whole systems' approach to encouraging behaviour change".[81] However a caveat is made: "The review of 'real world' initiatives suggests that it may be easier to replicate the smaller and less complex projects than those that are attempting whole system change".[82] Consequently, government interventions need to address multiple social levels, and ultimately address "society as a whole in order to achieve long-term normative change".[83] This means that policy-makers aiming to trigger pro-environment behaviour will need to have strategies directed towards, individual, organizational/group and the systemic/societal levels.

There exists a substantial body of research dealing with the core principals underlying pro-environment behaviour interventions among individuals.[84] However these small incremental changes may get diluted in complex systems and ultimately nullified unless whole systems change occurs.[85] From a systems perspective "change programmes involving whole systems invariably occur over long timescales".[86] Further, Darnton (*et al.*), although presenting evidence from a number of case studies of pro-environmental behavior change at the individual and group level[87] (such as West Sussex Real Nappy (diaper) Initiative, a program to increase the use of non-disposal nappies[88]) conclude:

> Unfortunately, there is insufficient evidence from this study to properly judge the effectiveness of the different policy approaches in provoking pro-environmental behaviour change, the level and extent of that change or the adequacy of that change in light of global targets. The limited evidence that is available suggests that it will be easier to replicate the more straightforward projects that the more complex "whole systems' approaches.[89]

**Lessons and Limits**

The lesson to be drawn from the work surveyed in this chapter is that a change in pro-environmental behavior sufficient to address the crisis of civilization described in chapter 2 of this book will require a "whole system" change. This conclusion was reached in our consideration of legal, economic and technological approaches considered in the last chapter. It is further confirmed in this chapter, through a consideration of the limited, but growing literature of conservation psychology and sustainable consumption. However, this raises a fundamental socio-political paradox: micro-social pro-environmental behavior change ultimately requires whole-of-system change, but whole-of-system change requires that complex systems be broken down into components. But those component sub-systems, networks, groups and individuals, can only effectively change if there is at least substantial whole-of-system change. A small group of socially responsible individuals may adopt sustainable behaviors, but their efforts may be "washed away" unless larger-scale social change occurs. Darnton has found in a study of the impact of sustainable development on public behavior in the UK that not only is public awareness and understanding of sustainable development low, but that people mistrust government information, believe the government is damaging the environment and that they are already doing all that they can to "save the environment".[90] It is difficult to be optimistic that societies will respond adequately to the environmental crisis that they face.

The epistemological vicious circle described above needs to be broken. The way which two of the present authors suggest that this could be done in *The Climate Change Challenge and the Failure of Democracy* [91] is for a special group of intellectuals, an ecological vanguard, to work at producing changes in the social system, groups and individuals. At present the environmental movement itself is a loosely structured attempt to do this. However focus needs to be made by these elites upon changing the "nervous center" of society, such as the universities and institutions of knowledge, as well as changes to various cognitive enterprises such as philosophy. This approach will be explored in the rest of this book.

# Chapter 5

## Wisdom Regained: Universities and the Institutions of Knowledge

> The manufacture of consent...was supposed to have died out with the appearance of democracy...but it has not died out. It has in fact, improved enormously in technique...under the impact of propaganda, it is no longer plausible to believe in the original dogma of democracy.[1]

### Introduction : The Problem of the University

Yesteryear universities met society's need for religious studies. Today they service society's needs as perceived by government and industry. These are predominantly the production of graduates to service the 'three C society', Capitalism, Competition and Consumerism. Should not the role of the universities be to offer us a vision of an alternative society based on peace, world health and environmental redemption?

The number of voices expressing concern about this state of affairs and the need to modify it increases daily. Some of us believe that the present economic system has to be changed because environmental damage is central to its successful operation.[2] The major theme of this book is the use of knowledge, intellect and leadership to prevent environmental catastrophe and to further the health and well-being of all society. This chapter will examine the role of universities in responding to the crisis of civilization. The second section, "The University Under Siege", will focus upon the present existential problems of the modern university and its resulting identity crisis, born from its present service to corporate and capitalist interests. The third section "The Idea of a University", will defend the necessity of a "liberal" university, but one that is ecologically informed. The fourth and fifth sections will outline in more detail what universities will need to do to contribute towards solving the environmental and socio-political challenges of our time. The fourth section, "Science-Saviour or Desecrator?" discusses the role of science in the promotion of a sustainable environment and the fifth section, "Environmental Education and the Life Support Systems of the Earth", the importance of an environmental focus as the nucleus of a contemporary liberal education. Existing universities may be so tight in the corporate grip that these types of changes may be no longer possible. If this is so, then new institutions, "Real Universities" will need to replace them by something of a process of natural selection.

The pursuit of the fundamental needs of society, environmental sustainability, health, peace and the eradication of poverty requires freedom of speech and expression in the face of the vested economic interests of corporatism, governments and often the media empires. In recent times, there has been criticism of academics for failing to speak out on important issues [3]

Is the university an institution capable of bringing about change within itself to contribute significantly to issues fundamental to the world? Ability to change requires insight, ability, resolve and open minds. All these attributes would be expected in those committed to academic research and teaching. Unfortunately, events over the past few decades and decisions within many universities are a bad omen for any meaningful contribution to the pursuit of environmental protection, peace and world health. For these have the potential to become contentious political issues that often sit uncomfortably with current economic ethos. For their progress they require protection, tolerance and freedom of debate in a climate of fair and caring management of staff resources. It is sad but necessary, therefore, to document the demise of university presses, the suppression of academic free speech by university administrations, the corruption of ideals and principles and the imbalance of education fomented by corporate gifts and sponsorship. Perhaps some of the foaming torrent of criticism presently directed at universities is related to these happenings?

**The University under Siege**

Criticism of universities comes from many directions. Governments sometimes blame universities indirectly for unemployment, saying that students cannot write or communicate. They complain that the universities are inefficient and do not produce graduates who can read or write. As a result universities are "reformed" repeatedly. Nor is industry and commerce satisfied, for they believe that the university should be more responsive to producing graduates who fulfill their immediate needs and skills for the job market. The press has frequent articles from vice-chancellors or members of their committees indicating the consequences of inadequate funding for their so-called "world class" status. The press, which is usually in tune with whichever government is carrying out incessant reform, preaches and exhorts "excellence", responsibilities and the global market.[4] Relatively silent are the students, who understandably worry about future jobs in a competitive world with high unemployment, and the academics who fear for their jobs and are too busy to say much anyway. The turmoil induced by criticism and change is perhaps greatest in Australia because financial constraints, administrative impositions and corporatization are incessant and there is no end in sight. Concomitantly, excellence has become a spin term that conceals inadequacy. Indeed it is hard to find the term on the home pages of Harvard or Yale. In Australia attempts are made to brand the university for marketing purposes with the use of slogans and images. There is a concentration on marketing rather than true contribution to humanity.[5]

In some other countries, for example the UK, universities have retained greater independence and funding has improved during the period of Blair government. In the US, there is opulence in the prestigious Ivy League universities and those funded by corporations, but poverty in the majority of others.[6]

Widespread public concern that the universities in many Western countries are functioning poorly is understandable for there is considerable evidence to that effect. Many books have been written detailing inequity, mal-administration, inadequate financing and teaching.[7] Universities as public institutions have always attracted criticism. Often it reflects much more than the discontent with change that occurs with successive generations of aging academics. Now however, it does not relate to denied utopian dreams of contribution to world happiness. It comes from the political left that believes universities have been taken over by economic rationalists and that university education is no longer equitable.[8] The university is under real threat as a centre for reflection and thought about society and the world. It has become merely a professional or vocational training centre. There is also vociferous criticism from the political right that regards much of university thinking as "leftist".[9] This is somewhat unexpected in view of the increasing cohabitation of the universities with national economic agenda. Consider for example these comments by British conservative journalist Paul Johnson:

> Universities are the most overrated institutions of our age. Of all the calamities that have befallen the 20th century, apart from the two world wars, the expansion of higher education in the 1950s and 1960s was the most enduring. It is a myth that universities are nurseries of reason. They are hot houses for every kind of extremism, irrationality, intolerance and prejudice, where intellectual and social snobbery is almost deliberately installed and where dons attempt to pass on to their students their own sins of pride. The wonder is so that so many people emerge from these dens still employable, although a significant minority -- as we have learned to our cost - goes forth well equipped for a lifetime of public mischief making![10]

Johnson, like many contemporary American conservative critics of the university, believes that the barbarians are already inside university gates, and those barbarians are the hard left radicals eager to push for the politically correct line.

For the editor of *Quadrant*, P. P. McGuinness, Australian universities are in decline and the idea of universities as places for a disinterested pursuit of knowledge, together with the idea of the university as advocated by John Henry Cardinal Newman, has long faded:

> [It] is probably honoured more in the breach than the observance by academics in the humanities and related areas these days, treated with contempt by the newer manageralists...dismissed as inimical to good

> administration and accountability by the bureaucrats who oversee funding, and it is totally foreign to all but one in a thousand of the student body.[11]

McGuinness has a low opinion of the humanities in Australian universities where the standard has always been "mediocre", lacking the first rate minds of the sciences. He has detailed why he believes this is so: that the pursuit of leftist thought, such as environmentalism, has scrambled the free-thinking ability of Australian academics and that the hope for free thought in Australia lies with independent thinkers who exist outside the university system. One can agree with this latter thought, but to equate environmental concern with "leftism" epitomizes the difficulty of addressing the environmental crisis in the present culture of economic rationalism.

The conservative Australian magazine *Quadrant*, has printed articles that attribute the demise of the universities to the political left. A satirical article by Andrew Irvine is a fictional correspondence between Gottlob Frege, (1848-1925), a professor of mathematics at the University of Jena, Germany, and the administration of the university.[12] Frege's work laid the foundations for the development of modern mathematical logic, yet, on today's performance criteria he would be made redundant. Irvine begins his article with the sentence: "What would have happened had the first wave of political correctness swept through the universities almost a century ago?"[13]

The answer seems to be that geniuses would have been sacked and scientific progress would not have occurred. The subtext of this article is that the left is guilty for suppressing intellectual progress. It is difficult to understand this conclusion. Certainly it is likely that Frege would be sacked today, but surely the most likely reason would relate to his failure to produce sufficient publications to prevent a reduction in departmental funding. Today such funding rates ten insignificant publications as ten times more worthy than one earth-shattering discovery described in one publication! Furthermore today, an important thrust of the university is to attract finance, and Frege's lack of student enrollments would further prejudice his retention. Why then does the left have to bear sole responsibility?

Australian universities, of all those in the West, have been subject to the most government control in order to influence research and opinion. In the 1940's Australia had an independent Australian Universities Commission to maintain full and free independence of university functions. These freedoms were gradually eroded so that within the past few years research grants awarded by independent panels have been struck out by ministers on the political advice of lay persons. This would seem to be beyond the bounds of incredulity so one case will be recounted.[14] The Australian Research Council (ARC) assesses applications for funding in the social sciences and humanities and the reports of its independent assessors are sent to the Minister of Education for approval. The process of assessment is diligent and adheres to international best practice. A right wing newspaper columnist attacked the award of certain grants and claimed that the ARC had fallen into the hands of "a club of scratch-my-back lefties" whose work was "hostile to our culture, history and institutions". The Minister was driven to veto some grants in 2004. Possibly impressed by his own power, in 2005 the Minister appointed lay persons to a grant-giving panel, they were a High Court judge, the editor (McGuinness, mentioned above) of *Quadrant*, the right wing magazine, and a television newsreader. The board of the ARC was abolished so that the Minister had unfettered control. McGuinness was asked to review applicants on the basis of the title and abstract of their research but insisted that he receive the full application and was qualified to judge them because of his knowledge of economics. Some applications approved by the ARC were cancelled, the Minister indicating that they were not in the national interest.

Fortunately this series of events did not proceed without protest from senior university figures, the Vice-Chancellor of Sydney University, and the Australian Vice Chancellors committee. The Minister's scheme was undermined by the possibility that academics would refuse to serve on ARC committees. Other examples of direct manipulation of research funding by the Australian government have been detailed.[15]

In general it is the academics in the humanities and social sciences that are attacked because their research suggests political solutions. It is often easier to denigrate the academic than to refute their research or teaching. This phenomenon is on public display in the United States. In his book *The Professors: The 101 most Dangerous Academics in America,*[16] David Horowitz identifies and criticizes academics who he asserts are anti-American, anti-semitic and left wing. The assertions were seen as McCarthy-like by "Free Exchange on Campus" a wide coalition of university organizations.[17] Professor Mark Levine, tenured Professor of Middle Eastern history at the University of California Irvine, one of the 101 "Most Dangerous Academics", whose academic record requires no defense, responds to Horowitz: "*The Professors,* and the kind of political and cultural discourse it represents, are dangerous to the functioning and purpose of the university, and to the larger notion of both free speech and civil debate that have long been cornerstones of American higher education, and through it, culture".[18]

Horowitz and his like present a threat to the untenured academic, an encouragement to silence. Who are "the like?" To the authors of this book who have looked from afar at this controversy, they are extreme right wing enthusiasts who carry their beliefs aggressively into all sections of American society. One example is Campus Watch who's role is defined as follows: "Campus watch will henceforth monitor and gather information on professors who fan the flames of disinformation, incitement and ignorance. Campus watch will critique these specialists, and make available its findings on the internet and in the media".[19] The foot soldiers of campus watch are the students. No-one would deny that feedback on the teaching abilities of academic staff is an essential part of the normal functioning of a university but is a student able to "identify key faculty who teach and write about contemporary affairs at university Middle East studies departments in order to analyse the critique of these specialists for errors or biases". [20]

Emeritus Professor Charles Birch believes that Australian universities have become elitist factories.[21] Students flock to courses which they believe will give them highly paid jobs. Birch, in contrast to most critics, would wish to see universities create interdisciplinary courses that address the problems of our age and which attempt to understand the world in a comprehensive and holistic way. Central to his vision is the rehabilitation of philosophy. These views are in tune with many of those presented in this book.

It would be possible to add even more criticisms, mostly justified, about the performance of universities. In Australia they are well documented in *Degrees Galore* by Frank Crowley, self-published because today's publishers mostly partner the ongoing economic ethos.[22] To them, his book was probably difficult to accept because it presented a blunt unpalatable truth. Perhaps it may have sullied their commercial relationship to universities. But the plain facts presented by Crowley lead us to acknowledge that the university as we knew it has been reformed, or destroyed, according to one's political opinion, and there is probably no going back. So is the volume of criticism justified or fair? Is it possible to list some positive academic advances of the past decade as a counterbalance?

The publicly funded university system that was irrevocably changed was certainly more distant from government control, much smaller and more inefficient than today. There was room for reform and in particular for the extension of higher education to many more students. But the expansion in student numbers has been accompanied by a disproportionate increase in the size of the bureaucracy and a relative reduction in the number of teachers. Academic staff are disillusioned, stressed and have a low morale. There is student overcrowding, financial stress and poverty. The only true positive is the increase in the number of students being educated. Others find positives in the successful gearing of research and teaching to economic output. Whilst this is important, it has become so dominant that the deployment of effort and resources to the more basic needs of humanity have been eroded.

While all these problems merit further discussion, their cause can be traced to a change in values. Of greater concern is the loss of freedom to communicate for this stifles a vital contribution to society by the universities. Worldwide, many universities established their own publication presses for the publication of books and monographs. It was reasoned that a university must be able to publish important material without having to suffer delays and rejections by commercial publishers. The publications by university presses were therefore subsidised.

In general, commercial publishers are beholden to the profit motive dictated by sales, and their book lists are dominated by novels, biography of the glitterati, sports persons and the powerful, and a scattering of newsworthy, extreme, unusual and New Age thinkers who can titillate the mind and sales. They are part of the market system ideology. It is possible to see how a book that could change the world might remain unpublished. With a few notable exceptions, it is probably fair to say that it is difficult to find a serious writer of non-fiction who has anything positive to say privately about publishers, for what they feel are unpredictable selections, delays in publication and poor distribution.

Yet today virtually all one hundred university presses in the USA and those in the UK and Australia operate as commercial publishers.[23] The demise of their non-commercial functions commenced within their own universities and was partially self-induced. They published scholarly work for a limited audience, usually to satisfy local productivity agenda and curriculum vitae. Often the material was highly specialised and had a scholarly readership of a few dozen copies.[24] When the universities became more financially accountable and rationalist, an important mechanism for publishing original work was lost. The situation has deteriorated further with contracting university budgets and with the rejection of material that constructively criticizes the financial ethos of the university system. If the universities are to reclaim their public voice, it will be necessary to revive an independent, non-commercial press.

Democratic systems have been characterized by freedom of speech with intellectuals able to make public comment and criticism. This freedom is under siege in academia. Indeed, it is irrational for politicians and ministers to make statements about the decline of university stimulated debate on issues of public importance, when government policy has gradually, though perhaps inadvertently, encouraged suppression of debate. In Australia, the eminent former politician and thinker Barry Jones has deplored the decline of the public intellectual.[25] He attributes this to super-specialization so that the academic finds it difficult to comment on issues outside his or her strict domain. There may be some truth in this, especially in science, but there are many other more cogent reasons for the decline. The most obvious are the overwhelming teaching and administrative duties resulting from more students and fewer staff, and the reluctance of the press and media to voice viewpoints that offer radical alternatives to the functioning of modern society. The world issues on which we should be expressing a viewpoint, poverty, global climatic change, environmental destruction and international conflict are all potentially political and an academic statement is therefore often regarded as political. Consequently the university may be subjected to pressure from government or its corporate friends. For example, critiques on genetically modified foods fall into this category. Often the university's response is to curtail comment.

Intellectuals should be leading or participating in debate but if they do, they are vulnerable and may be victimized by their own universities. The following examples occurred in Australasia, but there are examples from many other countries.

The new managerial systems in universities are averse to criticism from academic staff. The present system is increasingly modeled on industry and commerce where freedom of speech is severely limited. It is regarded as disloyal to make constructive criticism. A university staff member who was an office bearer in the university staff union, alerted members of the union to the lease by the university of an expensive corporate box at football matches.[26] The university intended to entertain potential sponsors in the hope of increasing funding for the university. The staff member had concerns about this expenditure, felt a duty to inform the membership and hoped for an informed debate on the issue. The university disconnected the staff member's email and telephone.

In another instance, a distinguished staff member held a workshop on the subject *"Why do universities matter?"* The contributions were integrated into a book that was accepted for publication by Melbourne University Press, but was subsequently rejected after intervention by the university administration.[27] The reason given for the rejection was that the book presented a 'traditional' view of the university. Presumably, university administration felt that it had moved away from the traditional university model and did not wish such views expressed. This was censorship. Clearly the book was worthy of publication for a well-known publisher then accepted it.

This and other examples illustrate the university view that staff should accept the present system, they should work within it and say little. On this basis, the universities have metamorphosed from an independent "traditional status" to conform with corporate practices. Good practice in universities surely means the delivery of educational outcomes of benefit to the student and the community together with a fulfilling occupational experience for employees of the university. This requires participation and debate in the development and managerial functions of a university and the provision of pastoral care for the needs of staff. Today it is difficult to find a university that sets such a good example to its students and the community. The so-called "excellence" that is trumpeted widely requires to be practiced internally. Unfortunately, in many universities, relationships between administration and staff are at best hopeless and at worst, appalling. The use of administrative power has suppressed important voices in community education and information.

The academic staff is also subject to outside pressures related to funding from governments and industry. The Hinchinbrook coastal resort development in northern Queensland Australia, adjacent to a World Heritage area, was controversial because in the view of many independent environmental scientists, it threatened the heritage values of the area. An academic in environmental law at James Cook University who was also president of a community environmental organization, provided legal advice to organizations opposing the Hinchinbrook development.[28] This academic was attacked by supporters of the development and was accused of using his university position inappropriately and they used freedom of information requests to obtain material to pursue him. The university supported his right to speak, but had concerns about his association with a community organisation. This displayed a conflict of interest by the university; it wished to be seen to encourage staff to establish community contacts, but presumably only non-contentious community contacts.

An analysis of instances of dismissal or suppression of an academic reveals that many occur because of their environmental advocacy. Brian Martin documents Australian examples in *Intellectual Suppression.*[29] In Adelaide, an attempt was made to dismiss Professor Manwell from his Chair after he wrote a letter to a newspaper criticizing some aspects of a fruit-fly spraying program of the Government Department of Agriculture. The process of dismissal was initiated by a complaint to the Vice Chancellor from another professor who had strong links with the government and its spraying program. In Canberra, tenure was denied to an academic in a Department of Human Sciences within a Centre for Resource and Environmental Studies. His "transgression" was to introduce interdisciplinary teaching that was seen to challenge the standard, rigid model for teaching. Sometimes the reprisal is petty and ridiculous; for example an academic philosopher who wrote a book on forestry was barred from using the university's Forestry Department library.[30]

Many of the cases described by Martin have in common hostility to environmentalism by the establishment and the exercise of power by scientists when their dominant paradigm is challenged. Local elites act to protect their power and status that is often linked to powerful outside interests. Examples of issues precipitating retribution are nuclear power, the use of pesticides, forestry practices and climate change science. The driving force for action may be the state, a profession or industry, and it is exerted through individuals in the bureaucracy or universities. The nuclear industry, a state creation, supports many scientists who become powerful and protect their position vigorously.[31] In the case of forestry, forest industries, governments and unions have worked together to utilize a cheap resource and preserve jobs with resulting damage to the environment.[32] A common series of events in the genesis of victimization is for a scientist to become involved with, or give advice to, a community group that is seen to oppose a vested interest. This gives credibility to the community group and the integrity of the scientist is attacked.

The influence of vested interests on freedom in the universities is not a sudden development in the age of economic rationalism. In the 1920's Veblen wrote about the effects of business within American universities.[33] Scholarly ideals were replaced by "the tastes and habits of practical affairs", and publicity and research results were influenced by sponsors. Business was influencing the universities in the same way that it influences all society. The corporate influence of today simply reflects its increasingly dominant influence in society and on government.

The suppression or dismissal of individuals have profound effects on most academics but the following quotation from C. Wright Mills written in 1963 before the recent curtailment of academic freedoms, illustrates that additional more subtle mechanisms operate:

> The deepest problem of freedom for teachers is not the occasional ousting of a professor but a vague general fear - sometimes known as 'discretion', 'good taste', or 'balanced judgment.' It is a fear which leads to self-discrimination and finally becomes so habitual that the scholar is unaware of it. The real restraints are not so much external prohibitions as control of the insurgents by the agreements of academic gentlemen.[34]

This mechanism reflects the operation of a university hierarchy intent on preserving its power and dominance. It is an instinctive behavior in mankind and blame for its debilitating effects cannot be attached to the universities. However it might be reasonable for intellectual leadership to show insight and provide constructive mechanisms to correct the problem. Solutions would be assisted by transparency and freedom of speech.

A further mechanism to limit the freedom of speech is the development of codes of conduct which allow the academic to speak only within their sphere of expertise. In Australia statements on other topics must be made as a member of the public.[35] Those academics who think laterally and apply their thoughts and intellect to many subjects can be penalized and silenced. This is humbug, for a good academic contributing to society must be able to use the power of thought and wisdom in many situations. An academic training promotes this. The code of conduct policy has further emphasized the compartmentalization of academic disciplines. Those advocating environmental solutions often fall foul of codes because their interests have to cut across many specialties if they are to offer holistic solutions to world problems.

Codes of conduct are being made part of the conditions of appointment and infringement of the code can therefore lead to dismissal. University management explains the need for a code in terms of responsibilities to others. In effect this means a responsibility to the university that wishes to vet, change or censor media statements. This raises two important issues. What is the responsibility of the academic to the community that provides more than half of the universities resources? Is this responsibility to be overridden by loyalty to a university management? In freedom of speech, a responsibility to others is already subject to the law, including defamation, and academics are no exception.[36] We must therefore ask who will be the arbiter when these so-called rules of conduct are transgressed: a vice chancellor appointed for business acumen but with a meager contribution to knowledge? Little wonder that silence for survival is now an academic imperative. However it is encouraging that freedom of speech is preserved to some degree in some countries that allow academics to be a critic and conscience of society. Such was the following case in New Zealand.

On occasions the direct suppression of academic free speech has resulted from the commercialization of public hospitals. The need to preserve commercial confidentiality is often used to suppress exposure of the failures in a hospital. This was illustrated in New Zealand when academic surgeons decided to speak publicly about the safety of patients in hospitals where they taught medical students.[37] They believed that the deaths of a number of patients should be investigated. Breaking the ranks of silence on such issues led to intimidation, attempts to discredit them, and private investigation to look for points of personal attack. Pressure was then put upon their employer, the university medical school, to silence them. Eventually an inquiry was held and their concerns were vindicated. It was academic freedom in their capacity as university employees that allowed them to speak out on this issue. [38] However it is sobering to realize that many universities may have used similar suppression if its own corporate links had been under threat.[39]

There are many concerns with commercial agreements between universities and companies. The number of agreements increases as government funding is reduced and the universities seek funding from the commercial sector. The academic staff member tries to uphold the concept of freely available scientific research data; industry does not, for reasons of competition and profit. The two viewpoints are often incompatible. The usual outcome is that the academic agrees to silence or it is imposed.

It can be argued that the universities transgress the spirit and perhaps the letter of Article 19 of the Universal Declaration of Human Rights. This states that everyone has the right to freedom of opinion and expression and that this right includes freedom to hold opinions without interference and to seek, receive and impart information and ideas through any media.[40] This freedom has become more restricted as the functioning of society has bowed to economic imperatives. As recognized by Martin and his colleagues,[41]an individual is only free to make some categories of public criticism if the position can be defended by force of power or by purchasing legal services. The restrictions on speech in public and academic life come at a time of growth in international human rights law.[42] It is dispiriting that academic institutions in wealthy, developed and democratic countries are so indolent in these matters.

There are a number of unprincipled decisions that lead to financial gain, which in the long term harm the image and focus of the university. They may also turn out to be counterproductive financially. The types of actions to which we refer are the manipulation or re-orientation of bequests using the most agile legal minds, so those funds can be used expediently for the crisis at hand. Such actions when they become known lead to university staff and others developing an aversion to making bequests and so are counterproductive in the long term. Therefore with bequests judgment is needed as well as rules. The intent of bequests should not be deviated from the wishes of the benefactor. Instead there is often room for explanation and maneuver at the time of the bequest. Universities also trade their assets for financial gain without consideration of the social, environmental or heritage considerations. Again, such actions lead to unwillingness to serve the university in many of its necessary tasks. Land given to a university to preserve its environmental heritage might be sold for development, within the donor's lifetime, causing enormous hurt and anguish.

Gifts and sponsorships present particular problems for universities. They are welcome and necessary if universities are to remain functional at a time of contracting government funds. Many of the worlds' greatest universities are great and remain so because of philanthropy. Endowed funds bequeath varying degrees of independence from government direction. There is, however, a problem when gifts or sponsorship are directed entirely to a specific purpose. Let us discuss one example.

In 1999, a major oil company donated AUS $25 million to a university for the establishment of a School of Petroleum Engineering.[43] It was probably the largest gift from industry to an Australian university. The Prime Minister of Australia commended this "far-sighted investment in Australia's knowledge and skills base".[44] Few would disagree because the proposed teaching would embrace a wide range of skills applicable to many industries. Nor would there be disagreement with the ethos of the oil company in supporting the university.

But the issue is much wider than this. Of all energy consumed commercially in the world, 88 per cent comes from fossil fuels. Oil is a fossil fuel responsible for a significant proportion of present greenhouse emissions and it is not too extravagant to suggest that its use will have to be seriously curtailed in the future if civilization is to survive in its present form. Furthermore, in general the oil industry has a deplorable record of world wide environmental damage from oil spills that continue unabated.[45] Of course, a School of Petroleum Engineering may play a role in improving the efficiency of extraction and utilization of oil and it is possible that it will direct its expertise to the use of gas, which is less contributory to greenhouse emissions. We must also accept that some petroleum products will continue to be required by humanity.

Balance is needed in university investment, endeavor and education. It is therefore appropriate to ask if AUS $25 million would be available from industry for a School of Alternative Energy or for greenhouse mitigation? The answer is "no." Such proposals do not have an immediate economic endpoint in the view of those who might donate. The combined funding of all academic environment departments in Australia is probably exceeded by this donation for the establishment of a School of Petroleum Engineering.

Knowing the threat of global warming to humanity it is difficult to see how such funding could be accepted without qualification. The solution to this imbalance lies with the university and the recognition of its duty to the community. Some minor principles of balance are already established in the university systems, though not necessarily for altruistic reasons. For example, when a research grant is received, a proportion of the grant is levied by the university for administrative and other university functions such as library maintenance. To achieve balance in the university's educational functions a proportion of donations, sponsorships and bequests should be sequestrated. Many would believe that a proportion of a AUS $25 million donation should support environmental initiatives in the mitigation of climate change and indeed assist the survival of non-economic subjects such as history and philosophy. The university's response to this suggestion is likely to be that it will lose the donation. But it should not be beyond the skills of the university's intellectual leadership to explain, persuade and accomplish a satisfactory solution.

The pressure to obtain sponsorships is sometimes opportunistic and unprincipled. The University of Nottingham, U.K., accepted 3.8 million pounds from British American Tobacco to finance an international centre for social responsibility. It is estimated that 120,000 people die from smoking each year in the U.K.[46] The company's director of corporate affairs, said "We are serious about demonstrating responsible behavior in an industry seen as controversial."[47] The company was attempting to buy some respectability and the university was prepared to facilitate this for the payment of money regardless of the potential effect on the image and standing of the university.[48]As a result, the editor of the *British Medical Journal*, Professor Richard Smith, resigned from his Chair at Nottingham University.[49]

The principle of balance in the distribution of finance has received little debate in the universities yet it is relevant to all income. In medical faculties, donations provide research funds for cancer, heart disease and childhood leukemia's whereas less fashionable causes of illness, suicide, depression and degenerative disorders of the elderly, to mention a few, are rarely supported. Yet the latter are responsible for much human misery. This imbalance is also perpetuated by the grant-giving bodies. It is a university responsibility to redress these imbalances to serve the needs of all humanity.

A slow realization is dawning in industry and commerce. Their lack of commitment to employees is reciprocated and performance deteriorates. As the universities follow global management trends with outsourcing and short term contracts to reap financial savings, they are suffering the same consequences. There is no collective support for the enterprise. There is disillusionment and despair that the university is no longer worthy of loyalty. The consequences are far reaching within and outside the university. The university is just another corporate body, which arouses cynicism about its motives. The universities talk of "excellence". It is difficult for any institution to espouse virtues when its own house is in such disarray! A legion of employees will testify to the increasing remoteness and inhumanity of university administrations with many instances of abject indifference, incompetence and even what is seen as personal malignancy. A university of the future, if it is to serve humanity successfully, has to succeed in the management of personal relationships, for local failure epitomizes the ills of the world, conflict, war, self-aggrandizement and a lack of concern for one's neighbor and environment. Furthermore it is the human fundamentals, world health, peace and environmental studies, that suffer most in this uncaring and pragmatic environment. In its values, the selfless personal performance of its employees and its commitments to all humanity, it has to be a cut above the rest.

**The Idea of a University**

So what should be the purpose of a university?[50] The idea of higher education was present in Greek society. There, led by Plato, the criticism by students of conventional knowledge in a series of dialogues raised perception to a higher level and enhanced personal development. Embryonic universities existed in medieval times and once more, structured discussions were used to examine knowledge. These institutions were independent, open to all students and functioned by means of a collaborative internal government. In 1845, Cardinal Newman set out his vision for a university that has remained a reference point ever since.[51] He believed that higher education should bring expansion of the mind and that learning should form "a connected view of things." He took this viewpoint because the industrial revolution was encouraging particular forms of knowledge in detriment to others. Thereafter, prior to World War II, teaching and vocational courses predominated; however after this war, the need for armaments and industrial production, together with their new technologies, stimulated a surge of scientific research. Knowledge became fragmented, a trend that continued despite the distinguished writings of Karl Jaspers (1883-1969).[52] He believed that knowledge is not absolutely objective and that truth is the present point in a dialogue intended to reach consensus. Furthermore, the university was about the proper development of modern society itself. To fulfill this role the university needed independence to interact with society without duress. Increasingly, with some exceptions, for example, universities in the US, this independence has been lost and with it the ability to modify society. So although it is accepted that the fundamental activities of universities are the advancement of knowledge, the passing on of this knowledge by teaching, and the provision of vocational training of doctors, lawyers and others, a leadership and reformist role in society is difficult.

The value of the university to the national interest in Western civilization was fully recognized as a result of two world wars, which demanded advances in science and technology and the training of skilled personnel. Universities became national institutions with the provision of national funding to facilitate national needs. Many universities with funding from endowments, bequests, and so on became, in effect, public because their expansion necessitated the acceptance of governmental funds. Other changes followed which are highly relevant to the present discussion. Initially the increased governmental funding did not affect the autonomy and independence of universities, but eventually it opened the door to political pressure and a succession of changes imposed by government. Numerous consequences have flowed from these changes which make universities the handmaidens of nation states and their economic and business needs.

Today, in Western countries, modern universities are a fading cornerstone of society, supporting national needs for both education and economic advancement. For some students they fulfill a thirst for knowledge, but for most, universities are ladders to jobs with more money and prestige. To parents they seem a solid insurance policy for the future of their offspring. For governments today, universities are an expense that needs to be justified and utilized for the good of the nation, with "good" interpreted as economic advancement. For government ministers, universities are captive arenas for management and reform, and for political correction and personal ministerial advancement by the demonstration of vigor and action. Their agenda is to wed the universities to the needs of industry and commerce and thereby serve national economic goals. For teachers and researchers, the "academics", universities offer a base for the pursuit of ambitions of personal research achievement, for many inherit the individualistic, competitive, egocentric missions so characteristic of our society. Teaching students is a necessary trade-off to accomplish this personal mission. For a minority in the university, the idealists and teachers who worry about the future of society, the university is now a contracting haven.[53]

In general, universities have provided for the needs of society.[54] Religious studies were important in universities when society was structured to this need. Now, different values, mainly economic, pertain. But the universities of yesteryear educated only a small minority of the people; usually they were the centers of elitism to sustain the small professional and ruling classes. Today, universities in Western society are cosmopolitan and educate millions. They are a major component of society, in education, in endeavor and in cost to the community as a whole. There are some voices that say that the role of this major resource, the universities, should be much wider than the education and training of individuals to serve economic needs. Professor John Niland goes some way to acknowledging this when he states

> [But] we must never lose sight of the fact that our bottom line is not really a dollar bottom line. Universities are for the long term, we prepare teachers and researchers of the future, we train the professions, we generate knowledge and we educate in the widest sense for a civilised society, an informed electorate and an innovative and flexible workforce.[55]

However, the issue is even more comprehensive than this. To explain the need for a much wider role for the universities, there are some statements and intents, agreed internationally, which are worthy of consideration for they define fundamental needs upon which humans can build a life of peace, well-being and fulfillment. There are declarations of human rights, agreed descriptions of health and consensus statements from the world's governments on the approaching environmental crisis.

The Declaration of Human Rights[56] was born of the horrors and deprivation of the Second World War. It was an emotional response to turmoil, bereavement and catastrophe. The Declaration was a collective cry for change. It obliged society, institutions, governments and individuals to work for its ideals; it was a modern epistle of Biblical intent; it was a clarion call to education. In its preamble it is stated:

> The General Assembly proclaims this Universal Declaration of Human Rights as a common standard of achievement for all peoples and all nations, to the end that every individual and every organ of society, keeping this Declaration constantly in mind, shall strive by teaching and education to promote respect for these rights and freedoms and by progressive measures, national and international, to secure their universal and effective recognition and observance, both among the peoples of Member States themselves and among the peoples of territories under that jurisdiction.[57]

As part of article 26 it states:

> Education shall be directed to the full development of the human personality and to the strengthening of respect for human rights and fundamental freedoms. It shall promote understanding, tolerance and friendship of all nations, racial or religious groups, and shall further the activities of the United Nations for the maintenance of peace.[58]

In 1998, the United Nations Educational, Scientific and Cultural Organization (UNESCO) made an important declaration on higher education for the 21st century.[59] The declaration made the point that higher education should be accessible to all on merit. It should be more student-oriented. It should promote national, regional, international and historic cultures as well as enhancing society's values and contributing to the development of education at all levels. There should be emphasis on higher education's ethical role, autonomy, and responsibilities. There should be equality of access and promotion of the role of women. Most importantly it stated that higher education should reinforce its service to society, especially in assisting in the elimination of poverty, intolerance, violence, illiteracy, hunger, environmental degradation and disease.[60]

The United Nations Environment Program has requested all educational systems to focus on all aspects of sustainability, economic, cultural, social and ecological.[61] Yet the universities participate uncritically in the global market with its undemocratic power structure and rules promulgated from the World Bank and the World Trade Organization that promote the needs of multinational companies, and they fail to address the needs of humanity as defined by democratic international organizations.[62]

With regard to health, in 1948 the World Health Organization defined health as "a state of complete physical, mental and social well-being, not merely the absence of disease or infirmity".[63] The use of the words "physical", "mental" and "social well-being" opened a vision of health that would come to embrace absence of war and poverty and an unpolluted and stable environment. It recognized that health is the product of political, social and economic conditions. The thinking of 1948 evolved to that of the Rio Earth Summit in 1992, which recognized that the health of humankind had to be seen in the context of the total human environment, including the health of the physical and living world itself (the biosphere).[64] This was the conservation medicine or ecological health perspective. It was seen that there was deep conflict between development on the one hand, and the stability of biosphere on the other, as evidenced by global warming, loss of biodiversity and worldwide pollution caused by a world economy that is to grow apparently forever.[65]

The universities give insufficient emphasis to the interrelated, fundamental issues of peace, poverty and the environment.[66] The reasons are complex. In the corporate university, they are not seen to have an economic endpoint or gain. And for the student who studies these subjects, what is their job expectation? Most importantly universities have had difficulty with the diffuse nature of the topics. They are uncomfortable with the advocacy that stems from research findings. The subjects cannot be categorized easily. They need to be approached through an "ecological" type of thinking. Ecology is the scientific study of the ways in which all living things interact between themselves and with their environment. In the 1850's, the scientist Alexander von Humboldt (1769-1859) believed in a "chain of connection" between all elements of earthly life including humans and that knowledge of that interdependence was "the noblest and important result" of all scientific enquiry.[67] He was referring to ecology, though it was not given this name until after his death. The word "ecology" was first used by a German biologist, Ernst Haeckel in 1866. It comes from the Greek, *oikos,* referring to the family household and its daily operations and maintenance. The term has evolved to refer to the "cosmic household".[68] Humboldt extended his ecological thinking to the political social and environmental problems of that time. The study of present world problems is best approached with this manner of thinking.

The Universal Declaration of Human Rights calls upon us to promote the principles of justice, freedom and the needs of humanity by teaching and education.[69] This is a call for a liberal or broad education that requires cultural, social, linguistic, literary and analytical abilities, with knowledge of history and the world today. What is meant by a "liberal" education? It is "concerned mainly with broadening a person's general knowledge and experience, rather than with technical or professional training" (New Oxford Dictionary of English) and "knowledge"means all knowledge with no emphasis on what is useful. A broad liberal education implies a breadth of knowledge across many topics with a grasp of their interrelationships. It confers the ability for self-learning, for analysis and criticism and the application of intellect to a wide range of issues - such as the world problems described in this book. It provides a depth of vision that allows science to be placed in the context of the needs of the human species.

The pendulum of university education has swung emphatically to vocational education and most courses and degrees are constructed with a job in mind, be it lawyers, doctors, engineers, computing, business studies or agricultural chemistry. Its role to provide a broad liberal education has all but disappeared because it is not seen to have economic relevance. The primacy of specialist education in universities exerts undue influence on learning in school and a narrow school education does not provide a passport to world citizenship. Despite this, there is an increasing recognition that a broad liberal education in many professions enhances their ability to serve the public. For example, it forms the basis of the doctor's maturity in supporting patients.

It is important for the universities to have an ethical, moral and professional component in vocational education; but mostly this has been surrendered to the professional organizations where it merges with the matters of salary, fees and conditions of service. This is not necessarily an ideal solution, for conflicts of interest may arise with important ethical questions being clouded by self-interest.

The university as we knew it has changed irrevocably and will not be resuscitated. The present institutions that call themselves, or wish to be "world class universities", may remain as servants of economic machines that produce educated persons for the job market. They will compete nationally and internationally and will be run as corporate enterprises with little respect for culture, ethics, the humanities or a liberal education. They will be funded and influenced by commerce and industry and particularly by multinational interests. The subjects fundamental to the health of humanity in its widest sense, medicine, environmental and peace studies, will require more autonomy and multidisciplinary inputs if they are to progress and contribute to humanity. As autonomous or semi-autonomous institutes they will have the opportunity to incorporate the necessary ethics, philosophy, values and commitment which are being discarded by the universities. Science will remain within the present university systems for it requires massive funding, but it will need to evolve as a professional organization so that scientists and the public can be protected.

The concept of human rights has encountered difficulties of interpretation since the Declaration was made in 1948.[70] Many governments have professed support but have often failed to act. Some governments have used human rights to politically attack countries that have different cultural concepts. Furthermore, in an aggressively capitalist world, human rights in democratic societies are being interpreted as a freedom to exercise rights with little or no responsibility. An exclusive pursuit of rights can cause conflict and hostility and therefore thwarts the aims of the Declaration.[71] As a consequence, in 1993, a Universal Declaration of Human Responsibilities was prepared by a large group of former leaders and statesmen of the world. It aims to balance the rights of the individual with the responsibilities to others.[72] In fact it built upon Article 29 in the 1948 Declaration that "everyone has duties to the community to which alone the free and full development of his personality is possible".

The Universal Declaration of Human Responsibilities seeks to commit humanity to the eradication of poverty, conflict and environmental damage. Article 7 said:

> Although every person is infinitely precious and must be unconditionally protected, the animals and plants which inhabit this planet with us likewise deserve protection, preservation and care... as beings with the capacity of foresight we bear a special responsibility - especially with a view to future generations-for the air, water, soil, that is, for the earth and even the cosmos.[73]

Hope must remain eternal that the present university system will reform to foster and teach these fundamental issues. New or Real Universities will be required with leadership to pursue the needs of humanity. They will be created by the interaction of the world's best minds using international communication systems but they will also develop a site for interaction and student encouragement. In the next two sections we will discuss how science and environmental studies fit into the conceptual framework so far sketched in this chapter.

**Science – Saviour or Desecrator?**

Science has blossomed since the Enlightenment when the critical use of reason replaced the religious superstition of the Dark Ages. It has become a cornerstone of Western culture. The quest for new knowledge has been overwhelmingly scientific. It has created economic success and the power that follows. Science is the basis for industrial and technological developments, whether the science is based in industry itself or in the universities. Science dominates the universities; its support requires expensive infrastructure and a large proportion of their budgets.[74]

The role of science in the promotion of one of the most fundamental needs of humanity, a sustainable environment, will be examined in this section. The performance of the universities cannot be discussed without contemplating the success and consequences of science. Advances in scientific knowledge are often breath taking. But the universities have allowed the dominance of science to suck the life-blood from subjects that are not judged to have an immediate economic outcome. The universities swim with the tide of science and technology and disregard some of their other responsibilities to society. In their pursuit of technological brilliance, they have failed to teach the place and role of science in a caring and concerned society. As a result, some suspicion and fear of science has arisen in the community.[75]

The almost God-like image of science is reflected in some attitudes to it. Criticism or revision of a scientific finding is permitted at a scientific meeting, or when a scientific paper is published, for both comply with the unwritten rules of the scientific community. But writers who criticise science may have difficulty in attracting a publisher, and critical journalism is often resented by the scientists. The independent critic of an official document based on science is often accused of misunderstandings and misinterpretations followed by slurs on his or her integrity, credentials and motives. It is the effrontery to criticise which produces this reaction not so much the content of the criticism. Scientific statements have become encyclicals, which cannot be criticized with impunity.[76]

To many of those without scientific training, science is the truth. This view is unjustified, for scientific findings may be "the truth" today, but may be revised or even rescinded tomorrow, when additional knowledge becomes available.[77] Therefore, "scientific truth" today is the consensus of knowledge today.[78] But this consensus may be fallible because it brings together widely divergent results, perhaps obtained by different experimental methods, each with its proponents. Such a consensus therefore includes different interpretations and opinions. For this reason an opinion expressed by scientists that knowledge can only produced and refined by scientific methods is seen by many as arrogant. The point is made by Satish Kumar, founder of Schumacher College, that there are other ways of knowing about the world, artistic, cultural, philosophic and religious, and that science must be looked at in the context of society, culture and the environment.[79]

The shifting sand nature of scientific knowledge provides ample opportunity for government and corporate manipulation. Decisions that are wanted are based on "sound scientific data" when this may be a fragile consensus. The precautionary principle becomes operative only when delay and inactivity suits the agenda. Often decisions are made without examining the scientific data in a global, cultural or environmental context.[80]

There is little recognition that the positive results of scientific discoveries that have given humanity health, longevity and comfort have also been used as to cause the global environmental crisis that threatens to wreak civilisation this century. Science is so omnipotent that it can now be used to solve the problems, clean coal, mirrors in space to reflect the sun's rays, or so it is thought by many. In an essay, important because it placed the needed responses on climate change firmly on the agenda in the US, we hear that technology will preserve our lifestyle:

> I am not proposing that we radically alter our lifestyles. We are who we are-including a car culture. But if we want to continue to be who we are, enjoy the benefits and pass them on to our children, we do need to fuel our future in a cleaner, greener way.[81]

There is overwhelming evidence that science will not save us in a world where biological resources and land and water are being consumed at a rate much than their renewal.[82] In a world where carbon dioxide is but one pollutant produced by a burgeoning population addicted to everlasting economic growth, the need is for a change in lifestyle, equalization of resources population limitation brought about by political, intergenerational, and philosophical change.

In many ways science has replaced God.[83] Darwinian science indicates that strands of protein in a primeval sea evolved by a process of chance and natural selection to a human existence for function, reproduction and biological fulfilment. According to Nobel Prize winner Francis Crick,[84] the soul is a mere assembly of nerve cells. We can think of the brain as a computer that will be superseded eventually. The thousands of scientists working away at new knowledge are merely the silicon chips of facilitation in this nihilistic journey.[85] There is no more to life than this, but science can make us comfortable, indeed very comfortable so let us enjoy this biological existence, or so the argument of the scientific materialist goes.

These viewpoints have consciously and subconsciously pervaded Western culture and have contributed significantly to the functioning of society today. Apart from the occasional talented scientific commentator or writer, a David Attenborough or a David Suzuki, science has failed to reveal and explain the infinite experience of creation whether the term is used in a biblical or biological sense.[86] The implications of scientific discoveries are generally ill-understood and need to be explained and taught in the context of the entire human condition. Science is incapable of giving meaning to human life, as society scrambles amongst a galaxy of New Age theories and thinking to seek signposts and a return to traditions.[87] Yet our relationship to science *is* changing. The multiplicity of scientific opinions on a given subject is not surprising to those who understand scientific methodology, but the majority of people are confused and insecure when they can no longer rely on firm scientific opinions. A pillar of Western society is showing signs of corrosion. The number of sceptics is increasing,[88] which may be good or bad depending on their integrity. Their role in the climate change debate deserves discussion

The role of skeptics in the climate change debate deserves further discussion. The consensus of the 2300 scientists preparing the IPCC reports is quite remarkable in the history of scientific review processes. Yet there are number of climate change skeptics who have the attention of the media. Their criticisms are answered but they continue to crusade with the support of the fossil fuel industry.[89] Their efforts have produced a form of schizophrenia in the political domain. Their pronouncements are favored by governments fearful of the implications of climate change—should these be correct-yet these same governments promulgate scientific resolution of climate change

Science itself is not always responsible for this scepticism. It is the way it is used by the powerful. The problem is compounded by the narrow outlook of some scientists. The rationale and philosophy of science is explained inspirationally by E. O. Wilson in *Consilience.*[90] The pursuit of science is reductionist; the building blocks of matter and life are broken down and analysed. The experts come to describe more and more about less and less. Expertise becomes fragmented and the scientist is often incapable of taking a broad view. But the scientist becomes honoured as an expert because of this confined but unique knowledge. Wilson calls this the "fragmentation" of knowledge.[91] For "fragmentation" the term "compartmentalization" is used in the present text. As a consequence topics such as environmental studies that encroach on and embrace so many other subjects are neglected. We are left with the anachronistic situation that a broad liberal education requires an understanding of scientific thinking. Yet often the scientist fails the community because of reductionism and a narrow outlook.

From the perspective of society, we must grapple with the fundamental problem of the permissiveness of a liberal scientific exploration of anything and everything, regardless of the likely use of discoveries for good or evil. But if we restrict our liberty to proceed down any path even when it seems irrelevant, significant advances could be jeopardized. Contemporary society objects to any limitation of freedom of inquiry. Yet in some instances it is possible to look at the potential ethical problems of a scientific investigation before proceeding. And it is also possible to define some essential tasks necessary for the well-being of humanity and harness science to solve them. Surely this was the dream of the Enlightenment. The problem now is that science proceeds at such a pace that society, its ethical and legal framework trail behind.[92] Why does it proceed at this rapid rate? Mainly because of competition between the egos of scientists who cannot resist the challenge and because science is a competitive economic force.[93] There is an urgent need to revise and control this situation.

Science was the kernel of the Enlightenment that would nourish the growth of knowledge. The failure of the Enlightenment to fulfil its promise can be attributed to the activities and attitudes of the proponents, of science. The scientists have adopted a reductionism to the exclusion of the whole. The dream of the Enlightenment is compromised by elitism, self-interest and individualism, all too often divorced from the care, concern and needs of humanity. The Enlightenment has been corrupted by its main proponents.[94]

Today the quest for scientific knowledge has become one of the most competitive aspects of human culture. To retain or aspire to a high standard of living, each advanced nation must spend heavily on science, or the competitive edge of industry will be lost. Basic research is strongly linked to economic success. Consequently a case can always be made for increased spending on science. But, as in any competition or race, only a few win medals; there are always more also-rans or losers. The explosion in knowledge as evidenced by scientific papers is almost exponential. There always will be scientific knowledge to pursue for we will never comprehend the Earth and universe in their entirety. Each question answered always raises many more questions. The case for more expenditure will always be urgent as competitive economies vie for dominance. Furthermore, we must accept that this race for knowledge has benefited the comforts of life but has added little to the happiness and peacefulness of humanity. It has fuelled economic growth so placing the Earth's environment under unsustainable pressure and its unforeseen consequences have caused the Earth and humanity many insoluble problems. The independent arbiter might conclude that the scientific revolution is a mixed blessing.[95]

We cannot turn back the clock. Some scientific and technological advances must assist the solution of problems we have created. We can perhaps curtail the gadarene rush to illusory success by controlling runaway science to provide us with an agenda for survival and well-being. This may seem a hopeless task given the ethos of individual freedom to pursue scientific goals and its linkage to power and money. At the moment, these issues are rarely raised in our citadels of higher education.

When we examine the role of science in society, a broad liberal education must expose the true nature of science and its mistakes as well as its successes. The precautionary principle must be understood. It must be appreciated that science has difficulty in accepting findings that diverge widely from current paradigms that tend to flow through the scientific community like a fashion. New theories or scientific findings that originate in unfamiliar cultures are often disregarded. We will now examine examples of these problems.

The scientific work of Vladimir Vernadsky remained unknown for cultural and political reasons.[96] Vernadsky (1863-1945) was born in St. Petersburg and was a professor in Moscow. His concept of the biosphere was unknown and unappreciated for half a century. He recognised the real implications of the Earth being a self-contained sphere. He realised that geology was not the science of the Earth's materials often thought of as "rocks", but a continuum of the transformation of elements by living matter. Today, when scientists observe and measure the flow of carbon, nitrogen and other elements through the hydrosphere, lithosphere, and atmosphere and through living things, they recognise that their interpretations of these flows in a holistic system depended upon Vernadsky's thoughts.

Yet when James Lovelock put forward his ideas on the biosphere in 1979,[97] he was not aware of the work of the person whom was his illustrious predecessor. Lovelock made a strong case for the biosphere being a single physiological entity, with everything having relationships to everything else. Vernadsky had remained unknown in the West ~~because he was a Russian and~~ isolated by culture, politics and language. In addition, his theories were framed in an era of increasingly mechanistic attitudes in science which separated biological and geological sciences. It was not fashionable to consider integrating life with the rest of the physical world. Both Vernadsky and Lovelock have a holistic view of the Earth. Both Lovelock and Vernadsky are scientists, but also wise men with a comprehensive understanding of all aspects of science and its place in the world. They were separated, not just by time but by political systems and worked in different scientific communities that seldom referred to each other's work.

The scientific study of the gastric brooding frog did not fit into current thinking and was dismissed out of hand.[98] This rare gastric brooding frog was discovered in Queensland, Australia in 1973. The female swallows the fertilized eggs that develop in her stomach. Over several weeks the stomach distends greatly as the many froglets grow inside it. The muscular wall of the stomach changes to resemble a uterus, and the digestion of the froglets is prevented by the secretion of prostaglandins, which suppress acid secretion. At birth 15 to 20 fully formed, but tiny, frogs leap from the mouth of the mother and swim away. These events and processes were studied at the University of Adelaide and scientific papers were prepared for publication. It was felt that the findings were so interesting and unusual that the journal *Nature* should be approached. The paper was rejected because it was thought to be a hoax. Presumably, in the mind of the reviewers and editor it did not fit into any preconceived notions of biology. The science was not at fault because the findings were later accepted for publication in the international journal *Science*[99] and in the leading gastroenterological journal *Gastroenterology*.[100]

The public values simple, concise opinions from scientists on issues of great public interest. Independent scientists should provide these. Current examples requiring scientific clarification are the controversy over genetically modified foods, and climate change resulting from global warming. Thousands of scientists are involved in these issues, directly or indirectly, yet it is regrettable that there are few who can present a considered, balanced opinion to the public. It is useful to reflect on the controversies over smoking and health a few decades ago. We now know that the research by distinguished academic scientists that linked smoking to cancer of the lung was negated by data and opinions promoted by vested financial interests.[101] These financial interests also controlled their own scientists and influenced many others. Similar machinations continue today with the debate over passive smoking. Reports are prepared by paid scientists to create the perception within the community that the issue is still controversial. On the contrary it is certain that environmental tobacco smoke causes disease in non-smokers.[102]

Similar nefarious actions are being repeated today. Energy companies play down the relevance of global warming, and the biotechnology industry fights a public relations battle on the safety of its products backed by its own scientific opinions.[103] Ben Santer was a climate modeller and lead author in a 1995 UN report on climate change that concluded that on balance human activities, such as the burning of fossil fuels were a cause of global warming.[104] The Global Climate Coalition, representing the interests of American oil and automobile companies attacked and damaged him by accusing him of “scientific cleansing“, that is, manipulating the scientific information.

Nor is it difficult for vested interests to manipulate the media. Commonly a scientist, expert in climate change, will make a consensus statement based on hundreds of studies. The television station will give equal time, "in the interests of fairness", to a single scientist, linked to an energy interest who disputes the findings. In many countries, organisations of industrialists who are committed to the production of fossil fuels have been formed to dispute the scientific findings on global warming. In Australia, the Lavoisier Group lobbies politicians and seeks media coverage from a handful of scientists. In a submission to a parliamentary Treaties Committee, the Lavoisier Group wrote: "With the Kyoto Protocol we face the most serious challenge to our sovereignty since the Japanese fleet entered the Coral Sea on 3 May 1942."[105]

Unfortunately the independent scientist is now an endangered species. The university has scientists with the ability, independence and integrity to analyze issues and report their conclusions openly to the public. Or at least they should have. The scientific community often passes the buck: "we do the work, it is society's role to decide what to do with it". Universities are increasingly dependent on funds from industry and commerce that often influence what scientists are prepared to say. Furthermore the universities are themselves increasingly restricting what their academic staff can say in public. Freedom of speech on scientific issues should be promoted and protected in any university worth of its name.

The hundreds of thousands of graduating scientists are employed predominantly in industry and government departments which is where scientific technology and instrumentation tend to be available. In industry, these scientists serve the economic machine; in government they provide regulatory public services. They are beholden to the employer who controls their freedom to speak in public. Furthermore, scientific endeavour is often harnessed to a government's particular opinion or need and there is subtle or overt suppression of uncomfortable findings. Occasionally, whistle blowers will expose dishonesty, but at the risk of severe consequences to their careers. In an editorial, *New Scientist*[106] draws attention to censorship in science. It is subtle, for it is often a failure to tell the whole truth. The editorial goes on to say "[c]orporate science has, of course, no choice but to serve corporate needs...."[107] This scenario needs to be remedied if we are to avoid the destructive misuse of science. The nuclear power, GM food and tobacco industries, are examples of this problem.

The university system, therefore, increasingly fails the community in the sphere of independent scientific opinion, more so than in other areas of opinion, because the independent scientist is unlikely to be found elsewhere. James Lovelock has maintained his independence, but not many others. Yet the conditions and facilities for team analysis of complex situations can function in universities when the topic is not controversial. We can take the example of meta-analysis.[108] In medical practice it is important to know if a new drug has a specific effect in a particular illness. Some studies of the drug show it does, some show it does not, and overall the results seem equivocal. There may be 250 scientific papers written on the subject, so how does the doctor make a decision to use the drug? In meta-analysis, all data from these 250 clinical trials is re-analyzed according to specific criteria, which are decided before the meta-analysis commences. By this means the results from a large number of patients can be analyzed statistically and true result produced. This is time-consuming work for medical scientists and statisticians which should not be done by industry wishing to sell a product or by governments not wishing to pay for it! Similar analyses are used in Evidence Based Medicine.[109] Here all available literature on a particular illness is reviewed and optimal treatment is recommended. Other methods for the examination of existing knowledge can be applied to the analysis of scientific issues that concern the community, but they are not, for political and economic reasons.

There are also a number of options which involve community participation and in which universities should involve themselves. The consensus conference is one such example and these have been held in several countries. In March 1999, one took place in Australia on the topic of genetically modified foods.[110] A widely representative panel of members of the public met to study written submissions from a wide range of experts who were then interrogated. The findings of the panel were reasoned and practical. Their recommendations included methods of regulation, the involvement of an ethicist in major decisions and a commitment of government to open public education on the topic. The conference was run by the Australian Museum, an appropriate independent organization, but this kind of function should surely fall within the remit of a university.

ProMED-mail is another important example.[111] It was established in 1994 as a project of the Federation of American Scientists and is now a program of the International Society for Infectious Diseases. It depends upon academic scientists from universities and national institutions who provide reports and analyses on outbreaks of emergent infective diseases of humans, animals and plants. This information is sent via the internet to its subscribers around the world. Calls for assistance are passed to appropriate experts. It provides a more rapid early-warning system than official channels. It is independent, which has advantages over official newsletters, for example those produced by World Health Organization, which are edited to take account of governmental sensitivities. The value of ProMED resides in the scientific judgment and expertise that can sift and analyze the information. Should not this type of service be provided by a university. But universities are not independent and conflict can arise when university monitoring impacts on a company that provides funds to the university.

The nuclear power industry and in particular its safety is a further example which enables us to analyse the contributions and problems of science.[112] The public fails to appreciate that electricity produced by nuclear power is very expensive compared to its generation from other fuels. The cost to the consumer is hidden by huge government subsidies to maintain a powerful industry. The advantage of nuclear power over fossil fuel is that it does not produce greenhouse gases. However, the nuclear industry has yet to demonstrate adequate safety in particular areas, for example, the storage of waste and is a significant health hazard.[113] It presents itself as safe by the suppression of facts about several serious incidents.[114] Even in the case of Chernobyl, a huge catastrophe in terms of radioactive contamination of large areas of the Northern Hemisphere, numerous deaths and a high probability of genetic damage to many future human generations, there is the culture of denial, excuse and reassurance.[115]

For decades the effects of radiation on human health have been based on the dose of radiation energy absorbed through the human skin. The survivors of Hiroshima received a single external dose of radiation of varying magnitude dependent on their distance from the explosion. The survivors were studied for the development of cancers in a "Lifespan Study". The number of cancers in these survivors was proportional to the dose of radiation received.[116] This finding has been utilized to reassure the public that low doses of radiation are without risk, a finding necessary to the continued survival of the nuclear power industry. This is because low doses of radiation due to leakage have occurred in many areas adjacent to nuclear power stations. Leukaemia and other cancers, which have occurred in people near to these sites, have been attributed by governments to other causes.[117] Yet for decades there have been alternative theories and evidence showing that low levels of radiation, especially when ingested, cause cancer of internal organs.[118] Because this information was inconvenient for governments, agencies and industry it has been suppressed and its proponents denigrated.

The failures of science are well illustrated by these events. National bodies such as the Atomic Energy Commission in the United States are extremely powerful in that they have government backing, support the salaries of numerous scientists and give large grants for scientific research. In itself this has a powerful influence on scientific viewpoints. The investigation of the health aspects of nuclear power installations has encountered difficulty or obstruction, and findings have not been easy to publish. Furthermore there has been much scientific support to continue the nuclear power industry in the face of a total failure to find a safe storage method for nuclear waste. If, as an independent scientist, one looks at the totality of the facts, one is driven to the conclusion that there must be many in the vast community of nuclear scientists who have extreme doubts about the prudence of continuing with nuclear power. Yet there is silence.

There are lessons for universities in these examples. The scientist is trained in a creed that promotes freedom of scientific exploration without limits, and with little fear of involvement in the consequences. As a result, adverse consequences are explored only when they occur. There always are some adverse consequences of a new technology, some can be foreseen, but are ignored in the scramble for development and commercial gain. In this regard there are interesting differences between medical and non- medical science and medical science. In general some non-medical science exercises liberty without responsibility, and adverse consequences may result. By contrast, in general medical science is subjected to checks and balances imposed by public demand and the ethics of patient care.

The correction of scientific misconceptions depends upon education. The universities have failed to educate those training in scientific disciplines, so that scientists are unaware of the history of science, of historical scientific mistakes, of the philosophical basis of science and its place in culture and society.[119] The university must guide the scientist away from the position of responsibilities only to himself and to an employer, towards a professional responsibility to the worldwide community. This involves the teaching of ethics, communication and a responsibility to speak out on important issues. And it requires leadership in public debate and the development of a scientific Hippocratic Oath.[120]

The precautionary principle is a most important concept that needs to be discussed in all curricula. All scientific findings are likely to be changed or modified by the advance of knowledge. Therefore, we never know when we have arrived at the truth. Often, we have to appraise those facts available to us and predict the likely truth in the future. This upsets many classical scientists. They have often told governments and politicians that decisions must be based upon scientific facts. Little wonder that politicians take advantage of this advice when it suits them. The precautionary principle is disregarded when it might hinder or prevent a government's need for development or economic growth.

In the state of South Australia, tuna farming was commenced in pens in the shallow coastal waters. Tuna was exported to the lucrative markets of Tokyo. In 1995, near to the pens, dead pilchards were washed onto the beaches in the hundreds of thousands and this kill spread around the coast of Australia. Scientific opinion indicated that a herpes virus was the likely cause. Pilchard is a vital part of the food chain in Australian waters and it can be assumed that the deaths had a serious impact on the marine ecology. In 1998 the kill was repeated.[121] It commenced in the approximate region of the tuna farms and spread along the coast as far as Western Australia. On both occasions, the supply of local pilchard was insufficient to feed the voracious tuna, and frozen pilchard had been imported from North America. Biological products brought into Australia are subject to analysis by the Australian Quarantine Inspection Service. Permission to import frozen pilchard was based on a process of risk analysis. Despite warnings from its own marine scientists, the Government of South Australia dismissed the danger of importation of herpes virus because it was not found in any of the imported samples examined. The import license was granted on the basis of a process of risk analysis which, by its very nature, is inappropriate because it discounts the vital precautionary principle The Government wanted no restriction on lucrative tuna production regardless of the ecological consequences.

There is an additional lesson to be learned from these events. A report from government scientists that raised the question of infection imported with pilchards, was changed at the instigation of the South Australian Government, with the agreement of the authors.[122] This event confirms the unacceptable practice of suppression of scientific evidence to allow government to act with short-term expediency to the detriment of long-term consequences. It also illustrates the failure of scientists to speak out on issues of importance to the community

The precautionary principle strongly promotes the adoption of prudent policy and measures before scientific confirmation. In the tuna example, infection of captive fish is a worldwide problem. Australia is isolated and its flora and fauna, including those of coastal waters, are highly susceptible to introduced species and diseases. This is why rigorous quarantine regulations should apply. The precautionary principle should have ensured that the farmed tuna were fed on local pilchards. If imported pilchards were used, it should have been assumed that they might carry infection. The imported feed should have been tested rigorously and perhaps put into isolated tanks with Australian pilchards to test whether they became infected. Indeed the precautionary principle might go so far as total avoidance of imported pilchard.

There is no excuse for environmental decisions based on ignorance of the scientific process and the precautionary principle. Australia is replete with disasters from imported species. The rabbit and fox were imported in the early days of the colony to satisfy the hunting urges of home-sick Englishmen. At the time the danger was not understood, but even by 1935 the lesson had not been learned. As a result the cane toad was imported into Queensland, Australia to rid the sugar cane of beetles.[123] Even though the federal government echoed concerns voiced initially by conservationists, economic considerations and pressures prevailed. The cane toad has now spread over thousands of kilometers and has had a devastating effect on local fauna.

As we have seen in this book, much of the scientific evidence relating to the future effects of various levels of greenhouse gas emissions is uncertain and probable to various degrees. After a long debate, the international community is accepting the precautionary principle on global warming. To await definitive scientific proof, may allow the Earth to be damaged irretrievably.[124] It follows that the precautionary principle should be understood by all. The scientific response of "we don't know yet" should indicate to individuals and governments that the precautionary principle must apply, whereas often they proceed as they wish to until proof of adverse effects is available. A broad liberal education for every graduate should include science, not necessarily practical laboratory science, but a study of how science functions and the many local, national and international examples of the use of the precautionary principle.

Many of the concerns about science and some of the insecurity and cynicism it creates relate to the scientific freedom given to scientists and the economic systems that employ them. This situation is reinforced by the operation of government research grants in many countries. The research budgets for individuals and institutions are competitive. For example, let us consider medical research grants. Applications are made by medical scientists to carry out research into any subject they wish, ranging from molecular biology of a rare disease to the social aspects of suicide. Scientific referees and committees judge the applications. The committees are balanced to cover a wide area of expertise but necessarily consist of individuals who have been successful in mainstream medical science. These individuals may also have their own applications under consideration by a different committee. Thus they have some self-interest in the system continuing to function in the same way that it has for the duration of their careers. The applicant and the application are ranked by scientific referees on scientific criteria and intellectual standing. Then a committee interviews the applicant.

In this system, the best competitive brains are often attracted to the most intellectually challenging frontiers of medical biology and as a result studies of genetic engineering in relation to disease may secure a hundred times more funding than youth suicide. This reflects the perceptions of scientific merit by the grants committee, their training and perhaps their subconscious value systems based on their scientific beliefs. In support of the system, members of that committee will explain that the results of molecular exploration of a rare disease may open avenues that will lead to the cure of other diseases; the discoveries may be universally applicable to many problems. In any event, in their view, the most able people must be supported, but "able" is judged under rules which favour the lines of thought and form of exploration most familiar to the majority of the judging committee. As a result, research budgets are spent on topics that bear no resemblance to the priorities for human advancement as judged by the community. Furthermore funding for laboratory research exceeds that for preventive and environmental health by a factor of at least ten. Even within the laboratory research sphere, many important diseases are overlooked for funding because they attract no applications or because the applications are judged inadequate.

This system of research funding is continuing the tradition of academic freedom of investigation. But it has to be asked whether this is appropriate in relation to the needs and well-being of society. Within the medical research funding policy, what is the magnitude of spending on some major or important ills besetting our society, human suicide, depression, substance dependency, paedophilia and road accidents? It will be found to be meagre in wealthy Western countries. As already indicated, the research funding and budgets applied to the major problems of the world is even more meagre.

Are we then to have research directed to the agreed needs of humanity? It will be argued by many that such research is unproductive because it reduces the motivation by personal choice and dulls the competitive entrepreneurial drive of the individual researcher. But history tells us that research directed to a single, overwhelming goal is often successful; recruitment of intellectuals to Los Alamos produced the atomic bomb. This method of operation continues to apply to a large proportion of the world's research on even more destructive armaments. It occurs when companies direct research into genetically modified plants and food for commercial gain with such speed that possible ecological damage is neglected. It has been applied successfully to the development of space programs. Is it not conceivable therefore that, with commitment, science could be harnessed to solve some of the world's major problems?

Public confidence in science has waned because of the adverse consequences of scientific discoveries and because of the behavior of some scientists. There is distrust and concern, which will have to be addressed by the scientific profession itself, and particularly by independent scientists in universities, if any remain. They will need to offer leadership to retrieve this situation. All professions and disciplines have been subjected to fraud, plagiarism and falsification but it is becoming apparent that science may have more than its share of these unfortunate events [125] and the public cannot comprehend why rules and regulations are not imposed.

Many scientists are worried that the rapid pace of scientific discovery has greatly outstripped ethical, moral and procedural responses. In the United States a survey of 2600 researchers revealed that 8 per cent knew of other researchers falsifying or plagiarising data. When 400 medical researchers were interviewed, more than half knew of colleagues who had cheated in clinical trials.[126] The occurrence of fraud is probably greater than is apparent because of cover-ups. Fraud occurs because of pressure to produce results, competition, personal ego and financial gain. Personal prestige and the standing of the institute, company or university are at stake to produce results and this often leads to the suppression of fraud. Fraud in clinical trials can often be traced to an overconfident investigator who knows that the drug works anyway, so why be diligent and honest? It is not surprising then that most scientists can recall episodes of fraud and falsification by their colleagues. One of the most spectacular examples of scientific fraud was perpetrated by a team of South Korean and American researchers led by Woo Suk Hwang. They claimed to have created human embryonic stem cell lines that matched the DNA of patients and the results were published in *Science*.[127]

The universities should set an example in these matters of scientific integrity. In the past the in-house mechanisms of personal discussion and unwritten rules did not prevent untoward events. What chance, then, of corrective action with the increased economic and performance pressures of today? This is an issue that requires appropriate national and international action because the results of scientific investigation have global effects. Universities must redouble their efforts to prepare students for these difficult issues. In addition, a regulatory system is required with individual registrations of scientists who adhere to a code similar to the way in which doctors should adhere to rules supervised by a medical board, or industry adheres to environmental regulations supervised by environmental protection agencies. In a regulatory system for scientists, scientific fraud could be reported to a board with the power to cancel registrations so preventing employment in public service or universities. This system would protect scientists working in industry for they would be obliged to speak out on developments detrimental to the community and would be protected. Of course there would be many difficulties with such a system, for example the co-opting of independent scientists and experts to sit on a board and judge their colleagues

An appropriate definition of a profession is as follows:

> A profession is composed of a body of knowledge, a substantial portion of which is derived from experience. A profession is responsible for advancing that knowledge and transmitting it to the next generation. A profession sets its own standards and it cherishes performance above personal rewards. And finally, a profession is directed by a code of ethics which includes the moral imperative to serve others. [128]

The medical profession espouses these ideals though it may fall short of them on some occasions. If science were a profession it would have a commitment to the entire community and an ethical intent. Like health care, science could be private in structure but public in purpose and this implies responsibility. The Universal

Declaration of Human Responsibilities lists scientists as professionals along with physicians and lawyers.[129] It states "Professional and other codes of ethics should reflect the priority of general standards such as those of truthfulness and fairness".[130] This has been transgressed so widely that it is difficult to consider science as a profession

This is an issue that would benefit from guidance at an international level from distinguished scientists, ethicists and others in a reformed university without economic political and managerial pressures. It needs a scientific Bill of Rights for the scientists and the public. The future of science utilised for the benefit of humanity is too important to be left to the human foibles of greed and personal aggrandisement.

Science is decreasing in popularity with students at school and university and enrolments in universities are falling.[131] Governments and industry bewail this because science feeds the economic machine. But the image of science is tarnished by the events described here. As a consequence, the student enters science for a job, with little thought that it will be a satisfying experience to serve humanity. An altruistic motivation survives in the scientific contributions to medicine and environmental studies but is sparsely evident elsewhere. The Enlightenment needs to be reborn if science is to serve the true needs of humanity. It is clear that the corporate-style universities cannot or will not participate in a rebirth.

In conclusion, it the duty of the universities to offer leadership by teaching and discussing the place and role of science in society. This should be part of the reawakening of a broad liberal education for every student. Each individual in the community needs to understand the workings, benefits and dangers of science. They must recognise the increasing control of important scientific discoveries by powerful corporations, discoveries that have implications for every person on earth. The public needs to be involved in discourse and decisions for their future is at stake. If the universities do not promote this discourse, it must be provided elsewhere by a Real University. The universities should also promote the independence of its scientists and encourage them to be public ambassadors of wisdom.

**Environmental Education and Life Support Systems of the Earth**

Science tells us that environmental damage, pollution and climate change are slowly eating away at our life-support structures. They behave as a cancer typically hidden but with emerging symptoms when growth in the number of cancer cells leads to the invasion of human organs. And like war, we do not wish to face the likely scenarios; and indeed, what's the point, because each of us seems impotent to influence the future?

In developed Western countries environmental degradation has resulted from economic growth, greed, and disregard of prudent policies. But in developing countries, particularly in Africa and some parts of Asia, numerous wars and poverty are the consorts of environmental damage. Here, war, poverty and environmental damage are tied in a vicious circle of misery.[132] The public has a poor understanding of the crisis and is confused by conflicting views. The commitment of governments is totally inadequate even in terms of those actions that they are prepared to initiate. Their funding often languishes as a budgetary afterthought, and recently in Australia the sale of public utilities has raised money for environmental repair rather than a recurrent, committed segment of the budget. We need to ask therefore why so little action is taken on these important matters? In what ways has environmental education failed and what responsibilities do the universities have in this regard? Educational systems have reflected the views of governments and society and have failed to resource environmental studies adequately. The universities' response to these needs is minimal. The fundamental reasons for the totality of these failures are buried in our economic beliefs, psychology and culture.[133]

During the 19th and 20th centuries, Western societies developed an interest in the environment by forming field naturalist and other groups, but environmental science did not register with the universities until the early 1970's.[134] Then simultaneously in most Western countries, there came a surge of public interest and concern about environmental issues sparked by books such as *Silent Spring* by Rachel Carson, first published in the US in 1962.[135] The quotation from poet John Keats (1795-1821) *Las Bell Dame sans Merci* immediately after the title page epitomizes the terrible environmental consequences of the indiscriminate use of pesticides and insecticides:

> Though the sedge is wither'd from the lake,
> And no birds sing.

Accompanying this quotation from Keats there is a prophetic quotation from Albert Schweitzer (1875-1965) "Man has lost the capacity to foresee and to forestall. He will end by destroying the earth".

Increasing environmental awareness led to the formation of many conservation societies that soon attracted a large and vocal membership. Governments and international organizations responded. In the United States the *Environmental Education Act*, 1970 was passed to educate the public about ecological balance and environmental quality. In 1972, the United Nations Conference on the Human Environment called for a greater involvement of universities in environmental problems which was the diplomatic way of saying that there wasn't any significant involvement.[136]

In Australia, between 1972 and 1978, seven universities developed academic departments for the study of the environment. Mostly these departments emphasized postgraduate studies and soon they were successful in attracting a large number of students. Graduates became eagerly sought and employed in governments and industry. However during the 1990's a number of ominous clouds slowly appeared on the horizon and by the late 1990's a storm of economic rationalism decimated the number of postgraduate students in one university from 200 to 7 over a period of only two years. Other universities had a similar woeful performance. Thirty years of progress was demolished. Why? The specific reasons related to the imposition of fees on students in a milieu of economic rationalism. But the climate that allowed this to happen without regret had evolved with the changing values and commitments of society and universities over two decades. It occurred not only in Australia but also in other Western countries.[137]

The demise of academic departments of environmental studies in Australia occurred when the Commonwealth government instituted full fees for postgraduate students. These fees discouraged potential entrants who already recognized that an environmental qualification was not a passport to personal economic success. This contrasted with the surfeit of applicants prepared to pay university fees to enter professions that offered rich monetary rewards. Those wishing to do environmental study have an altruistic view of society and indeed environmental skills are as necessary to society as medical ones, but there comes a point at which the personal sacrifice is too great. Potential students may be prepared to be insecure and impecunious to serve humanity in developing countries but they are not prepared to take out loans to pay university fees. Furthermore, the university treatment of environmental studies as a non-economic and dispensable topic was recognized by the student community. In an attempt to alleviate this situation, some universities have instituted a modicum of protection of places without fees in the hope preserving environmental studies for more enlightened times. This is no more than an attempt to survive.

Some seeds of failure were also sown when universities around the world first recognized a need to establish environmental studies. Efforts were made to include them in the curriculum, usually by inserting them into existing structures. This created difficulties. Often, there was no defined focus or financing for environmental studies. Finance, when available, was shared between a multitude of departments, often science departments, to provide components to the course. Administration was by a multidisciplinary committee with little commitment to the issue. Those teachers contributing part of their time became frowned upon from within their own departments and they were often seen as letting down the common cause of that department. Sometimes segments of environmental studies were in departments in different faculties and their small size put them at a disadvantage in the competition for funds. As a result the subject never reached the cohesion and multidisciplinary status that it needed to succeed, and it often failed to attract and retain first-class academic staff.

Often the most long-established universities had the most difficulty. In recent years, in Edinburgh University, Scotland, the Vice Chancellor, Sir David Smith, a person committed to environmental education, spent his time there activating the subject at all levels in the university.[138] When one of us (David Shearman) visited the university in 1998, he had left and environmental studies had contracted. A Center for Climate Change and a science department covering environmental studies, though excellent as separate departments, failed to provide a cohesive undergraduate course. The representation of environmental studies in other well established universities is also poor. In the University of Adelaide, some resources were provided for the development of environmental studies and numerous committees sat and deliberated. The responsibilities were apportioned largely to science departments and their existing programs and indeed it seemed that some departments even changed their names in order to be able to capture resources and students.

By contrast, some young universities were able to achieve a more appropriate model by creating a new structure suitable for environmental studies. Ideally a "school" of environmental studies would embrace both undergraduate and postgraduate education and research and would have core staff, structures and budget. As in a faculty of medicine, the undergraduate course would be non-specialized in order to produce a good all-round, non-differentiated graduate who is able to specialize in one of many areas at a postgraduate level. To produce a graduate, a faculty of medicine brings together many disciplines, including the physical and biological sciences, communication skills, clinical disciplines, ethics and psychology and preferably also a wider knowledge of the world including its religions, cultures and value systems. A similar situation should apply to the undergraduate course in environmental studies. Thus, a "school" of environmental studies would include representation from the biological and physical sciences, meteorology, economics, the health sciences, psychology, philosophy, politics and an understanding of the role and workings of international and local environmental organizations. At a postgraduate level this ideal "school" would have a structure that would include supervisors for a wide spectrum of higher degree study including the areas mentioned above. It would not be confined to the biological sciences. The aim would be to produce graduates who could assist government, industry, legal frameworks and environmental medicine and be leaders, educators and activists in the area. Like medicine, environmental studies require a modicum of committed activists and thinkers to offer enthusiastic leadership.

The similarities between the needs of environmental studies and a school of medicine are worthy of further emphasis. One provides care for the health of all the components of the human body in a holistic manner, the other attends to all aspects of the health of the world's environment which in turn is essential for human health. There are other analogies with medicine; for example the encouragement of personal insight into health and the environment. It is unacceptable for a doctor to smoke; this is a personal responsibility to self, family and community and a responsibility to set an example. Our environmental consciousness is less mature. A professor of environmental studies was seen by one of us driving his four-wheel drive over fragile sand dunes, cursing at the tracks of motor cyclists who had preceded him. He thought his work to be so important that he could disregard environmental rules! The teaching of environmental studies begins with a personal responsibility for the environment.

From this discussion, it is apparent that environmental studies require an identity that recognizes its interdisciplinary needs. If we were to select one mechanism that would be likely to damage its evolution it would be a forced marriage to a discipline with different aims and needs. In Australia this was the outcome chosen in a majority of universities by the dominating managers who believed that large departments would be more efficient financially.[139] Geography was selected for the mergers for it was in decline. The origins and traditions of environmental studies and geography are radically different.[140] The links of environmental studies to social and political movements and the function of advocacy fit uneasily with geography. It is difficult to understand the reasoning behind such inappropriate mergers and one is driven to the conclusion that they reflect only the ease of administration. Economic rationalism triumphed over academic rationalism, and it signaled the failure of universities to establish environmental studies as an interdisciplinary specialty.

There are however much wider issues that impinge upon the viability of environmental studies. Perhaps the most important is the conflict in society between the increasingly dominant economic agenda and humanity's need for a sustainable environment. The two opposing views can be summarized as follows. One viewpoint says that unlimited economic growth will produce health, wealth and wellbeing for increasing numbers of people. The world market will promote these aims by freeing constraints and fostering democracy. There will be more choice and opportunity for all, for the wealth created will trickle down to the poor in both developed and developing countries. Increasing knowledge and technology geared to the market will not only improve our lives but it will enable us to solve the problems we create on the way, for example global warming, repair of soils and production of fresh water.[141] As we master the Earth's environment and then the space of the universe, there will be no limit to progress. In effect, this seems to be the consensus in The World Bank, the OECD, the World Trade Organization and the world's leaders in industry and economics.

Most major schools of economic thought tend to see the economy as independent of the environment and believe that progress, that is economic growth, will lead to a cleaner environment. The origin of this spurious theory probably resides with Simon Kuznets.[142] He suggested that the gap between the incomes of poor and rich increases as average income rises but beyond a certain threshold, the gap begins to narrow.[143] Some economists have applied this hypothesis to economic development and the environment. They have theorized that pollution in poor countries will increase as average income rises. Once incomes have reached a certain level, money will be spent on cleaning up the environment.[144] There is little evidence for this and it has to be of concern that economists should give birth and nurture to such beliefs. To them "cleaning up" the environment seems limited to litter and grime. They fail to grasp that a native forest felled remains felled, an extinct species remains extinct and a hotter earth will remain so. There is a naïve assumption amongst politicians that the primacy of wealth and economic development is necessary for environmental solutions. On the contrary, the surging economies of the west have caused the environmental crisis and one might ask, when today the greenhouse emissions of China approach those of the US at what point will there be sufficient wealth in the world to solve this problem?

The alternative viewpoint to that of "growth" is perhaps best introduced with the words of David Orr:

> If today is a typical day on our planet earth, we will lose 116 square miles of rainforest or about an acre a second. We will lose another 72 square miles to encroaching deserts, the results of human mismanagement and overpopulation. We will lose 40 to 100 species and no one knows when the number is 40 or 100. Today the human population will increase by 250,000. And today we will add 2700 tons of Chlorofluorocarbons to the atmosphere and 15 million tons of carbon. Tonight the Earth will be a little hotter, its waters more acidic, and the fabric of life more threadbare. By year's end the numbers are staggering: the total loss of rainforest will equal an area the size of the State of Washington; expanding deserts will equal an area the size of the State of West Virginia; and the global population will have risen by more than 90 million.[145]

This statement was made in 1991 and now 16 years later little has changed. There are fewer tons of chlorofluorocarbons discharged into the atmosphere because of an international treaty of singular success, but as we have seen, many more tons of carbon are discharged and most other measurements of the health of the environment have continued to deteriorate. Continued economic growth in its present form is incompatible with a sustainable environment.

David Orr summarizes what he defines as "modernity", that is, "the cult of growth" and its opposing views, by six related propositions:

> The world is assumed to have no biophysical limits; the ecological worldview sees the biosphere as finite with the inference that economic growth is unsustainable.
>
> Humans are seen as malleable, rational consumers; the ecological worldview sees them as biosocial organisms.
>
> There is no unity of man and nature; nature is a resource to be dominated and used. The ecological worldview believes that unless balance is achieved between man and nature, human tenancy of the earth will be short.
>
> There are no social limits to growth, and society is infinitely adaptable to technological change; the ecological worldview is that societies are fragile

> and there are social and political limits to social change.
>
> The modernist has unending faith in scientific and technical progress, which are seen from an allegedly neutral viewpoint; the ecological worldview is at least cautious about scientific and technological progress and denies that science is value neutral.
>
> The modernist believes that economic growth will create a politically stable society; an ecological worldview believes that while economic growth may produce political stability in the short term, in the longer term social chaos will result because of ecological decline.[146]

This is a debate between two fundamental issues, one that is anthropocentric, promoting the use of nature for mans needs, and the other recognizing the need for a partnership with nature to a varying degree. The debate is one sided for the "growth lobby", the corporate sector, commerce and financial institutions, have firm control of information services and pay little attention to analyses of environmental problems. Certainly some major media publications are grossly unbalanced and seem to be propagandists for the ethos of growth. It is a matter of conjecture whether this omission is through ignorance of its importance or whether it is deliberate. These omissions of intellectual debate fall within the remit of the universities.

The universities are not seen to have an independent voice in this debate for the following reasons which we have detailed earlier in this chapter. They have adopted policies that emphasize "economically relevant" subjects and neglect studies important to the wellbeing of humanity. Their cynical and poorly resourced commitment to environmental studies falls into this category. They have shown increasing discomfort with environmental studies having social and political connotations that are necessary for an informed debate about the reform of society. The increasing restrictions on free speech in universities clash with the advocacy role of environmental education and the financial and philosophical links between the universities and the corporate world present an increasing number of problems. There is an official view that environmental aims, policies and management have been successfully absorbed into mining, industry and commerce. This leads to the universities believing that the need for advocacy is diminishing. Corporate leaders have become an integral part of the university system, for their companies act as sponsors and they become council members and chancellors, from where they influence university policy regardless of the environmental record of the companies they control. For all these reasons the university environment is not conducive to debate and the advancement of environmental knowledge.

The need for multidisciplinary environmental studies is further emphasized by a consideration of future environmental threats. We can see the effects of climate change, pollution, erosion, land clearing and species loss and our efforts to reverse them have only meager results at the most. The Kyoto protocol to reduce greenhouse gas emissions by just over 5 per cent of 1990 levels by 2012 is unlikely to be met yet it represents only a small fraction of reductions needed if the climate is to be stabilized . Why is it that when the damage to humanity and the Earth from global warming is potentially so great there is denial by many of those empowered to do something about it? Why do we have so little concern for our descendants in a hundred years time? Let us put it this way. Would it be fair and reasonable for each of us to die leaving a credit card debt of $15,000 or $10,000 or even more to each of our children? They would have to pay the debt or allow compound interest to accumulate during their lifetime and then pass the debt to their children. It would seem outrageous. But this epitomizes our response to climate change, for we have not yet recognized intergenerational equity. Our descendants will pay for the tremendous storms, sea level rises with the submergence of islands and lowlands, spread of diseases and other disasters.

In the development of the human brain and our commitment to others, like all other species, we develop commitment to our children. Some of us, by our behavior, show that we are committed to others, by acting as nurses, doctors, care workers and teachers. In families, commitment may extend to grandchildren with finance bestowed for education by those who can afford it. But the buck stops there. There is no commitment beyond. Subsequent generations and the state of the world in which they will live, is beyond our comprehension.

There are similar problems of comprehension and commitment in our lifetimes. This is why health education has difficulty in changing behavior. Smoking at the age of 18, the threat of lung cancer at age 50 or 60 means nothing, for 50 or 60 seems so old, what does it matter? Neither is global warming seen as a perceived threat for the same reasons. Environmental studies in universities require an extra dimension to address this future problem. Its solution as we have outlined in this book will draw upon the disciplines of psychology, behavioral science and the techniques of health education and disease prevention. Again the breadth of this interdisciplinary study resembles that used in schools of medicine.

There are three further aspects of environmental education that are pertinent to this discussion. These are the relationship between science and academic environmental studies, the need for environmental education to be universal and the concept of biophilia.

Environmental studies are often regarded as part of science with the advantage that they can wear the cloak of recognizable economic outcomes. However they are not part of science, they are multidisciplinary and science is a tool within environmental studies. Yet when environmental studies are separated from science in university budgets they suffer even more from financial strictures. Some universities proclaim their environmental expertise in compendiums that list the environmental projects in numerous disciplines ranging from biotechnology to engineering; they are the modernists with a blind faith that technology will solve all problems and especially save us from the ecological limits of the Earth.[147] The demise of broad liberal university education has inhibited consideration of complex issues. Specialization compartmentalizes each subject with scientists thinking only as scientists, economists only as economists and so on. There are few minds and little encouragement to bridge the gulfs.

In stating that science ought to be considered as a tool, it is important to understand the great disparity between the speed of discovery and technological change and the snail's pace of human institutions to deal with the legal, environmental and social aspects of these advances. This issue was discussed in the previous section, in our analysis of the role of science. David Orr[148] points out that thirty-five years after *Silent Spring,* we use more synthetic chemicals than ever. Twenty-five years after *The Limits to Growth,*[149] economic obesity is still the goal of governments everywhere. Twenty-two years after Amory Lovin's prophetic,[150] and as it turns out, understated projections about the potential for energy efficiency and solar energy, we are still using two to three times more fossil fuel than we need. Twenty-years after Wendell Berry's devastating critique of American agriculture,[151] sustainable agriculture is still a distant dream. The same is true today. Human society is addicted to a rush into new technology not recognizing that most of the ecological, economic, social and psychological ailments which beset contemporary society can be attributed to knowledge acquired and applied before we've thought it through carefully.[152] We rushed into the fossil fuel age only to discover problems of acid rain and climate change. We rushed to develop nuclear energy but still don't know what do with the radioactive waste. Nuclear weapons were created before we had time to ponder the full implications. High input, energy-intensive agriculture is also a product of knowledge applied before much consideration of its full ecological and social cost.[153] More recently genetically modified foods have defined a change in procedures in agriculture from which there may be adverse consequences of great ecological importance although we cannot predict them.[154]

Economic growth is driven in large measure by new knowledge that often results in environmental damage, social disintegration, unnecessary costs and injustice. The fundamental problem is the relatively slow rate at which we collectively learn and assimilate new ideas. This has little to do with the fast speed of our communications' technology or with the volume of information available, but has everything to do with human limitations of our social economic and political institutions. Indeed, the slowness of our learning, or at least of our willingness to change, may itself be an evolved adaptation. Short circuiting this limitation reduces our fitness. And even if humans were able to learn more rapidly, the application of fast knowledge generates complicated problems much faster than we can identify them and respond.

We cannot foresee all the ways in which complex natural systems will react to human-initiated changes at their present scale and velocity. We are playing catch-up with the consequences of fast knowledge but we fall further and further behind. We are on the slow train of correction, rehabilitation and mitigation, whereas the express train of new technology roars along leaving passengers behind and having no destination in mind. Fast knowledge has created an unassailable power structure because so many depend upon it. This structure suppresses or ignores any alternative viewpoints that might slow the reckless speed of the express train. The benefits are here and now, the costs later, always a tempting offer in a finite human life.

In conclusion, science is a tool within environmental studies, to be utilized as needs be, along with many other tools. The discipline of environmental studies espouses this philosophy in its practice and teaching. Many philosophers and thinkers believe that the increasing divorce of our civilisation from its inherent environment and ecology is a fundamental cause of our lack of resolve to act on the environmental crisis.[155] During human evolution, in relative terms, many hours were spent in hunter gathering, a few minutes in agrarian society and a mere second in industrialization and the growth of the free-market economy. Our long hunter-gathering existence was in general a state of ecological balance with nature. A similar state exists today in many of the indigenous peoples of the world who have complex societies that link family, health, rituals, ancestors and the physical and living environment. Our mental attachments to the environment remain with many of us because of environmental experiences. This orientation to nature has been called *biophilia,*[156] which has been defined as the urge to affiliate with other forms of life. Since the agrarian revolution we have moved away from biophilia with increasing speed. Yet it is necessary for us to correct this as part of our environmental education. As Stephen Jay Gould has said "We cannot win this battle to save species and environments without forging an emotional bond between ourselves and nature…for we will not fight to save what we do not love".[157]

Education needs to be reformed to include this experience of nature, not only in childhood but also throughout life. We cannot save, preserve or rejuvenate nature until we appreciate and respect it to the full. This emotional response conflicts with scientific thinking, which is dry and unemotional. It breeds discomfort between the environmental scientist and the conservationist. Yet how can we study music without an emotional response to the works of Mozart or Berlioz or art without a response to the colors of Van Gogh? We must recognize that science is part of our culture and as such elicits emotional responses to its findings. The intellectual rules of science dictate against the involvement of feelings, but inevitably they intertwine, and some would recognize this. Is it possible to remain emotionally distant when the creative components of the universe are displayed?

We recognize the bonding to land by many indigenous peoples so that loss of their land leads to disintegration of their society and to disease. Westernized peoples often recognize their bond to place. There is grieving and depression with change of house or country and anger and deprivation when those places to which we are attached, have their environment destroyed by pollution or development. Many of us recognize in ourselves the need for attachments to other, non-human forms of life with gardens, pets and through travel to natural wonders of the world. Some of us recognize our total dependence on the earth's ecology for food, medicines and life itself. We can even speculate that the present increasing mobility of populations is a cause of stress, illness and social breakdown, whether the mobility results from the moveable job market or from refugee status; there is no longer a bond or attachment to earth, so necessary for our mental stability.[158]

This is why the reform of education must include wisdom as well as knowledge and it must recognize feelings, ethics and morals.[159] This removes environmental studies from the scientific sphere into a system of broad education that includes philosophy, culture, history, practical skills, community and family structures and more. This has led educationalists such as David Orr to say that all students should have a comprehension of the basic laws of thermodynamics, the basic principles of ecology, energetics, the limits of technology, how to live well in a place, sustainable agriculture and forestry, environmental philosophy and ethics, as well as others.[160]

In contrast to technological knowledge, David Orr points out the value of what he calls "slow knowledge".[161] Slow knowledge is accumulated during evolution in the process of cultural maturation. It involves how to do practical things within the limits of social and environmental contexts, yet retain harmony and social cohesion; it is the careful conservation and increase of knowledge over many generations. It constructs a society on the basis of wisdom rather than cleverness and it recognizes that the careless application of knowledge can destroy all knowledge, for example by nuclear or biological warfare. Indeed education, that is knowledge only, without wisdom, may allow people to become greater and greater destroyers of ecology.[162] As David Orr says: if one listens carefully, it may even be possible to hear the Creation groan each year - when another batch of smart degree-holding, ecologically illiterate, Homo sapiens who are eager to succeed is launched into the biosphere.

Education alone is insufficient to mitigate the environmental crisis. It has been thought that if people are educated, environmental problems will resolve, but they worsen because of the poverty of millions in developing countries and the greed and consumerism in "developed" and educated countries.[163] This is not to say that there is not an awareness of environmental issues among young people assessed in the Asia Pacific region.[164] However, society seems unable to act on this knowledge because of the dominance of economic considerations. Indeed, today, education itself can be implicated in the environmental crisis because it is wedded to the idea that human progress is equivalent to economic progress. Education, requires a committed environmental component.

There is a strong case that environmental education needs to be universal. The United Nations Environment Program argues a need for environmental responses that involve the whole of government :

> Institutions such as treasuries, central banks, planning departments and trade bodies frequently ignore sustainability questions in favour of short-term economic options. Integration of environmental into the mainstream of decision-making relating to agriculture, trade, investment, research and development, infrastructure and finance is now the best chance for effective action.[165]

In effect, this recognises that environment ministries in governments manage national parks and cuddly creatures, whereas major decisions that affect the environment are made in other, more powerful, departments. It is so difficult for governments to hold sway against the economic driving forces that even the educated fail the environment. Al Gore, former Vice President of the United States and author of *Earth in the Balance,*[166] proposed making "the rescue of the environment the central organizing principle for civilization".[167]The formidable barriers in his way stopped him giving effect to his beliefs during his Vice Presidency. His record of cooperation and compliance with big business resulted in a Green candidate running for the Presidency in 2000. There is a history of governments denying environmental problems and disregarding environmental laws. The United States Congress has dismantled environmental laws in the interests of free trade. After three years in office, supported by an overwhelming mandate, former British Prime Minister Tony Blair admitted that environment had taken a back seat in his list of priorities. If the environmental agenda is to move forward it seems crucial for there to be a profound change in attitudes throughout governments. The Bush Administration has removed hundreds of environmental regulations.[168]

The inadequacies in environmental education are throughout the entire educational system. The United Nations Environment Programme's *Global Environmental Outlook - 2000,*[169] expresses the view that environmental education, like mathematics should be part of the standard educational curriculum. Yet environmental education today is marginalized from mainstream education policy, has no coherent plan for progression from kindergarten to university has a low profile in examination subjects and its teaching has problems within the traditional culture of schooling. In rectifying these omissions it is essential that the universities offer leadership by developing strong undergraduate courses in environmental studies and including these in entrance requirements so that school education will follow. The seeds of environmental concern and education are often sown in childhood, but are adversely affected by the competitive consumer outlook perpetrated through parents, relatives, television and schools. Education has to counteract these overwhelming interests.

There are those who believe that all education is environmental or that it should be seen in this light. David Orr has defined the educational needs to achieve ecological literacy.[170] We need to recognize that all education is environmental education because humans are merely one part of the natural world. Education must be seen as self-development and self-mastery rather than mere subject mastery. Knowledge carries responsibilities, hence the myth of the freedom of knowledge must be exposed and rejected.[171] There is only genuine knowledge if its effects on communities and the environment are uncovered. There is a need for more concrete thinking to bridge the gap between ideas and reality, rather than the high level abstractions of the intellectuals.[172] These thoughts necessitate a profound reform of education.

Three components of environmental education are important in school and university. "Education about the environment" essentially has involved science but needs a wider focus to put science in context. "Education in the environment" traditionally involves planting trees by children and field trips for university students but essentially it should be about experiencing nature and humanity's relationship with it. "Education for the environment" involves value judgments and is of vital importance. Its utilization of philosophy, history and cultural studies brings an understanding of the social and environmental movements and how these are based on different value systems in each culture. Learning involves multidisciplinary lateral thinking instead of working within a single discipline. The aim will be to provide and foster personal and practical contact with ecological systems.

A lack of relationship with non-government organizations is a further impediment to the university contribution to environmental studies. Such links can be established by a variety of mechanisms. For example, the concept of university open learning can be applied to millions of voluntary workers in non-government organizations and for reciprocity, non-government organizations could provide a vital resource for community studies and implementation of simple measures being researched and developed by universities. Such arrangements do sometimes occur on an ad hoc basis but there is no university ethos for them and no such role in university mission statements. Frequently the academic is uncomfortable working or associating with non-government organizations, which they see as non-academic, unstructured and amorphous. Such reactions often smack of elitism. A historical study of the environment movement describes how scientists often used their expertise to provide leadership on conservation.[173] However, these prophets were often harassed by government and bureaucracy.

**Conclusion**

To summarize the university's role in environmental education: from the magnitude of the world's environmental problems, one would assume that environmental studies would have a major place in teaching and research. There would be faculties or schools of the size of law or medicine. But as in much of society, university leaders are in a state of denial regarding worldwide environmental problems. Even those who grasp the magnitude of the problems, fail to warn or act. There is a failure to speak to government and industry for the issue is too difficult politically and self-preservation is paramount. The leadership failure is almost universal. Even when a teaching and research program exists, it is heavily attuned to scientific explanations and solutions. Yet science and its discoveries are partly responsible for the crisis. It is an ecological crisis that requires an understanding of philosophy, values and history in the context of the ecology of the Earth. Environmental studies has been severely reduced in the face of budgetary restraints and there has been little attempt to protect or promote recruitment. An acceptance by community and university of the environmental crisis should lead to the reform and enhancement of environmental education with a multidisciplinary approach and more resources. But even this represents an interim solution, one which falls far short of the necessity for total reform of environmental education.

Leaders in universities could also signal that they understand the importance of environmental issues by implementing four measures. University buildings, grounds and properties should be managed according to best environmental practice in terms of energy, water use and waste management. Management should include the education of all staff and students in the implementation and progress of the measures; the system should be managed by a representative group of the university. Some universities are moving in this direction. University land should not be sold opportunistically to manage budgets when this land has important value for its biodiversity, “open space”or conservation. Examples of this expediency are now legion, with universities behaving like marauding property companies. University investment should be ethical in environmental terms, as well as social. In Australia, both university and staff superannuation investments include many companies that engage in major environmental damage often of worldwide significance. There is an opportunity to remedy this for one fund is used by all academic staff. The involvement of industry leaders in university councils and management bodies is now common, as part of the ethos of linking to industry and attracting funds. When university councils include leaders of companies, they should have impeccable ethical credentials. Furthermore, Councils should always include people with knowledge of environmentalism and community health. These small reforms would signify that university leadership has acknowledged its responsibilities to the community.

While such developments within the university would represent a major advance in university thinking, they cannot be regarded as a solution to the problems created by the runaway express of growth and development. The slow train may become more efficient in picking up more passengers and in mitigating damage but in the express train as in the economy, it will be business as usual. More goods will be produced more efficiently but economies will continue to expand and environmental repair will never catch up. An appreciation of the failures in environmental education provides an opportunity to recognize the need for a complete reform in all education. Education equips the individual to join the world labor force; but it does not create a “calling” to a better society. Environmental education today suffers most from this attitude. At best it may provide ecocrats to work for the market, to smooth the bed sheets of an expiring Earth. The complete revolution in education must ensure that ecological thinking permeates our culture and education - anything less will offer only mitigation.[174]

Of all the failures to be acknowledged by the university, the most abject is their performance in environmental education. Leaders of universities and other academics have failed to speak out on a world crisis which they continue to document scientifically yet deny in their words and actions. This abdication of responsibility has been filled by the independent scientists working alone outside universities and other institutions like Rachel Carson and James Lovelock, by individualistic environmentalists working via the media like David Attenborough and David Suzuki and by independent Institutes such as World Watch.

Science in its reductionist march has shown reticence in accepting Gaia but has recreated the concept under the name of Earth Science Systems, another specialty. It is surely time to move forward: environmental studies should now move to reflect the concept of the unity of nature, a tenuous thread in the history of science.

# Chapter 6

## Philosophy, Religion and the Environmental Crisis

What is the use of studying philosophy if all it does for you is to enable to talk with some plausibility about some abstruse questions of logic, etc. & if it does not improve your thinking about the important questions of everyday life?

- Ludwig Wittgenstein (1889-1951)[1]

Generally speaking, the errors in religion are dangerous: those in philosophy only ridiculous.

- David Hume (1711-1776), Scottish philosopher[2]

In the seminars of philosophers it is easier to find objections, refutations and counter-examples to any proposed definition of morality - and of such associated concepts as 'the good' and 'right conduct'- than it is to secure general agreement about what any of these fundamentally important notions mean. This is not to claim that we do not know what 'right' and 'wrong' mean, it is instead an indictment of contemporary philosophy, which has allowed the term ' academic' in the phrase 'it's only academic' to mean 'empty and futile'.

- A.C. Grayling[3]

**Introduction: Philosophy, Death and Resurrection**

This chapter will consider the contributions that philosophy and religion should make in dealing with the crisis of civilization, and especially the environment problems detailed in this book. We discuss philosophy and religion together in one chapter because they both deal with ultimate questions and certainty with questions of value and goodness and the reasons for human action. There are also, of course, important connections between philosophy and religion and many thinkers have attempted to show that there is or is not, a rational philosophical grounding to religious thought.[4] Philosophy and religion have also attempted to articulate *weltanschauungs* or world views, comprehensive views of the place of humanity in the cosmic scheme of things.[5]

In this chapter, based upon the philosophical work of one of the authors (Joseph Wayne Smith) we will attempt to chart a direction for philosophy and religion to take if humanity is to deal with the enormous challenges which face it in the 21st century. In short, we believe that academic philosophy has lost its way and in the attempt to construct itself into a rigorous "scientific discipline" has lost sight of many fundamental issues that are of much more importance than the rigorous logical resolution of various standard "puzzles" and paradoxes. Philosophy cherishes reason and rational justifications of positions by the construction of deductively sound arguments in accordance with the canons of formal (symbolic) logic.[6] We doubt whether philosophy in this sense, as a quasi-applied mathematics, is theoretically coherent and of value to humanity as an aid in dealing with the crisis of civilization as far as the discipline is practical: as an adjunct to the sciences, allegedly resolving their "conceptual" problems. With respect to religion, we very much doubt whether an attempt to rationally ground or criticize religious beliefs is at all a worthwhile activity in the light of the urgency of issues facing humanity. Religion, we argue, even in its "fundamentalist" form, which many non-philosophers and theologians accept, may be of considerable practical use in mobilizing people to radically change their consumption patterns and to address core environmental issues. This chapter then will not be explicitly discussing climate change as such: the ramifications of the conclusions reached here for climate change studies will be made in detail in the concluding chapter.

### The Case Against (and For) Philosophy

Ernest Gellner (1925-1995) a British philosopher and social theorist said about contemporary American philosophy:

> If the several thousand or more of professional philosophers in America

> were all assembled in one place, and a small nuclear device were detonated over it, American society would remain totally unaffected ... No one would notice any difference, and there would be no gap, no vacuum, in the intellectual economy, that would require plugging.[7]

Why should this be so, when throughout the recorded history of humanity, fundamental questions have been asked about human existence and the writings of philosophers have influenced the course of civilization? Because philosophy in the universities is dying from self- inflicted injuries and in this state of weakness its demise is easily expedited by the economic rationalists who seek to "streamline" the modern university.

In the scramble to conform to economic rationalism the universities have tacitly allowed many non-vocational subjects to wither. Perhaps they are seen as unimportant to what is now called "the real world". Students soon recognize this message of official disapproval, enrolments fall, the subject becomes uneconomic to teach and is abandoned. Many of these academic subjects are categorized as "the humanities": examples are history, philosophy, literature, languages both ancient and modern, music and visual art. Here, philosophy, as representative of these subjects, will be discussed to assess its relevance to the fundamental needs of humanity. A case will be made that philosophy and like subjects are vital in a broad education that in turn facilitates study of fundamental human needs of peace, health and a sustainable environment. But to do so, academic philosophy as it is practiced in Anglo-American and Australian universities must undergo a fundamental change.

In the British and Australian university systems, philosophy as a discipline has been contracting for the last two decades. In Britain, entire university philosophy departments have been closed. The situation in Australia has not been as bad. Funding cuts have meant that departments have not been able to employ new staff, and many have retired or have been made redundant. The situation in continental Europe is different. France especially takes philosophy very seriously; it has not succumbed to the attacks of the economic rationalists. But in many countries, philosophy is regarded as not economically viable in today's ethos of economic rationalism.

By contrast, this academic, tunnel view of philosophy has not been accepted by the public. Their thirst for philosophy is quenched at café meetings and by best-selling books on philosophy such as *Zen and the Art of Motorcycle Maintenance*[8] and *Sophie's World*[9] and others, though these may suffer the disdain of the academic philosopher. Humans have an inherent need to explore the fundamental nature of knowledge, reality and existence. This is philosophy. We can see the stars and planets in the night sky and by radio astronomy we can discover many more which are light years away, but what does this mean? What is the significance of humankind in the cosmic scheme of things? This is a philosophical question.

To ask, "What is philosophy", is to already ask a philosophical question, as introductory textbooks tell us. Philosophy is best understood by studying "philosophy" by plunging in the deep end. This, it may be thought, would make it impossible to explain the meaning of philosophy to the non-philosopher or non-academic. A philosopher friend remembers his female high school librarian saying to a first year class that "there are no books in this library under the classification of "100" [philosophy], because we have no need for any of them". His mind drenched with surging adolescent hormones immediately thought that philosophy must be about hard-core X-rated sex. Well - no, not exactly. Philosophy can certainly investigate questions of sexuality and eroticism. However, philosophy does not have a fixed subject matter at all, although there are distinctive philosophical questions about the nature of knowledge and reality. As teenagers, haven't we all thought about the meaning of life, what is the point of it all? Is life absurd or does it have a meaning? Some of us continue to ruminate on this throughout life and many great works have been written on this topic.[10] In doing this we are philosophizing. Associated with such questions is the issue of the existence of God. Is there a God? Did God create the universe, or did it evolve by purely physical processes? How do we know that God exists or doesn't exist? Are there arguments for this? These questions too, are philosophical ones.

A brief examination of the history of philosophy explains its relevance to all knowledge. Philosophy in Western civilisation arose in ancient Greece, about 2,500 years ago, although there are traces of reason-based argument in other traditions such as the Jewish religion that gave us the Bible. What characterizes philosophy is a concern with arguments and reason rather than an appeal to authority. Philosophers such as Socrates (470 – 399 BC), Plato (427 – 347 BC) and Aristotle (384–322BC) believed that by reason, rather than by faith, knowledge about the world and about beauty, good and evil could be reached. The Greek Philosophers were the first people in the Western world to think systematically about the nature of the universe – whether the world was finite or infinite, whether there were "atoms", indivisible building blocks of matter and the like. These philosophers discussed such questions with few facts because, with limited technology, they could not practice experimental science.[11] But many of the questions which they asked did not require scientific technology; e.g. what is the "whiteness" common to all white things and how can the "whiteness" be at distinct places at the same time (the problem of universals).[12] Socrates is famous for giving us the *Socratic method* that can still be used in debate even today. We have seen it best in courtroom dramas on television, where questions are asked and inconsistencies are revealed in cross-examination.[13] The Socratic method is a common teaching tool in law schools in the United States. It involves a dialogue between two speakers with one speaker questioning another speaker and asking them, with respect to a thesis that they accept on a topic, whether they accept or reject various assumptions. Frequently the person questioned is required to deduce various consequences from these assumptions, which are then shown to be false or contradictory. The questioned person may be led to contradict him/herself or come to see that they cannot cogently defend the originally accepted proposition.[14]

Science as we know it was once called "natural philosophy". Many scientific and mathematical theories have developed in response to philosophical questions. Look at the night sky. Who has not wondered about whether the universe was finite or infinite? But what do these terms mean? If the universe was finite what would happen if one stood near the edge of it, and threw a spear, or could one fall off, and where to?[15] Philosophers speculated about this for hundreds of years until a group of mathematicians conceived a finite universe as having a "curved space" like the surface of a ball, unlike the more familiar space of inside of a box. Advanced science generates many profound questions. For example, what is time? What is matter? What is energy? In general, the more complex the scientific theory, the more difficult the philosophical implications become. So philosophical analysis partners scientific discovery and it is the neglect of this partnership that separates science from humanity and contributes to the perception of scientific elitism.

Since its inception, at least in the West, philosophy has been concerned with the rational justification of world views, comprehensive views about the nature of everything in the universe, and even things which do not seem to have a physical existence such as numbers. We speak of six apples - but what is this "six" of which we speak?[16] Where is the number six? We can also say, correctly, that we *know* many things. We *know* them rather than merely *believe* them because we have good reasons, arguments and evidence.[17] In daily life we seldom have problems with any of these questions. However if we are in a court of law, or practicing science in reflection about the foundations of law, science and other fields of knowledge, questions about justification and knowledge – epistemological questions - become very relevant. Epistemological questions are questions about the truth and validity of whole systems of thought. For example, we all know that there are laws of nature, such as the laws of a science called thermodynamics. Thermodynamics is a branch of science concerned with the laws of energy transfer, especially heat. Anyone who has ever wanted to save money on their home heating bill, has, whether they were aware of it or not, been thinking about thermodynamics. One of the most basic laws of nature, the second law of thermodynamics, says that a system that receives no new energy will run down. For example, we never see the ash from a burnt cigarette suddenly reconstruct itself into the old cigarette.[18] Things fall apart in time unless energy is used for work to repair them. We know this on the basis of experience. If this was not so, no housework would ever need to be done! But how do we know that our experience is true – are these claims true for all parts of the universe and for all times? How can we reason from the *known* to the *unknown* without reasoning in a circle, assuming what we have to prove? This is the famous problem of induction.[19]

Philosophy is about reason and argument. Every assumption must be questioned. For every claim there must be reason and evidence. Nothing can be accepted on faith or on the basis of custom or tradition. Even personal experience is questionable unless it is *objective* - being in principle open to independent assessment and evaluation. This sounds like the scientific method and in a sense it is. Science took this methodology from philosophy at the dawn of the modern age because many early scientists such as Rene Descartes (1596-1650) were primarily philosophers. Descartes, for example, wanted to obtain a secure foundation for both science and religion, so he introduced a principle of methodological doubt. One must doubt everything until one can find something which cannot be doubted - which for Descartes this was his own existence - I think, therefore I am. This was Descartes secure foundation for all human knowledge.[20] Philosophy must call everything into question. Descartes' system, like most in philosophy was a failure and was refuted by his critics.[21] Nevertheless his approach stimulated philosophical thought and constituted an advance in knowledge.[22]

In summary then, philosophy attempts to explain the universe, humanity's place in it, the purpose and meaning of life and its values. It analyses all aspects of knowledge, science, medicine, mathematics, political and social theory, law, history, economics and any other subject. It takes the building bricks of knowledge and attempts to place them into a structure, a framework, a pattern or the architecture of civilization. With the demise of philosophy, the universities have become the manufacturers and marketers of bricks. Thought, wisdom and meaning, the philosophy of life have been discarded. This narrow focus should be an embarrassment to universities at a time of public thirst for philosophy with a philosophy café movement in many cities and overflowing public lectures. Humankind looks for meaning in science, art, education and public policy. The Greek word for philosophy means "the love of knowledge and wisdom". The two are wedded.[23]

To return now to the demise of philosophy in the universities, we have seen that university administrations believe that it lacks substance and relevance. However, it is also dying from a self-destructive illness. This will be discussed now. It should be recognized that intellectual battles are an integral part of healthy debate, but battles can become too bloody and are then counterproductive and destructive. Unfortunately, destructive forces have existed in many university departments to the detriment of universities, students and philosophy, and have provided a convenient excuse for budgetary restrictions.

The history of philosophy is turgid with battles and dissent as humanity struggles with the meaning of life. Since the 1960's, perhaps due to the increasingly rapid changes in society, the rationality of philosophy and of science itself have been questioned. Great doubt has been raised about scientific methodology and whether scientific progress can be said to have occurred at all, despite technological advances.[24] In the philosophy of physics, philosophical doubts arising from investigations into quantum mechanics exist about the commonsense understanding of reality, causation and existence.[25] Even in mathematical logic, seemingly self-evident assumptions have come into question.[26] In short, at the deepest levels of almost all academic subjects from mathematical logic to history, a feeling is emerging that the subject is reaching its end or limits. John Horgan argues this point for the physical sciences.[27] It has been noted by others that the philosophical enterprise may have come to the end and that philosophy may have exhausted itself.[28]

The arguments leading to this "death of philosophy" scenario have arisen within the Anglo-American philosophical establishment, through a questioning of orthodox models of the nature of science, logic and language. In addition French "postmodernist" and "deconstructionist" writers have also signaled the end of philosophy in the sense of being a master discipline supplying a foundation to human knowledge.[29]

The loss of faith in philosophy as capable of investigating the nature of reality and science and criticizing scientific claims has led philosophers in various directions. Some believe that the theory of knowledge, should be "naturalized" that is, replaced by a study of empirical science, logic and linguistics. Others believe that philosophers should not be preoccupied with natural sciences such as physics, but rather they should look towards sociology and history for answers to their questions; they should address issues of everyday importance. Some others believe that philosophy is best abandoned completely.[30] But few of them though who believe that the game is "not worth the candle" have resigned their positions to make way for younger blood; their working conditions, compared to the factory floor or the dole queues, are just too good! This brings to mind the words of one of the philosophers, regarded as a precursor to postmodernism, Friedrich Nietzsche (1844-1900), the German writer and philosopher who said:

> Experience teaches us that nothing stands so much in the way of developing great philosophers as the custom of supporting bad ones in state universities. It is the popular theory that the posts given to the latter make them free to do original work; as a matter-of-fact, the effect is quite contrary. No state would ever dare to patronize such men as Plato and Schopenhauer. And why? Because the state is always afraid of them - it seems to me that there is need for a higher tribunal outside the universities to critically examine the doctrines they teach. As soon as philosophers are willing to resign their salaries they will constitute such a tribunal. Without pay and without honors, it will be able to free itself from the prejudices of the age. Like Schopenhauer it will be the judge of the so-called culture around it.[31]

In fact, this has happened, philosophical genius is often more fertile outside the universities.

Postmodernism and in particular deconstructionalism been said by more orthodox philosophers to have resulted in philosophy's loss of faith in its traditional tasks. It is important to discuss the role of postmodernism and deconstructionism as mechanisms undermining philosophical and scientific objectivity particularly in the universities of the United States. Modernism is a term applied to the scientific and technological changes and the accompanying cultural developments that commenced with the Enlightenment. Postmodernism claims that modernism and the thoughts that supported it have collapsed. Many American conservative educationalists see postmodernism, together with multiculturalism, as a threat to the objectivity of the university system. The key to the solution of all these problems lies with philosophy itself. The so-called "cultural wars" (to be defined), seen so acutely in the American university system, are a contemporary version of the skeptical debates that philosophers have engaged in for more than 2500 years.

Postmodernists are philosophers who proclaim the end of the modern age of philosophy, scientific and moral progress and all the grand projects of Enlightenment.[32] Postmodernism can be seen as an expression of a loss of faith not only in God and the ideas represented by God, but in "man" "himself" as postmodernist feminists observe.[33] Rather than relying upon logical argument and scientific evidence, postmodernists use rhetorical strategies such as parody and debunking to challenge the thoughts and institutions that dominate society.

Deconstructionism is a "subspecies" of postmodernism that primarily proclaims the death of language, logic and reference, at least as traditionally conceived. It also challenges hyper-rationalism of disciplines like traditional Anglo-American Philosophy. The movement arose in Paris in the late 1960s and has as its key figures Jacques Derrida, Michel Foucault and Roland Barthes. It has had a major impact upon the humanities and social sciences in America, Britain and Australia. The critical legal studies movement, a deconstructionist movement, has had great influence in American legal studies departments.[34]

While mainstream analytic philosophy, as practiced in British and Australian University departments, has exerted a diminishing influence upon both intellectual and daily life, French philosophy has had an important influence on our culture, media, and social policy. Many people who know little about academic philosophy may have heard of at least some of the ideas of a French philosophy. The great concern of the ordinary person about the meaning of life and why indeed there is a world at all, does not rate highly in the research agenda of orthodox Anglo American- Australian philosophy. Once it would have been dismissed as "unscientific" or "meaningless" , now it is merely regarded as "unimportant" compared with scientific topics such as the strong program of artificial intelligence, the creation of thinking machines. In general, analytic philosophy in Britain, Australia and the United States has been largely concerned with trimming the edges of the scientific/ technological model of the world.[35]

French philosophy however has a different starting point. It believes that the physical sciences are not the sole operators of "Truth". Instead, the mind, consciousness, language, social discourse and the personal world are the starting points. These points are reflected in the writings of deep ecologists and indeed many other philosophers. This is not an anti-scientific stance, it simply agrees that science, even the science of ecology, cannot answer all the questions and that there are other ways of knowing.[36] This "other" way includes value systems and religious systems - and we will have more to say about religion below.

Much recent French philosophy is concerned with the philosophical meaning of language or discourse. Perhaps the most famous French philosopher, one who through the schools of postmodernism and deconstructionism has exerted a major influence upon American academia is Jacques Derrida. However, in general, the scientific based philosophies in Britain, Australia and the United States regard Derrida's work as contradictory and meaningless. Derrida's style is more at home in the field of the great philosophical play writers and novelists, than in mathematics and physics. There, as in daily life, language is full of double meanings, metaphor and allusion. Derrida has to be understood as one would understand writers and poets, with an eye for hidden meanings. In fact, Derrida challenges the standard theory of the way words are thought by philosophers to obtain meaning, namely the idea that they do so by "corresponding" to some nonlinguistic object. Derrida has another theory of meaning, namely that the meaning of words is determined by their relationship to a network of other words in our language. However Derrida also addresses concepts of relevance to the world order, for example sovereignty, the rock upon which state politics has been built. Sovereignty is being "deconstructed" by military interventions and global economic systems and these events merit further analysis

Derrida's name is associated with the idea of *deconstructionism*. This is a method for showing that a sequence of words has accepted certain assumptions about language and so it has undermined itself in some way. How deconstructionism works varies from one article to the next depending upon the nature of the work. Let us choose an example, orthodox rational economics, the free market economy. Our understanding of "the free market" depends upon understanding the opposite, a non-free market economy. Deconstructionism involves showing that the distinction between the free market and non-free market is a false one, perhaps by showing that the free market is not really free at all, but is based on monopoly.

Deconstructionism, then, is primarily about questioning fundamental assumptions. Although mainstream philosophers like to think that they are free and open thinkers, most academic philosophers operate within a particular paradigm and seldom criticize the core assumptions of that paradigm. Derrida's work has been taken up by feminists and cultural theorists, many of whom are fundamentally dissatisfied with the present state of philosophy and society. Is it then any wonder that the orthodox philosophers have reacted so strongly against Derrida and deconstructionism? Philosophers like to set up problems in an either-or fashion whereas deconstructionism, like Eastern philosophy, says "perhaps both" or "perhaps neither" or maybe the whole question is not worth the effort. This directly threatens the security of employment and prestige of philosophy that depends upon endless dialogue with past ideas within an accepted framework![37]

The most easily misunderstood viewpoint of the deconstructionists is that history is a "fiction". When history is written, it is influenced by the culture and language of the writer. The history of World War II written by Japanese historians is vastly different to that of Western writers. Australian history written by the descendents of British settlers denies the genocide and injustice perpetrated on Aboriginal peoples. In writing history we favor our own concerns and justify our actions. Some historians have accepted a need for change and are seeking to rewrite history using a less biased approach. Clearly this process can go too far. One consequence of the acceptance of the idea that all history is fiction is that horrific events such as the Holocaust also become a fiction. In this scenario, the revisionist tales of contemporary neo- Nazis will have just as much validity as the historical work of Jewish scholars. But if we have good reason to believe in the existence of any historical event at all, then surely it is the Holocaust. It is clear that although history may be inadequate in reflecting the truth, to assert that it is a "fiction" is to overstate the case. Clearly the idea of a "fiction" is in need of deconstruction.[38] Alarm has been expressed at the historical viewpoint of the deconstructionists, but this questioning of history is relevant and each case must surely be judged on its merits.

A number of American conservative philosophers and public commentators have seen postmodernism and the movements associated with it such as cultural studies, feminism and multiculturalism as destructive of Western values of truth, freedom equality and justice. Multiculturalism has become part of this debate. The "cultural wars" debate in campuses in the Unite States has included multiculturalism because ethnic diversity has been used to attack the Western tradition, or so a group of conservative philosophers believe. This is because multiculturalism implies an acceptance and tolerance of widely different lifestyles, beliefs and practices and so questions the validity of our modern liberal lifestyles.[39]

Allan Bloom concluded that: "The crisis of liberal education is a reflection of the crisis at the peaks of learning, an incoherence and incompatibility among the first principles with which we interpret the world, an intellectual crisis of the greatest magnitude, which constitutes a crisis of our civilization".[40] Bloom believed that the American university has begun to decompose. Presumably American society will also dissolve, as the values of the West are abandoned and its vital traditions ignored. Although regarding Allan Bloom's book "ill considered", Bill Readings entitles his own treatise *The University in Ruins.*[41] For Readings the wider social role of university as an institution is now up for grabs. It is no longer clear what the place of universities is within society nor what the exact nature of that society is. One of the key factors responsible for this, Readings says, is the rise of the global economy and the decline of the power of the nation state. The national cultural mission that universities once had has gone. So what remains of the contemporary university is "post historical" in the sense that the institution has outlived itself. It is now survivor of a forgotten era.

Roger Kimball sees the crisis of the humanities produced by the deconstructionists, feminist and postmodern movement as constituting a "war against Western culture".[42] In this war Hollywood movies and comics replace

Shakespeare and other traditions; Star Trek and Seinfeld replace Chaucer and Keats and in Afrocentric studies, Beethoven has become an African American. Conservative critics allege that this is a war against the intellect.

The conservative challenge has been met to some degree by Lawrence W. Levine who notes that

> The 'traditional' curriculum that prevailed so widely in the decades between the World Wars and whose decline is lamented with such fervor by the conservative critics, ignored most of the groups that compose the American population whether they were from Africa, New York, Asia, Central and South America, or from indigenous North American peoples. The primary and often exclusive focus was upon a narrow stratum of those who came from a few northern and West European countries whose cultures and mores supposedly became the archetype for those of all Americans in spite of the fact that in reality American culture was forged out of the much larger and more diverse complex of peoples and societies. In addition, this curriculum did not merely teach Western ideas and culture, it taught the superiority of Western ideas and culture; it equated Western ways and thought with 'civilization' itself. This tendency is still being championed by contemporary critics of the university.[43]

Levine's defense of multiculturalism is well taken. However he does not deal with the real concern raised by the conservative critics of the "postmodern" university, namely that the postmodernists and cultural studies movements have abandoned the quest for objectivity and truth. Arguably one can accept multiculturalism, in the sense of acknowledging the existence of ethnic and racial pluralism and diversity without abandoning *completely* the Western ideas of truth and objectivity.[44] A book by one of the authors (Smith) *The Unreasonable Silence of the World*, deals with this issue in detail.[45]

**Where Now, for Philosophy?**

Philosophy has always been a battleground for ideas. With evermore rapid and frenzied changes in the world's social scene, the questioning of our values and concepts has become more searching and in many instances more radical. However, philosophy has been too busy with postmodernism and intellectual summersaults, and one should be critical that many current philosophers have aped the scientific and technological specialists to produce a jargon and phraseology unrecognizable even to those with high intelligence. This is irresponsible. The phenomenon of control of language or perhaps we should say control by language is discussed by John Ralston Saul[46] who points out correctly that many philosophers have joined the corporatist language makers, so rejecting their need to be understandable to all humans.

Unfortunately, the battlegrounds of philosophy have not been limited to the spoken and written word, they have also consumed academic departments around the world in sordid maneuverings. James Franklin has described the political battles in philosophy at the University of Sydney.[47] The Department of Philosophy split in two in 1973 between Marxist and mainstream philosophers who were generally to the political right. David Stove and David Armstrong, for example, were associated with *Quadrant*, a conservative monthly magazine and with the Australian Association for Cultural Freedom and were strongly opposed to the "left". In 1993 the schism was repaired in body if not in mind. Schism was avoided at Flinders University in South Australia where the philosophy department was led by the late Brian Medlin, founding Professor of Philosophy, one of the leaders of the Adelaide Vietnam protest movement. With the exception of one staff member, Marxism was embraced particularly in its Maoist form, from the 1960's until the late 1980's.

In both Sydney and Flinders, experiments were made in changing philosophical practice, and in recognizing its political dimension. Franklin saw these bold experiments, such as group assessment, as oppressive and believes that these left-wing philosophy departments were anti-learning. Certainly, the pursuit of a very narrow and dogmatic Marxist line by both departments, guaranteed that they would ultimately become irrelevant when their brand of Marxism died in the 1980s. Interestingly the Flinders radicals then began to teach "bourgeois philosophy." However for all their faults these departments could be viewed in their historical context as reactions against a particular style of philosophy. The radicals believed that philosophy should be relevant to the struggles of daily living and not a mere academic, scholarly discipline. But perhaps the most interesting aspect of the events is that they were tolerated by the university administration and by colleagues, although some of the departmental machinations and events are almost unbelievable and may best be told elsewhere.

The philosophy against which the Sydney and Flinders radicals were revolting unsuccessfully is still with us. It is a form of philosophy that Socrates would also have found repulsive. This philosophy is one concerned with the specialist investigation of clearly defined problems. It believes that philosophy should be a scientific discipline concerned primarily with the physical sciences. Moral philosophy is tolerated provided it is given a "scientific basis" . Metaphysics is acceptable provided that it is "scientific". Politics and the social sciences are regarded as non-studies. It is this type of limited technical philosophy that both the Sydney and Flinders radicals unsuccessfully opposed. Sadly these departments replaced one set of dogmas with another. For some unknown reason philosophers seem to be extremists, never settling for the happy and middle ground.

How then can philosophy be renewed when it is terminally ill? This is the key question which authentic philosophers must answer today. Its demise is sought by economically rationalist governments and is also fomented by forces of internal self-destruction. Why resuscitate the academic philosophy patient? The free spirits of philosophy have left the university body and function well beyond the gates, like Pierre Ryckmans in Australia. Ryckmans draws the parallel that: "the instinctive and universal awareness that the quest of the philosopher is as ancient and as essential to the human endeavor as the primeval occupation of the peasant".[48]

Culture has sustained us since humanity first settled down to sow, plant and harvest, and it is not by chance that the same word is used - to *cultivate* our gardens and to *cultivate* our minds. So philosophy lives and will continue to live outside academia. The present convulsion caused by the postmodernists should be viewed in the same light as the recurrent philosophical altercations that have occurred since the time of Socrates. The human mind will always pursue philosophical thoughts, even if there is no university discipline of philosophy.

Philosophy is a marker of the activity of the cultivated human mind; it should be part of academic endeavor and provide key input into other disciplines. In the University of Glasgow medical school, students studied philosophy as an elective subject.[49] These happenings reflect a re-entry of the humanities into the curriculum of some medical schools.[50] Its presence and activity is a measure of the health of the university, and its present collapse says much about the current sickness and debility of academia.

Having read this discussion on the role of philosophy, how would you react, if you are a parent, to the news that your son or daughter, having gained entry to an Ivy League University or medical school, had decided to study philosophy? Such a change might be seen to be as devastating as entry into a convent or monastery to the non-religious parent! You would ask: "What is the use of philosophy?" To which job would your offspring aspire, and is this adequate recompense for all your financial deprivations as parents? You might also be anxious that philosophy is no longer regarded as part of a "real world" created by an economically orientated society. Your discomfort would certainly be increased further by knowledge of contemporary philosophy teaching in the universities. However, in defense of your daughter or sons choice you should recognize that philosophy is the cornerstone of a broad liberal education which will develop and hone the powers of analysis and criticism so necessary in a world of information explosion. It will foster the open mind that can contribute to many walks of life. But in an academic course that traces human thinking from Plato to artificial intelligence, the student will participate in an analysis of contemporary issues ranging from health-care to nationalism. Will we embrace postmodernism? Yes, we will have a debate, postmodernism versus conservatism, during which your offspring will be able to make up their own mind. Furthermore we will address questions raised by the New Age Movement, neo-paganism as well as the relationship of postmodern spirituality to religion, medicine and the environment around us. Although university departments of philosophy are in decline, some schools of medicine, environmental studies and science are recognizing its importance and are introducing it into their teaching.

To lose philosophy and history from human life is to lose identity. It is a 1984 scenario that George Orwell predicted. In his totalitarian society the citizens had only vague memories of society before the revolution. The masses inhabited derelict cities with only slender clues as to what had gone before. It was a society stripped of collective roots or values. All activity was devoted to the glory of the Party. Orwell's predictions should have died with the collapse of communism, but to many people, unfettered capitalism is leading to similar ends. The universities' embracement of economic rationalism has led to the neglect and demise of cultural roots, history and philosophy. Our society led by the universities is losing touch with what has gone before. Society is becoming like a collection of anthills of many different sizes and shapes but all having the same intent. They work away at accumulation and growth with increasingly sophisticated practical abilities but with no insight or wisdom.

Philosophy provides an opportunity to analyze our culture, to recognize its non- sustainability and to look for alternatives. The philosophers and thinkers of deep ecology have done this but their writings remain closeted because philosophy is also part of our Western culture and works within it. Alternatives to present Western society are threatening for it is presumed that they will result in "turning back the clock", a return to primitive life and even cave dwelling! The response of those living at the time of feudalism might have been similar and democracy could not have been envisioned. It is equally difficult for us to imagine an alternative to our capitalist consumer society.

Despite the demise of academic philosophy, it is still taught and encouraged occasionally in some universities and your son or daughter should not be deterred if they wish to study it. Eventually higher education will have to be reformed to serve the entire needs of the world community. If humanity is to progress in happiness and fulfillment, philosophy will contribute to the many issues that confront us all. Philosophy will participate in the medical ethics debates and decisions in medical schools and health services. It will be required in all the professions, in politics, economics and science. Its dispersal will be a strength not a weakness, for like ecology, peace studies, democracy and poverty, everything is linked. Philosophy must not be compartmentalized; it is part of every aspect of human life.

Having defended the importance of philosophy as a cognitive enterprise which develops the critical faculties of students it is important to recognize one of the major limitations of modern philosophy, be it analytic philosophy or "French" philosophy: there is a self-limiting tendency of philosophers to focus in published works on technical problems and foundational issues rather than broad problems of human concern. For example, in the leading journal *Environmental Ethics*, a large number of articles from the first issue of the journal to the latest issue focus upon the foundational issue of whether an environmental ethics is possible and what its basis should be.[51] The approach of most of these philosophers is to adopt an inferential view of ethics, where moral propositions are deduced from general principles, and the general principles are in turn supported by "intuition".[52] This approach to ethics, and to philosophy itself has insuperable problems that renders the approach of these environmental philosophers untenable.[53] As Kenneth Sayre has noted, the discipline of environmental ethics has responded to the environmental crisis by giving us an exercise in ethical theory. But in the light of the possible self-destruction of industrial civilization and the collapse of civilized order, what is needed is a critical examination of those values, ideologies and doctrines that have led to this dire situation and an outline of an alternative world view, norms and values that would achieve an ecologically sustainable society. This book is a small contribution to that task and there are of course other important works which attempt to do this.[54]

Comments have been made in the pages of *Environmental Ethics* about why environmental ethics and philosophy has not had a significant impact upon public policy decisions.[55] David Jones had said that environmental ethics needs to embrace the power of "storytelling" to motivate human action.[56] The authors of this book have spent considerable time lobbying politicians on environmental matters, as well as two of us working on environmental law related issues. "Storytelling" could be a useful rhetorical tool in some respects but it requires more than the skilful telling of "stories" to achieve goals in the public arena. Donald Brown is correct in our opinion in observing that environmental ethics (and philosophy) needs to be applied to particular issues of public concern than it has been.[57] Too much of environmental ethics has been concerned with highly abstract foundational and metaethical debates without addressing specific environmental issues of concern such as global climate change. There are more papers written on postmodern feminist approaches to the environment, than there are on the ethical and philosophical issues associated with global climatic change, which can easily be confirmed by a Google search.[58] Hence, environmental ethics and philosophy have in general failed to critically examine the ethical and philosophical assumptions of the scientific, economic, legal and political arguments made in various public policy controversies. Environmental ethics and philosophy, because of an obsessive pursuit of the general and abstract, has made itself irrelevant to the major controversies of the day.

This situation is entirely curable. It means that philosophers need to spend less time addressing the "eternal" or "perennial" questions and more time working with experts in other disciplines such as psychology, medicine, climatology, environmental science, law, geography and so on, to address with first hand expert empirical data environmental and socio-political issues of concern. Almost all the major philosophical journals fail to do this. As has been said previously, if all the copies of these journals and their authors within them were destroyed the loss to society would be negligible.[59] For philosophy to become relevant to society once more, it needs to be engaged in the controversies of nations wherein the philosophers live.

John Gray's *Straw Dogs: Thoughts on Humans and Other Animals*[60] is an example of the sort of direction that environmental philosophy should *not* go. The book is very well written and free of the jargon, logical symbolism and technicalities which alienates non-philosophical readers from approaching philosophical texts. The book also sets out in almost breathtaking fashion to slaughter as many sacred cows as possible; on page four of the text Gray concludes that "in the world shown us by Darwin, there is nothing that can be called progress".[61] Darwinism presumably refutes the humanist belief in progress - which is of course to make the large and dubious assumption, never justified by Gray that Darwinism exhausts all there is to say on this subject.[62] The great bulk of humanity is moved only by self-interest, Gray believes and it "seems fated to wreck the balance of life on Earth - and thereby to be the agent of its own destruction".[63] And Gray continues pursuing other subjects: "Science will never be used chiefly to pursue truth, or to improve human life. The uses of knowledge will always be as shifting and crooked as humans are themselves".[64] Scientific research in neurophysiology shows, Gray believes, that our idea of free will is an illusion.[65] Finally, philosophy itself, upon which Gray bases his mediations "is a subject without a subject matter, scholasticism without the charm of dogma".[66] Truth, Gray says "has rarely been of the first importance to [philosophers]".[67] That, of course, must apply, self-referentially to his own work as well.

We cannot attempt to address point by point Gray's cynicism here, although in the above footnotes we have given an indication of how a response should proceed. It is clear that Gray's worldview ultimately holds out no hope of resolving the crisis of civilization because humans are in the end just a highly destructive animal species. It is possible that accepting such reductionist worldviews could lead to a self-fulfilling prophecy.

The concluding chapter of this book will examine the issues of values, ethics and global climatic change as this chapter has been focused upon the mere general issue of the importance of philosophy in responding to the environmental crisis. We turn now to look briefly at the relevance of religion as a response to this crisis. Again we approach this issue from a practical, commonsense position. Although there are great philosophical theological and empirical/scientific difficulties facing "fundamentalist" religious traditions-especially Judaism, Islam and Christianity - and many philosophically- inclined religionists would like to see a major metaphysical reform to their traditions, there is little hope of the great bulk of religious people who are "fundamentalists' or literalists turning away from conventional faith in the short term. It is the "short term"- a matter of decades - which we have seen is a crucial period of time for addressing central facets of the environmental crisis such as global climatic change.

**Religion and the Environmental Crisis**

> O Lord, how manifold are your works! In wisdom you have made them all; the earth is full of your creatures. Yonder is the great and wide sea with its living things too many to number, creatures both small and great

> … All of them look to you to give them their food in due season. (Ps 104:24-25, 27)

> Keep two truths in your pocket and take them out according to the need of the moment. Let one be "For my sake the world was created". And the other: "I am dust and ashes".
>
> - Rabbi Simcha Bunam[68]

If we are to address the relevance of religion to the environmental crisis, it is first necessary to understand, at least with respect to the Judeo-Christian position, some of the debates in theology that form a background to the environment argument. Again, as with our account of philosophy in the first part of this chapter, the following outline and overview is written for those who are not specialists in theology.

Some social theorists believe that the modern era is a secular one, and that for the West the era of "Christendom" has passed.[69] Christianity has, it is argued, been beating a hasty retreat in cultural and philosophical influences since the time of Enlightenment.[70] In the 17th and 18th centuries, under the influence of rationalist criticism, the Old and New Testament began to be subject to searching examination and criticism.[71] The former self-evident character of the Judeo-Christian conception of God was rejected in the light of the advancement of science and historical understanding.[72] This examination of the historicity of the Scriptures and the idea that in cases an allegorical interpretation must be made, was embraced by some early Christians such as Origen (? 185-254 AD). Biblical narratives, according to Origen, may not in all cases be literally true because the point of the Scriptures was not to record history but to present moral teachings. As Origen put it: "The Scriptures contain many things which never came to pass, interwoven with the history, and he must be dull indeed who does not of his own accord observe that much which the Scriptures represent as having happened never actually happened".[73]

Hermann Samuel Reimarus (1694-1768) and David Friedrich Strauss (1808-1874) questioned the divinity of Christ and regarded the Scriptures as mythopoeic rather than biographical.[74] Reimarus and Strauss' work would influence the "Higher Criticism"[75] (a primarily German school of biblical analysis based upon linguistic and textual appraisal) and the contemporary consensus among liberal Protestant theologians that "so far as the biblical historian is concerned... there is scarcely a popularly held traditional belief about Jesus that is not regarded with considerable skepticism".[76] As Hoffman and Larue put it:

> Biblical codes reflect older Mesopotamian regulations. Biblical psalms and Wisdom literature exhibit links with Egyptian writings. Biblical cosmological concepts parallel beliefs held by Israel's neighbours. The recovery of religious literature belonging to Canaanites, Assyrians, Babylonians and Egyptians have demonstrated that some biblical concepts may have been borrowed from other cults, and that festivals sacred to Jews and Christians have roots that reach back into pagan religious celebrations. In other words, the Bible, rather than proving to be a product of divine revelation, is clearly a human product, limited in what it contains by its setting in time (first millennium before the common era, and the first two centuries of the common era) and space (the ancient near eastern world).[77]

The idea that the Jewish and Christian (and Islamic) religions have their origins in Egypt was argued for by Godfrey Higgins (1771-1834), Gerald Massey (1872 -1908) and Alvin Boyd Kuhn (1881-1963).[78] The Hermetic Books of Egypt are the most ancient books in the world. The Seventh Book of Hermes, entitled, *His Secret Sermon on the Mount of Regeneration* has striking parallels with the sayings of Jesus, the sayings belonging, of course, to a much later era.[79] There are many other crucified saviours in other religious traditions, including Odin in Norse mythology,[80] and other man-gods (such as Zoroaster, the Persian God and Krishna), have been viewed as having a "virgin birth". [81] Not even the claim that the Jewish people gave the world monotheism stands up to historical scrutiny; early chapters of the Bible retain elements of polytheism which ancient Judaism had and the ancient Egyptians believed in one omnipotent, omniscient God-creator, Ra, the solar god, of which the other gods are only aspects of.[82] Harpur concludes that

> the entire Christian Bible, Creation legend, the descent into and exodus from Egypt, ark and flood allegory, Israelite "history", Hebrew prophecy and poetry, and the imagery of the Gospels, the Epistles, and Revelation are now proven to have been transmitted from ancient Egypt's scrolls and papyri into the hands of later generations who didn't know their true origin or their fathomless meanings.[83]

Archaeological research into ancient Israel also shows, according to Finkelstein and Silberman in their book *The Bible Unearthed*[84] that many of the key historical events depicted in the Pentateuch and the Deuteronomistic History, although containing the Jewish spiritual legacy are myths constructed as part of the national-building ideology and religious/political agenda of King Josiah of Judah during the late 7th century BCE. There is, in the considered opinion of Finkelstein and Silberman, no scientifically convincing evidence for the patriarchs, Exodus, the conquest of Canaan and the United Monarchy of David and Solomon. For example, the "Solomonic" gates, artitectural remains from cities allegedly fortified by Solomon according to 1Kings 9:15 have been now seen by a number of archaeologists to belong to 10th century BCE rather than Solomon's time, if he existed at all.[85] On the question of the origin of the Israelites, Finkelstein and Silberman conclude on the basis of archaeological surveys of Israel that the "process...[was]... the opposite of what we have in the Bible: the emergence of early Israel was an outcome of the collapse of the Canaanite culture, not its cause...most of the Israelites did not come from outside Canaan - they emerged from within it."[86]

The historicity of the New Testament, has as we have said, been open to varying degrees of doubt since at least the 17th century. The early church fathers were aware of the many Scriptural inconsistencies and made considerable efforts to reconcile diverging Scriptural accounts of events.[87] However, many leading Protestant theologians came to believe that the major "contradictions" within Christianity were between the belief in miracles and the supernatural on the one hand, and the naturalism and materialism of modern science on the other. For example, Rudolf Bultmann, a leading 20th century Protestant theologian, proposed that the worldview of the Bible was obsolete and that Christianity was in need of demythologizing whereby hermeneutical methods of exegesis the Christian doctrine is reformulated for a modern world.[88] Bultmann's own approach was a reformulation within the framework of the existential theology of the German philosopher Martin Heidegger (1889-1976).[89] Other theologians sought alternative philosophical foundations to their reformulations.[90]

This "death" of the traditional God[91] has led to a wide range of social and metaphysical reworkings of Christianity including feminist, Marxist, "dialectical" theism, "new age" and cosmic-Christ perspectives.[92] There has also been attempts of reformulating Judaism and Christianity in the light of the charge made by Lyn White, historian from the University of California, who argued that Christianity is at the historical root of our ecological crisis.[93] Science and technology, which have given humans mastery over nature, are products of Western culture. The philosophical roots of Western culture are Christian attitudes and principles and Christianity is arrogant towards nature and anthropocentric, seeing nature as having no reason for existence except to serve human beings. This has encouraged an attitude of exploitation.

White's thesis has generated considerable debate.[94] Some researchers have alleged that there is a statistically significant correlation between lack of environmental concern and Biblical literalism.[95] White's thesis can be restated for other religious traditions,[96] including Eastern traditions, [97] and there has been many attempts to derive an environmental ethics from traditional religious sources such as the Qur'an.[98] Others have sought to build alternative spiritual, metaphysical or philosophical systems.[99] Others still, have sought to show that Christianity, through the concepts of stewardship and frugality, can deliver an environmental ethics.[100] For example, Preston Bristow, is a creationist and an elder at the First Congregational Church of Woodstock, Vermont - but is also a conservationist. In his response to White's thesis published in the creationist magazine *Answers in Genesis*[101] he gives a response to White which would not be out of place, in a slightly revised form in the journal of *Environmental Ethics.*[102] Rather than encouraging exploitation, the teachings of Jesus support stewardship of a creation which God saw was good (Gen 1-2), neighborliness, frugality, non-violence and care for the poor (Colossians 1: 15-17; Luke 12: 42-48)[103]

Two best selling books - Richard Dawkins *The God Delusion*[104] and Sam Harris, *The End of Faith*[105] - have put the case that although the questions which religion deals with are philosophically meaningful and important to human survival, religious beliefs are irrational and religion has a negative impact on most aspects of human existence. Religion, as Harris puts it, leads us "inexorably to kill one another".[106] Surprisingly enough, especially so given that Richard Dawkins is an evolutionary biologist, neither author discusses the issue of religion and the environment. The case though can be made consistent with the atheistic critique of Dawkins and Harris that fundamentalist religions such as Christianity encourage a destructive and exploitative attitude towards the environment. James Watt, President Reagan's First Secretary of the interior, for example, has been *misquoted* as saying to the US Congress in 1981: "After the last tree is felled, Christ will come back". Although it appears that Watt did not say this,[107] many Christian fundamentalists hold to the idea of "dispensationalism", that the return of Jesus is so close that there is no need to conserve the environment. The bestselling Christian fiction of all time is the *Left Behind* series by Tim LaHaye and Jerry Jenkins, [108] dealing with life on Earth after the "Rapture", the time when the righteous are taken directly to heaven leaving the ungodly to deal with hell on Earth.[109] An influential part of the "religious right", a core part of President Bush's constituency, as Jon Carroll puts it "is actively praying for environmental degradation"[110] and the end of the world.

The theological and philosophical debates briefly outlined above have not had a significant impact upon the beliefs of fundamentalist Christians.[111] Further, as Richard Koch and Chris Smith point out in their book, *Suicide in the West*, as far as the growth in religious beliefs go "[The] Fundamentalists are winning".[112] To suppose as Richard Dawkins does that the weight of scientific and rational argument will change people's core religious beliefs, is in our opinion, not only unsubstantiated by empirical and scientific evidence (eg even Marxism and communism in the former USS and China did not bring about an end of religion), but is contradictory with his basic argument that religious beliefs are *irrational*.

This means, with respect to the environmental crisis, that it is hopeless to expect to convert Fundamentalist to some alternative "naturalistic" or "spiritual/metaphysical" position. People can and do live quite happily with modern technology *and* the belief that the world was created by God less than 6,000 years ago.

The great naturalist and evolutionary biologist E. O. Wilson believes that it is necessary for secular scientists to put aside the theological and philosophical concerns of the "culture wars" and to make a truce with fundamentalists and evangelical Christians.[113] In his latest book entitled *The Creation: An Appeal to Save Life on Earth,*[114] Wilson the secular humanist constructs his book as an "impassioned letter" to a hypothetical Southern Baptist Pastor. Whereas both Richard Dawkins and Sam Harris are combative and want to stamp out religion, Wilson more realistically accepts that basic metaphysical differences need to be put aside so that both camps can work to save the Creation. As he puts it: "religion and science are the two most powerful forces in the world today…[and if]…religion and science could be united on the common ground of biological conservation, the problem would soon be solved".[115] As Wilson puts it in conclusion:

> When you think about it, our metaphysical differences have remarkably little effect on the conduct of our separate lives. My guess is that you and I are about equally ethical, patriotic and altruistic. We are products of a civilization that rose from both religion and science-based Enlightenment. We would gladly serve on the same jury, fight the same wars, sanctify human life with the same intensity. And surely we also share a love of the Creation….For however the tensions eventually play out between our opposing worldviews, however, science and religion wax and wane in the minds of men, there remains the earthborn, yet transcendental, obligation we are morally bound to share.[116]

A hopeful sign is the Evangelical Climate Change Initiative which is an initiative by 86 US evangelical Christian leaders, including the presidents of 39 evangelical colleges, pastors of "mega churches" and the leaders of various church aid groups such as the Salvation Army.[117] This is an important development because as E.O. Wilson has observed there are 30 million members of the US National Association of Evangelicals, and if only one per cent of its members became religious conservationists, there would be 300,000 additional conservationists who may have considerable enthusiasm and passion.[118]

Apart from evangelical Christianity, Pope Benedict XVI on 1 September 2006, the First Creation Day, instituted by the Catholic Church, launched an appeal to preserve the "Creation, God's great gift". The "Creation…is exposed to serious risks from choices and lifestyles that could provoke its degradation". Environmental degradation "renders unsustainable the existence of the poor of the earth in particular". Thus: "In dialogue with Christians of various confessions, it is necessary to make a commitment to the care of creation by not squandering its resources and by sharing them in solidarity".[119]

Bartholomew I, Orthodox Christian leader, said in a 1997 speech that harming the environment is a sin: "To commit a crime against the natural world is a sin…to cause species to become extinct and to destroy the biological diversity of God's creation…to degrade the integrity of the Earth by causing changes in its climate, stripping the Earth of its natural forests, or destroying its wetlands… to contaminate the Earth's waters, its land, its air and its life with poisonous substances - these are sins".[120]

The attempt to change the view point of literalist or fundamentalist Christians who at present do not think like Bartholomew I, would be a major contribution to the quest for human survival on this endangered Earth. Even in the light of the theological concerns raised in this section, we still reject as implausibly idealistic the program of Richard Dawkins, which seeks to replace religion with the light of scientific reason and humanist morality. The theological and philosophical disputes discussed here will not be resolved in the near future, for these issues have engaged the best minds for centuries and no end of debate is in sight. The hour is too late for the energies of such minds to be continually channeled into these scholastic debates - as fascinating as such matters are. E.O. Wilson is correct in calling for an end to the cultural war between science and religion so that both camps can focus their individual attention upon dealing with the environmental crisis. In this chapter we have gone further and proposed that

E.O. Wilson's idea should be applied to philosophy as well.

**Conclusion: Philosophers Have Only Interpreted the World in Various Ways; the Point, However, is to Save It**

The above paraphrase and inversion of Karl Marx's famous aphorism from *Theses of Feuerbach (1845)*[121] serves as the conclusion to this chapter. We began our discussion by offering a defense of the importance of philosophy as an academic discipline in the university and in the affairs of life. Unfortunately academic philosophy in the US, Britain, Australia and Canada (but much less so in continental Europe), was made into a type of scientific, technical discipline in the 20$^{th}$ century, under the influence of various philosophical movements such as logical positivism and analytic philosophy.[122] Philosophy was to address problems primarily in the physical sciences and even then, philosophy was to be confined by the reductionistic agenda of showing that all reality (universals, matter, mind, space, time numbers and abstract essence etc.) could be explained within the confines of physics.[123] Little attention was given to defending this methodological assumption, something extraordinary given the foundational debates about the fundamental nature of reality in physics itself.[124] For these physicalist philosophers, the world of ordinary objects - of land masses, vegetation, climatic systems and the like-did not *really* exist. Only the constituents of the latest physical theory (e.g. string theory[125]) did. But, if this is what philosophy is about, then there is little reason for the "ordinary object" tax-paying public to support a discipline which essentially debunks them - and the reductionist philosopher can hardly object since she/he is hardly a *real* thing and neither is funding! Philosophy came to be seen by university administrators and funding "razor gangs" as an esoteric, even extremist discipline, that made no important contribution to economic and social life.

Contrary to the view of the technical, analytic philosophers, we believe that a socially and environmentally relevant philosophy, engaged with the real world problems of the day, rather than eternal or perennial questions, is what is needed today.[126] There is nothing wrong with undergraduate students receiving background training on the great problems in the philosophy of logic, metaphysics, and religion, as we argued earlier in this chapter. However, it is irresponsible to pursue postgraduate and academic research in esoteric areas of philosophy that we have very little contact with the "real world" and indeed, which typically concludes that it does not exist.

Many philosophers, of course, are aware of the limitations of academic philosophy and many have sought to address pressing issues of concern, such as the environmental crisis, through ethics research - in particular outlining systems of environmental ethics such as deep ecology and ecofeminism. Although this is an advance on physicalist reductionism, it has its own problems. The idea of this research program is that once the fundamental ethical framework is devised, then this program could be used to deal with the ethics of a whole array of environmental problems, such as global climatic change. However the discipline of environmental ethics has not moved very far from first base and from the late 1970's until the present day, philosophers are still largely preoccupied with defending the foundations of their systems. No doubt this practice will continue until an environmental collapse spells the end of civilized order- and the end of this type of philosophy.[127]

Against the current of contemporary philosophy we believe that there is an alternative. Philosophy should be an interdisciplinary critical and analytic cognitive enterprise concerned with addressing problems of life to obtain "wisdom".[128] At the present point in the human story, global climatic change and its wide value and ethical ramifications should be at the top of the philosopher's research list. As we will see in the concluding chapter, global climatic change, comes near the bottom of the citation list. Academic philosophy will continue to be largely irrelevant to the affairs of life until academic philosophers fundamentally change their approach. "Real philosophy" will be done by others, typically outside the university.

# Chapter 7

## Conclusion: Values, Ethics and the Ecology of Sustainability

It is one of the great illusions of progressive thought that there is always something to be done; that there is a solution to every problem that faces us, that we can discover that solution by reason, and that the solution consists in doing something, either politically or individually.[1]

Presidents of the United States tend to propose solving our problems by "growing" our economy. The problem of global warming cannot be solved by "growing" our economy fueled by fossil- fuels, unless some new unforeseen technology for recapturing the carbon is created and made economically feasible very rapidly. Carbon emissions must be reduced; the only way to do that is to use less carbon-based fuel, as long as there is no recapture technology in place. The only way to do that, as long as the economy is overwhelmingly dependent upon fossil-fuels, is to shrink the economy... Right or wrong, we will not do it! No U.S. president is likely to suggest it. If a President were politically foolhardy enough to suggest shrinking the economy, the U.S. populace would drive him or her out of office, even if he or she could correctly say that the number of jobs would not shrink.[2]

It comes down to this: modern civilization has no future. It confronts the same lethal combination of ecological collapse and inner decay that has extinguished previous civilizations.[3]

### The Limits to Growth and Global Climatic Change

This book has argued that the problem of global climatic change is but one part of a series of converging environmental and social crises which threaten human civilization. It has been shown that economic and technological "solutions" to the problem of global climatic change fail because these alleged solutions ignore fundamental problems with the dominant value system of the modern world, which are laid bare by the three quotations opening this chapter. This concluding chapter will expand on a theme developed in the previous chapter, which discussed philosophy, religion and the environmental crisis. Here the focus will be upon the role of ethics and values in dealing with global climatic change. In particular, the chapter will examine the presently limited - but growing field of the ethics of climate change - but from a "limits" to growth perspective, as outlined in this section.

Let us begin with an example. French researchers at the Climate Mission at the Caisse des Depots have stated that cattle are much greater contributors to global warming than previously thought because of their methane and nitrous oxide production.[4] Bovine gas consists of primarily methane and nitrous oxide which are volume-for-volume, more effective than CO2 at trapping heat. France's 20 million cows account for 6.5 per cent of France's greenhouse gas emissions. Bovine belches emit 26 million tonnes of greenhouse gas emissions into the atmosphere, and their feces produce another 12 million tonnes. French oil refineries produce, by contrast, 12 million tonnes of greenhouse gas emissions. One wonders what carbon trapping process would be proposed to deal with bovine belches and flatulence! As this example clearly shows, dealing with the full extent of the problem of global climatic change will involve an examination of the rationality and sustainability of the modern Western consumer lifestyle (in this example, "burger culture") and global capitalism.[5]

There is a large literature arguing that an ecologically sustainable society is incompatible with a growth economy.[6] Writers in this, "limitationism" research tradition generally take the view that modern capitalism (and socialism/ communism) is based upon continuous economic growth (defined as an increased level of matter-energy throughput in the economy) - and this, for various thermodynamic and ecological reasons is not sustainable.[7] Human industrial activity is constrained within various "limits to growth" and some writers in this field believe that "overshoot" has already occurred, meaning that humanity is essentially living on" borrowed time" at present consumption levels because of its use of the "ecological capital" of the future.[8] A "steady state"[9] or "conserver society",[10] which involves much simpler, largely self-sufficient lifestyles is advocated.[11] This limitationist or "limits to growth" framework is generally accepted by workers in the ecological health paradigm, because the very point of an ecological approach to human health is to situate human beings as complex bio-social organisms within their cultural-physical environments. Human beings may well use technology, but they are still biological organisms, subject to ecological limits.

The IPCC acknowledged the existence of this debate in its brief discussions of "Global Future Scenarios".[12] The database contained 124 scenarios from 48 sources. The scenarios range from the most pessimistic (social and/or environmental collapse) to optimistic views of unlimited economic prosperity from a high-tech future. No choice is made by IPCC among the scenarios and it is not clear what the analysis hoped to achieve. Forty eight sources is a rather small number to consider and the IPCC sample at that is biased towards more conventional sources who believe that economic growth and technological innovations will save humanity. Further, even given this bias, it would have been helpful if some sociological meta-analysis had been conducted examining *why* such a wide range of possible futures had been identified by futurologists. Does this mean that there is an inherent indeterminary in such future predictions due to an underdetermination of the theory by data - that there is simply not enough empirical evidence to know even with a degree of probability?[13] For example, Stephen Haller, in his book *Apocalypse Soon?* claims that our "best science can give us no assurance that doomsday is either likely or unlikely", because predictions of global catastrophe based upon existing models of global systems are "uncertain".[14] This mode of thought has led to philosophers and other theorists exploring the application of various decision principles for policy making under conditions of uncertainty, [15] such as the precautionary principle.[16]

Andrew Simms in his book *Ecological Debt: The Health of the Planet and the Wealth of Nations*[17] argues that the evidence about global climatic change does favor a scenario. Simms argues that the problem of global warming cannot be adequately dealt with by high tech fixes, but requires cuts in consumption and a change in lifestyle, at least in industrialized societies. He is critical of the consumer lifestyle of superfluous shopping, of greed and the accumulation of luxury items and a "throw away mentality. A "fast food" lifestyle and "junk culture" based upon exponential economic growth is a recipe for social disaster: "Economic globalization that depends on fossil fuels for its life blood is... socially divisive - increasing risk and instability - and undermining the basic security needed for viable livelihoods and stable communities".[18] In particular, Simms rejects the technology/efficiency argument which underlies many of the scenarios in the climate change debate. He quotes calculations made by the astrophysicist Alberto di Fazio. At present there is almost a "total correlation" between the size of the global economy and historically recent increases in CO2 emissions. With the global economy doubling every 17 years, technological optimists are faced with being on a technological treadmill, continually struggling to keep up with the pollution consequences of economic growth. As Simms puts it: "To make the planet fit for human life, CO2 in the atmosphere was converted by natural processes into fossil fuel reserves over the course of 180 million years. According to di Fazio, humanity is converting fossil fuels back into the atmosphere ' a million times faster'". [19]

The IPCC report does recognize that there is an extensive sociological and cultural anthropological literature showing that human lifestyles and behavior cannot be explained solely by economic rationality and in terms of the changing relative prices of commodities.[20] Social and cultural variables overlay economic ones. From the perspective of cultural anthropology, lifestyles are expressions of personal identity, attempts by individuals to position themselves in the socio-cultural universe.[21] The IPCC then concludes that attempts to intentionally change consumer lifestyles are "bound to fail" because of the deep socio-psychological roots of consumer culture. Social transformation will require acts of personal transformation and more.[22]

The IPCC makes sketchy suggestions about how the "barrier of consumption-based identity" could be overcome. One suggestion is that non-Western cultures (e.g. East Asia) view the self as a socially interdependent entity and that "intercultural communication" may aid in breaking down Western individualism. The phenomenon of rising individualism in East and South East Asia with rapid industrialization and commodity wealth, is not considered. The need for environmental education is mentioned as is "creative democracy".[23] These themes are not developed in satisfactory detail.

In general, philosophers have also failed to appreciate the ethical and philosophical consequences of the environmental crisis from a limits to growth perspective. The preoccupation of the few philosophers working in the field of ethics and climate change[24] has been to apply general moral theories such as utilitarianism (which in general defines social welfare as the sum of individual utilities), human rights theory and other theories, to an array of questions generally associated with questions of distributive justice.[25] The basic problem of global climatic change for these philosophers is about the distribution of responsibility for solving the problem. How much should "the poor" contribute as a fair allocation of costs?[26] What is a fair allocation of greenhouse gas emissions for nations, over the transitional period to long-term allocations, and over the long-term?[27]

The ethics of global climate change is significantly complicated by the question of intergenerational justice and the somewhat larger question of why posterity matters: "What is the correct principle for evaluating the requirements of both present and future inhabitants of the world?"[28] Do future generations have "rights" and if so, what are they?[29] The general problem facing this field of research is that the main families of ethical theories used to address the issue - such as utilitarianism,[30] contractarian theories,[31] rights theories,[32] Kantian theories[33] and communitarian theories[34] all face major objections,raised by competing theorists, the objections in themselves not always presupposing the correctness of the critic's own ethical position.[35] For example, one of the most respected books in this field, *Reasons and Persons* by Derek Parfit,[36] can be read as a major exploration of the problems facing utilitarian responses to philosophical problems of population theory and intergenerational ethics, problems which Parfit explores with fine and clear analysis, but which he admits in his conclusion that he does not solve.[37] As another example, the communitarian Alasdair MacIntyre concludes that there is no reason to accept the existence of human rights: "the truth is plain: there are no [ natural or human] rights and belief in them is one with belief in witches and unicorns".[38]

A deeper aspect of the problem of the search for the truth in ethics relates to the super-ultimate justification of moral theories. On what basis should one accept *any* moral theory? According to Bernard Williams, we "would be better off without morality," [39] for moral claims are not irrational to deny and moral claims do not make the strangest claims on us.[40] Not only do various fundamental ethical principles common to all or most moral theories, lead to contradictions,[41] but morality itself seems to conflict with personal survival values. For example, the demands of morality to help those in need according to one line of thought, "makes the pursuit of practically all sources of personal fulfillment morally impermissible."[42] According to Shelly Kagan in *The Limits of Morality*, morality requires doing those acts not otherwise forbidden that lead to best results overall.[43] These are acts not favoring one's own interests but of the world as a whole to produce the best results. Kagan is concerned with human conduct, but other theorists may wish to extend the analysis to the non-human world, so that moral actions produce the best results for humanity *and* nature. On this account, most of our actions are morally wrong because time and money could be better spent.[44] This means, self-referentially that writing books such as *The Limits of Morality* is immoral, since the time, effort and resources could be better spent, say saving lives in the "poor countries". It also means that if there are no limits to the sacrifices that moral agents must make to pursue the good, then people will respond by saying "to hell with morality" and may make even less sacrifices than they would if the demands of morality were not so severe.[45]

The application of moral theory to the problem of global climatic change generates its own related set of problems. Many moral philosophers, be they human rights theorists or utilitarians, have adopted a cosmopolitan justice approach to the distributive justice questions of global climatic change,[46] based on the idea of interconnected obligations, with things happening in one part of the world affecting elsewhere.[47] Cosmopolitan accounts of justice reject the idea that nation-states and communities have ethical standing.[48] Individuals are the basic unit of morality and moral theories should be egalitarian, impartial, universal and individualistic.[49] According to Charles Jones, cosmopolitanism underlies modern liberal political theory.[50] Cosmopolitans have tended to accept the view that there should be equal *per capita* rights to the atmosphere[51] and that the developed world bears the lion's share of responsibility for environmental repair as "the present global distribution of wealth is the result of the wrongful expropriation by a small fraction of the world's population of a resource that belongs to all human beings in common".[52] Paul Harris says that much of the greenhouse gas emissions in the United States "comes from arguably frivolous and certainly nonessential activities, whereas most of the emissions of the world's poor are due to activities necessary for survival or achieving a basic living standard."[53] Henry Shue says that the developed countries have an obligation to "shoulder burdens that are unequal at least to the extent of the unfair advantage previously taken." [54]

There are various problems with the cosmopolitan approach. It assumes an "us" and "them" world, where the developed nations have through colonialism, imperialism and now globalization,[55] exploited an underdeveloped "victim". This view ignores the role that global corporations, especially the international financial sector, has played in the story, which complicates the politically correct plot considerably.[56] Also, while we do not underrate the exploitation which colonialism produced, there is some merit to the conservative argument that colonialism may have helped developed poorer nations in *some* respects, even if the net results are negative. As well, although consumerism in the developed countries has, as Harris rightly recognizes, a frivolous aspect, development in the West did produce the medical and agricultural technology which has allowed the populations of the poorer countries to rapidly expand over the last two hundred years. People are alive who would not have been otherwise alive, even if they emit less greenhouse gas per capita than individuals in the West. Ethics requires that these Parfit-style benefits[57] based on identity, also be taken into account. Further to this, "few environmental risks are not induced or exacerbated by the pressures of rapid economic development."[58] Unless one believes that there is something racially special about Europeans, the starting of the industrial revolution in Europe, rather than Asia, Africa or Australia, was a matter of historical circumstance. The problem of global climatic change would have occurred in say Asia if circumstances were different. The important thing to note is that this problem, along with all the other environmental problems constituting the environmental crisis, is a problem of modernity and development rather than of the immorality of the West. Tracing the genesis of this, establishing causality and assigning responsibility will be more complicated that the standard ethics literature assumes.

The cosmopolitan idea of equal per capita emissions faces economic and efficiency objections. As MacCracken puts it:

> With not all energy being from carbon, with different needs for energy, and with different locales providing different opportunities for non-carbon energy, the difficulties of an equity based accounting scheme could be insurmountable. In addition, basic economic theory teaches that arbitrary and inflexible distributions of responsibility (e.g. requiring every nation to have the same per capita emissions) would tend to raise costs and create inefficiencies, only some of which could likely be overcome with an international permit system.[59]

Cosmopolitan supporters assume that the equal emissions share for everyone will involve the developed world contracting their greenhouse gas emissions, while the developing world increases their carbon emissions "to catch up".[60] This however takes a limited view of the environmental crisis centered only around relative differences in per capita greenhouse gas emissions. From a limits to growth perspective, *both* the developed and developing world have unsustainable social systems, the developed world through productive and consumption aspects of the environmental impact equation, and the developing world through excessive population growth.[61] The time for "catching up" is over, and is, in any any case nonsensical from the limits to growth position, for if the approach to development taken by the West is flawed and a mistaken road, why should others be encouraged to go down that road? [62] Ted Trainer points out that Ladakh near Tibet is a society where people have only hand tools and a low average GNP per capita, but they have an extremely culturally rich life:

> The Ladakhis are kind and generous. They have extensive community

support systems. They look after each other and value their old people, they have rich spiritual life, a relaxed lifestyle, and robust and sustainable food producing systems despite fiercely cold winters and a short growing season. Their production is labor-intensive, yet the pace of work and life in general is relaxed, with much time for ceremonies and religious observance. No one is isolated or lonely, they do not waste but recycle everything, they have no interest in power, domination or competition. They are very conscious of their dependence on nature, they are multi-skilled and practical, and they live simply. There is no crime and no poverty and no drug problem and no social breakdown. Above all they are notoriously happy people.

A strong case could be made that the people of Ladakh have a far superior culture to that of rich western countries. It is quite disturbing to ask of the Ladakhis "What development do they need?" The traditional Ladakh villages are in my view, more or less satisfactorily developed. A few possible technical changes suggest themselves, such as what to do with improved infant health care and perhaps the introduction of more tree crops with windmills. But they do not need supermarkets, television, freeways, cars, throwaway products, and packaged imports, higher incomes or a higher GNP. In fact it is precisely the coming of these things, the penetration of Western economic forces, that is now rapidly destroying the ancient culture of Ladakh.[63]

There is thus a major disagreement about the principles and values by which responsibility for greenhouse gas emissions should be divided between nations[64] and it is reasonable to suppose on the basis of past performances that the philosophers will not reach some sort of consensus before civilization falls. Most importantly, again from the limits to growth position, this entire debate begs a fundamental question about the nature of economic development and ecological sustainability. The poor "victim" nations on this view, must be compensated by the affluent West, so that these nations can develop as well, but in more sustainable ways, and become ecologically politically correct versions of the West. This means that the poorer nations must be still connected to the West through the global economic system. However, as Trainer, echoing the works of other radical development theorists concludes, globalization has been a disaster for the poorer nations:

> It forces the poorest and weakest to compete with the very strongest for resources and markets. It prohibits the Third World governments from regulating, protecting, assisting or intervening. Thus, it prevents them from taking control of their own development. Development is only of whatever corporations want to develop and unless governments assist this their country will be boycotted by investors and banks and the World Trade Organization. Globalization enables the corporations to come in take all the resources, markets and businesses they wish. Globalization forces poor countries to compete against each other to export resources to rich countries as cheaply as possible. Globalization enables the corporations to destroy businesses and livelihoods, by taking the sales and putting local people out of work. Globalization allows the corporations to take over industries meeting basic needs such as water supply, and maximize profits by jacking up prices and cutting supply to the poor. If governments resist they come under massive attack from rich world institutions; e.g., they will be cut out of trade and cut off from loans.[65]

According to this position, a sustainable strategy for poor nations and the rich, is to live within limits,[66] minimize economic connections with the global financial system[67] and maximize local and national self-sufficiency.[68]

Hershel Elliot in his book *Ethics for a Finite World*[69] presents a critique of conventional ethics from a limits to growth perspective which the present authors find instructive. Ethics should move away from its present a priori methodology and its basis in thought experiments and "intuition" and become more closely integrated with the social and natural sciences, because moral theories are about how human beings can live on earth and so factual matters are highly relevant to ethics.[70] We are in agreement with this orientation and believe that the failure of moral theory to solve its fundamental problems is an indication of something fundamentally wrong with Western ethics. In this section, with respect to the problem of global climatic change we have argued that the conventional approaches to dealing with the moral problems in the field do not appear to be successful. It is necessary in a situation of conceptual perplexity to move outside of the system of buzzing confusion and to take a fresh look at the matter. We suggest that too much intellectual energy by fertile and highly creative minds have been devoted to the articulation of general ethical theories which upon analysis inevitably give rise to unending philosophical debates about their problems and paradoxes. Meanwhile the world "burns".[71] Talents and energies need to be redirected.

**More Appropriate Research Problems**

The more appropriate research problems arise when one moves outside of the paradigm of liberalism and conventional growth-based values centered around orthodox economic theory.[72] If the ecological crisis is as bad as the evidence presented in this book suggests it is, then it is crucial for both the developed and developing world to adopt a new economic system not based upon economic growth as well as accepting a fundamental change in the present value system, a system which is enmeshed with consumerism, materialism and self-interest.[73] Simpler lifestyles, typically involving voluntary simplicity, local economic self-sufficiency and a greater degree of community co-operation, will be necessary. But the problem is, as we have detailed in other works, that the modern world has moved in exactly the opposite direction needed for creating ecologically sustainable societies. Economic globalization has moved all societies far away from local economic self-sufficiency and immigration and multiculturalism has further fragmented communities. A fundamental questioning of the basis of capitalism is now seldom conducted even by writers on the "left" who also accept that capitalism is the "only game in town".[74]

Global climatic change is, as Stephen Gardiner describes it, a "perfect moral storm", where a "perfect storm" is defined as "an event constituted by an unusual convergence of independently harmful factors where this convergence is likely to result in substantial, and possibly catastrophic, negative outcomes."[75] Global climate change presents, Gardiner argues, a worse situation than Garrett Hardin's "tragedy of the commons,"[76] when intergenerational aspects of global climatic change are considered. Future generations are not here to assert claims and have no bargaining power. The present generation, despite some altruistic actions, is still dominated by self-interest. This same egoistic inertia grips each generation, g1, g2, g3, resulting in continuing degradation of the environment.[77]

One aspect of this intergenerational tragedy of the commons is the institutional inadequacy of democratic institutions. As Gardiner says: "It is doubtful whether such institutions have the wherewithal to deal with substantially deferred impacts".[78] Present democratic institutions operate on a short time horizon. The benefits of continuous greenhouse gas emissions are on present voters, but the costs are on people not yet in existence. Barry Holden in his book, *Democracy and Global Warming*[79] argues that the problem of sacrifices now, for long-term gains can only be effectively dealt with by democracy - in fact global democracy. We have addressed this debate in another book, *The Climate Change Challenge and the Failure of Democracy.*[80] We do not intend this book to be the last word on this problem, but rather one of the first dealing with the frightening consequences of modernity's endgame. It seems to us that Gardiner's intergenerational tragedy of the commons problem is a challenging, unsolved and vitally important problem which has the implication of bringing the democratic approach to ecological problems into question. This is a much more worthwhile field for creative philosophical minds than the present preoccupation with the mechanics of highly abstract ethical theories, and the perennial debates that they generate.

Surprisingly little has been written on how humans can make a peaceful transition to a sustainable society. Without a solution to this problem humans, if they survive, will still make a transition to sustainable lifestyles, but they will do so because of the remorseless working of things[81] under the iron glove of ecological necessity.[82] Humanity's time left for choosing to make a "soft landing" is fleeting.

---

## Notes

### Chapter 1

[1] Sir Crispin Tickell, "Religion and the Environment," Lecture delivered to *The Earth Our Destiny* Conference, Portsmouth Cathedral, November 30 2002 at http://www.crispintickell.com/page18.html.

[2] J. Dimbleby, "The Coming War," *The Observer*, October 31 2004 at http://observer. guardian.co.uk/comment/story/0,6903,1340066,00.html.

[3] "King Sees 3 C Rise as Nearly Inevitable: Stark Warning Over Climate Change," BBCNews.com, April 14 2006 at http://www.heatisonline.org/.

[4] According to the dust jacket of J. Lovelock, *The Revenge of Gaia: Why the Earth is Fighting Back - and How We Can Still Save Humanity*, (Allen lane, London, 2006).

[5] J. Lovelock, "The Earth is About to Catch a Morbid Fever that May Last as Long as 100, 000 Years," *The Independent*, January 16 2006 at http:/www.comment.independent.co.uk/commentators/article338830.ece7, also cited in *New Scientist*, January 21 2006, p.13.

[6] Lovelock, as above. See also on the likelihood of the collapse of Civilization, J.W. Smith, G. Lyons and E. Moore, *Is the End Nigh? Internationalism, Global Chaos and the Destruction of Earth*, (Avebury, Aldershot, 1995); J.W. Smith and G. Sauer-Thompson, *Beyond Economics: Sustainability, Globalization and National Self-Sufficiency*, (Avebury, Aldershot, 1996); J.W. Smith and G. Sauer-Thompson, *The Unreasonable Silence of the World: Universal Reason and the Wreck of the Enlightenment Project*, (Avebury, Aldershot, 1997); J.W. Smith and G. Sauer- Thompson and G. Lyons, *Healing a Wounded World: Economics, Ecology, and Health for Sustainable Life*, (Praeger, Westport, Connecticut and London, 1997); J.W. Smith, G. Lyons and E. Moore, *Global Meltdown: Immigration, Multiculturalism and National Breakdown in the New World*

*Disorder*, (Praeger, Westport, Connecticut and London, 1998); J.W. Smith and G. Sauer-Thompson and G. Lyons, *The Bankruptcy of Economics: Ecology, Economics and the Sustainability of the Earth,* (Macmillan, London and St. Martin's Press, New York, 1999); J.W. Smith, G. Lyons and E. Moore, *Global Anarchy in the Third Millennium? Race, Place and Power at the End of the Modern Age,*(Macmillan, London and St. Martin's Press, New York, 2000).

[7] Lovelock, cited note 5. See also Michael McCarthy, "Environment in Crisis: 'We Are Past the Point of No Return'," *The Independent,* January 16 2006 at http://news.independent.co.uk/environment/article338879.ece .

[8] Lovelock, cited note 5.

[9] M. Milliken, "World has a 10-year Window to Act on Climate Warming-NASA Expert", *Common Dreams News Center,* September 14 2006 at http://www. commondreams.org/headlines06/0914-01.html.

[10] J. Hansen, "Is There Still Time to Avoid Disastrous Human-Made Climate Change?" Keynote Address to the Third Annual Climate Change Research Conference, Sacramento, California, September 13 2006 at http://www.climatechange.ca.gov/events/2006_conference/presentations/2006-09-13/2006-09-13_HANSEN. PDF.

[11] Professor Peter Smith, a professor of sustainable energy at the University of Nottingham, said in a speech at the British Association Festival of Science in Norwich in September 2006, that the world has only 10 years to implement carbon-clean new technology before climate change reaches a point of no return. The present level of carbon dioxide in the atmosphere is 380 parts per million(ppm). Professor Smith has said: "The scientific opinion is that we have a ceiling of 440 parts per million (ppm) before there is a tipping point, a step change in the rate of global warming. The rate at which we are emitting now, around 2 ppm a year and rising, we could expect that that tipping point will reach us in 20 years time. That gives us 10 years to develop technologies that could start to bite into the problem". See A. Jha, "Energy Review Ignores Climate Change 'Tipping Point,' "*Guardian Unlimited,* September 4 2006 at http://www.guardian.co.uk/science/story/0,,/1864802,00.html.

[12] J. Hansen (*et al*), "Global Temperature Change", *Proceedings of the National Academy of Sciences,* vol. 103, no.39, September 26 2006, pp.14288-14293, published on-line www.pnas.org/cgi/doi/10.1073/pnas. 0606291103; F. Pearce,

"One degree and We're Done For", *New Scientist*, September 30 2006, p.8.

[13] Hansen (*et al*), as above at p.14288.

[14] Hansen (*et al*), as above at p.14292.

[15] Hansen (*et al*), as above at p.14292.

[16] *The Bulletin Online*, "5 Minutes to Midnight," at http://www.thebulletin.org/minutes- to- midnight/timeline.html.

[17] As above.

[18] As above. See also S. Keen, "A Secular Apocalypse," *Bulletin of the Atomic Scientists*, January/February 2007, pp. 29-31.

[19] N. Oreskes, "The Scientific Consensus on Climate Change," *Science,* vol. 306, December 3 2004, p.1686.

[20] These reports are available at http://www.ipcc.ch/pub/online.htm.

[21] IPCC, Working Group I, Fourth Assessment Report, *Climate Change 2007: The Physical Science Basis*, at <http://www.ipcc.ch >; R. Harrabin, "Consensus Grows on Climate Change" at http://news.bbc.co.uk/2/hi/science/nature/4761804.stm.

[22] IPCC, as above, p.7

[23] IPCC, Working Group I, Fourth Assessment Report, Climate Change 2007: The Physical Science Basis (Summary for Policymakers), at http://www.ipcc.ch .

[24] As above p.7.

[25] S. Rahmstorf (*et al.*), "Recent Climate Observations Compared to Projections," *Science*, February 1 2007, doi: 10.11261/science.1136843; I.M. Howat, (*et al*), "Rapid Changes in Ice Discharge from Greenland Outlet Glaciers," *Science*, vol. 315, no. 5818, March 16 2007, pp. 1559-1561.

[26] IPCC, cited note 23, p.8.

---

[27] As above p.10.

[28] As above.

[29] IPCC, Working Group II, Fourth Assessment Report, *Climate Change 2007:Climate Change Impacts, Adaptation and Vulnerability (Summary for Policymakers)* at <http://www.ipcc.ch >, p.7.

[30] M. Warren, "Science Tempers Fears on Climate," *The Weekend Australian*, September 2-3 2006, p.1. The alarmist scenarios referred to include a one meter rise in sea levels by the end of the 21st century. See chapter 2 for a further discussion of "alarmist scenarios"of abrupt climate change.

[31] J. Giles, "US Posts Sensitive Climate Report for Public Comments", *Nature*, vol. 441,May 4 2006, pp.6-7, doi:10:1038/441006a.

[32] S. Knight, "Scientist Issues Grim Warning on Global Warming," *Timesonline*, June 15 2006 at http://www.timesonline.co.uk./article/0,, 2-2134760,00.html.

[33] Cited from Knight, as above. See further P. Baer and T. Athanasiou, "Honesty About Dangerous Climate Change" at http://www.ecoequity.org/ceo/ceo_8_2.htm.

[34] J. Salliot, "Kyoto Plan No Good, Minister Argues", *Globe and Mail*, April 8 2006, p.A 5.

[35] Editorial, "Kyoto in Crisis," *New Scientist*, 8 July 2006, p.3.

[36] As above.

[37] As above. However on January 31 2007, Chancellor Angela Merkel of Germany ended state subsidies for the German coal mining industry, effectively shutting it down: R. Boyes, "Germany to Bury its Coal Industry," *The Australian*, February 2007, p. 10.

[38] F. Pearce, "Kyoto Promises are Nothing but Hot Air," *New Scientist*, June 24 2006, p.10.

[39] As above. On the climate forcing potential of various greenhouse gases see J.T.

Houghton (*et al* eds.), *Climate Change 2001: The Scientific Basis*, (Cambridge University Press, Cambridge, 2001).

[40] D. Freestone and C. Streck (eds.), *Legal Aspects of Implementing the Kyoto Protocol Mechanisms: Making Kyoto Work*, (Oxford University Press, Oxford, 2005).

[41] A. Meyer, "The United States has it Right on Climate Change in Theory", at http://www.opendemocracy.net/debates/article-6-129-2462.jsp; "Greenhouse Gas 'Plan B' Gaining Support," *New Scientist*, December 10 2003 at http://www.newscientist.com/article.ns?id =dn4467.

[42] "Greenhouse Gas 'Plan B' Gaining Support," as above.

[43] www.climatecare.org cited from the *New Internationalist*, July 2006, p.14.

[44] UK Meteorological Office, *The Greenhouse Effect and Climate Change: A Briefing from the Hadley Centre*, (UK Hadley Centre for Climate Prediction and Research, Devon, 1999), p.5. See also Pew Center on Global Climate Change, *Beyond Kyoto: Advancing the International Effort Against Climate Change* at http://www.pewclimate.org/docUplands/Long%2DTerm%20Target%2Epdf. For a critique of the Kyoto Protocol on the basis of its inadequacy to deal with the problem of global climatic change, see. S.M. Gardiner, "The Global Warming Tragedy and the Dangerous Illusion of the Kyoto Protocol," *Ethics and International Affairs*, vol. 18, no.1, 2004, pp. 23-39, and D.C. Victor, *The Collapse of the Kyoto Protocol and the Struggle to Stop Global Warming,* (Princeton University Press, Princeton, 2001).

[45] L. Dayton. "Kyoto 'Too Late' to Stop Warming," *The Australian*, January 5 2006, p.3.

[46] "Kyoto Protocol Won't Solve Drought Crisis: Cambell," ABC Newsonline, October 13 2006 at http://www.abc.net.au/news/newsitems/200610/s1764365.htm.

[47] "Australia Pushes for New Kyoto Deal at UN Climate Talks", ABC Newsonline November 13 2006 at http://www.abc.net.au/news/newsitems/200611/S1786756.htm; S. Lewis and D. Shanahon, "PM Pushes a New Kyoto",*The Australian*, November 1 2006, p.1. Australia is suffering, at the present time, more from the impacts of global

warming than any other country, according to the Australian Bureau of Meteorology in its 2006 climate statement. Weather events in 2005 were "highly unusual and unprecedented in many areas", withnothing similar in recorded history back to 1900. North Australia had record rainfall,but southern areas had experienced an on-going drought. Australia's average temperature has risen by 0.9 C over the last century, compared to a global average of 0.7 to 0.8 C. The water problems already facing south and eastern Australia are "very likely" to increase by 2030. River flow from Australia's Murray-Darling basin, crucial for Australia's existing agriculture, could fall by10-25 per cent by 2050: Australian Bureau of Meteorology, Drought Statement-Issued January 4 2007, "Driest Year on Record in Parts of South Australia",at http://www.bom.gov.au/announcements/media_releases/climate/drought/20070104.shtml; Australian Bureau of Meteorology, *Annual Australian Climate Statement 2006,* January 3 2007 at http://www.bom.gov.au/announcements/media_releases/climate/drought/20070103.shtml.Climate change research conducted by Australia's CSIRO indicated that under worse case scenarios from climate models, many areas of southern Australia by 2070 will experience average temperature rises of greater than 6 C and a fall in rainfall levels by 40 per cent, with some regions having spring and summer days 7.1 C warmer and a 60 per cent reduction in spring rainfall: A Wahlquist, "More Heat,Less Rain on Way for States," *The Australian*, November 6 2006, p. 4. From 1910 to 2005, South Australia, the driest state in the driest continent on Earth,experienced an average temperature increase by 0.96 C (with 0.10 C per decade). South Australia's average maximum temperature since 1950, has increased by 1.2 C (0.21 C per decade): R. Suppiah (*et al.*), *Climate Change Under Enhanced Green- house Conditions in South Australia*, (Climate Impacts and Risk Groups, CSIRO Marine and Atmosphere Research, June 2006).

[48] "Howard, Beazley Trade Blows Over Climate Change," ABC Newsonline, October 312006 at http://www.abc.net.au/news/newsitems/200610/S1777938.htm; United Nations Framework Convention on ClimateChange (UNFCCC), *GHG Data 2006:Highlights from Greenhouse Gas (GHG) Emissions Data for 1990-2004 for Annex I Parties*, (2006), at http://unfccc.int/files/essential_background/background_publications_htmlpdf/applications/ pdf/ghg_booklet_06.pdf.

[49] "Out of Control Over the Limit," at http://www.smh.com.au/news/environment/out_of_control_and_over_the_limit/2006/10.

---

[50] *GHG Data 2006*, cited note 48; "Kyoto Vows Forgotten as Gas Emissions Grow,"*The Australian*, October 31 2006, p8.

[51] "Blair Sees Wider Climate Deal After Kyoto," ABC Newsonline, January 282007 at.http:/www.abc.net.au/news/newsitems/200701/s1834807.htm.

[52] D. M. Smith (*et al.*), "Improved Surface Temperature Prediction for the Coming Decade from a Global Climate Model," *Science*, vol. 317, August 10 2007, pp. 796 - 799.

[53] J. H. Kunstler, *The Long Emergency*, (Atlantic Books, London, 2005).

[54] L. Grant, *The Collapsing Bubble*, (Seven Locks Press, Santa Ana, 2005). The population issue is not deal with in detail in this text having been discussed by the present authors elsewhere. See D. Shearman and J. W. Smith, *The Climate Change Challenge and the Failure of Democracy*, (Praeger Press, Westport, 2007).

[55] J. W. Smith, *The High Tech Fix*, (Avebury, Aldershot, 1991).

[56] J. Kekes, *The Nature of Philosophy*, (Basil Blackwell, Oxford, 1980).

**Chapter 2**

[1] A. Gore, *An Inconvenient Truth: The Planetary Emergency of Global Warming and What We Can Do About It*, (Rodale, Emmaus PA, 2006), p.8.

[2] Tony Blair, Foreword to H.J. Schellnhuber (*et al* eds.), *Avoiding DangerousClimate Change*, (Cambridge University Press, Cambridge, 2006), p.vii.

[3] As above. British climate scientist Mike Hulme has said that although climate change is a reality, "the discourse of catastrophe is in danger of tipping society

on to a negative, depressive and reactionary trajectory": "Stop Spreading Warnings of Chaos and Catastrophe," *The Australian*, November 7 2006, p.11. It is not "scare mongering" if a threat has a real scientific basis, even if that threat is "unlikely". See further F. Pearce, *With Speed and Violence: Why Scientists Fear Tipping Points in Climate Change*, (Beacon Press, Boston, 2007).

[4] G.M.Tabor, "Defining Conservation Medicine", in A.A. Aguirre (*et al* eds.), *Conservation Medicine: Ecological Health in Practice* (Oxford University Press, Oxford, 2002), pp. 8-16, cited p.9., and G.M.Tabor (*et al*), "Conservation Biology and the Health Sciences: Defining the Research Priorities of Conservation Medicine", in M.E. Soule (*et al* eds.), *Research Priorities in Conservation Biology*, 2nd edition, (Island Press, Washington DC 2001), pp. 155-173; D.J.Rapport, "Ecosystem Health: An Emerging Integrative Science", in D.J.Rapport (*et al* eds.), *Evaluating and Monitoring the Health of Large Scale Ecosystems*, (Springer, Heidelberg, 1995), pp. 5-34: M. Pakrass, (*et al*), "Conservation Medicine: An Emerging Field", in P. Raven and T.Williams (eds.), *Nature and Human Society: The Quest of a Sustainable World*, (National Academy Press, Washington DC, 1999) pp. 551-556; P.R. Epstein, "Climate, Ecology, and Human Health", *Consequences*, vol. 3, 1997, pp.2-19; Norris, "A New Voice in Conservation Medicine", *BioScience*, vol. 51, 2001, pp.7-12.

[5] R.S. Osfeld, (*et al*), "Conservation Medicine: The Birth of Another Crisis Discipline", in A.A. Aguirre (*et al* eds.), *Conservation Medicine: Ecological Health in Practice* (Oxford University Press, Oxford, 2002), pp. 17-26, cited p.17.

[6] United Nations Environment Programme, *Global Environment Outlook 3*, (Earthscan, London, 2002).

[7] Consult www.unep.org and www.unep.net.

[8] *Global Environment Outlook 3*, cited note 6, pp. 301-318.

[9] As above. See also C. Arden Pope III (*et al*), "Lung Cancer, Cardiopulmonary Mortality, and Long-Term Exposure to Fine Particulate Air Pollution", *Journal of the American Medical Association* vol.287, no.9, 2002, pp.1132-1141; A.J. McMichael, *Human Frontiers, Environment and Disease: Past Patterns, Uncertain Futures*, (Cambridge University Press, Cambridge, 2001).

[10] World Meteorological Organization, *Antarctic Ozone Bulletin*, No.3 2006 at

http://www.wmo.ch/web/arep/06/ant-bulletin-3-2006.pdf; "2006 Antarctic Ozone Hole is Most Serious on Record", at http://www.wmo.ch/news/news.html; "Ozone Layer Hole Reaches Record Size", at http://www.abc.net.au/news/newsitem/200610/s1754959.htm. Severe Ozone depletion was also observed during the 2004-2005 cold Arctic winter. See M. von Hobe (*et al.*), "Severe Ozone Depletion in Cold Arctic Winter 2004-2005", *Geophysical Research Letters*, vol.33, 2006, L17815, doi:10.1029/2006GLO26945.

[11] "Record Loss During 2006 Over the South Pole", European Space Agency News at http://www.esa.int/esaCP/SEMQBOKKKSE_index_0.html. The record size ozone hole arose as follows. In the southern hemisphere winter the atmosphere mass over Antarctica is isolated from air exchange with warmer mid-latitude air due to winds known as the polar vortex. Very low temperatures result and polar stratospheric clouds form which contain chlorine. During the polar spring, sunlight strikes these polar stratospheric clouds leading to the release of chlorine, which when released, breaks down the ozone molecules. L.Dayton, "Record Ozone Hole Despite Cuts in CFCs", *The Weekend Australian,* October 21-22 2006, p.5.

[12] *The Montreal Protocol on Substances the Deplete the Ozone Layer* (1987) as either adjusted and/or amended in London 1990, Copenhagen 1992, Vienna 1995, Montreal 1997 and Beijing 1999, UNEP Ozone Secretariat, United Nations Environment Programme at http://www.unep.org/ozone/.

[13] Multisectorial Initiative on Potent Industrial Greenhouse Gases, *MIPSGGS Newsletter,* September 2006 at www.mipiggs.org.

[14] "Warming Gas Fear", *New Scientist*, September 30 2006, p.6.

[15] Millennium Ecosystem Assessment, *Millennium Ecosystem Assessment Synthesis Report* March 23 2005 at http://www.millenniumassessment.org/en/Products. EHWB.aspx.

[16] Report of the Secretary-General, *We the Peoples: The Role of the United Nations in the Twenty-First Century* at http://unpan1.un.org/intradoc/groups/public/documents/UN/UNPAN000923.pdf.

[17] Millennium Ecosystem Assessment, as above, note 15 at p.12.

---

[18] J.R. Mendelson III (*et al*), "Biodiversity: Confronting Amphibian Declines and Extinctions", *Science*, vol.313, no5783, 7 July 2006, doi: 10. 11261/science.1128396 at http://www.sciencemag.org/cgi/content/full/313/5783/48.

[19] As above.

[20] As above.

[21] J. Harte(*et al.*), "Biodiversity Conservation: Climate Change and Extinction Risk," *Nature*, vol. 430, July 1 2004, doi: 10. 1038/nature02718; J.A. Pounds, "Widespread Amphibian Extinctions from Epidemic Disease Driven by Global Warming," *Nature*, vol. 439, January 12 2006, pp. 161-167, doi: 10.1038/nature04246; M.B. Araujo and C. Rahbek, "How does Climate Change Affect Biodiversity?" *Science*, vol.313, no 5792, September 5 2006, pp.1396-1397, doi: 10.1126/science.1131758. A somewhat dissenting view is given by G.M. Hewitt, "Genetic and Evolutionary Impacts of Climate Change," in T.E.Lovejoy and L.Hannah (eds.), *Climate Change and Biodiversity*, (Yale University Press, New Haven and London, 2006), pp.176-192. Professor Hewitt argues that sudden drops in temperature, such as those associated with a new ice age, would be more destructive than present global warming trends. See L.Dayton, "Refuge from the Ice Next Time," *The Australian,* November 8 2006, p.24.

[22] "Water Crisis", World Water Council at http://www.worldwatercouncil.org/index.php?id=25.

[23] Millennium Ecosystem Assessment, cited note 15, p11; F.Pearce, "The Parched Planet", *New Scientist*, February 25 2006 pp.31-36 and *When the Rivers Run Dry: Water- The Defining Crisis of the Twenty-First Century*, (Beacon Press, Boston, 2006); S.Postel, "Self-guarding Freshwater Ecosystems", in L.Starke (ed.), *State of the World 2006*, (Earthscan, London, 2006), pp.41-60.

[24] The United Nations World Water Development Report 2, *Water: A Shared Responsibility*, (UNESCO, Paris and Berghahn Books, New York, 2006).

[25] As above, pp.89-90.

[26] As above, p. 90.

[27] As above, p. 99.

[28] As above, p.204.

[29] WWF Freshwater Program, *Rich Countries, Poor Water* (2006) at http://wwf.org.au/publications/rich-countries-poor-water/.

[30] As above p.6.

[31] As above p.3.

[32] Report of the Senate Environment, Communications, Information Technology and the Arts Reference Committee, *The Heat is On: Australia's Greenhouse Future*, (Parliament of the Commonwealth of Australia, Canberra, November, 2006).

[33] WWF Freshwater Program, *Rich Countries, Poor Water* cited note 29, p.11

[34] As above, p.15.

[35] P.S. Corsel (*et al*), "Cost of Illness in the 1993 Waterborne Cryptosporidium Outbreak, Milwaukee, Wisconsin", *Emerging Infectious Diseases*, vol.9, no.4, April 2003, pp.426-431.

[36] L.R. Brown, *Plan B 2.0:Rescuing a Planet Under Stress and a Civilization in Trouble*. (W.W. Norton, New York, 2006).

[37] See L. Brown at http://www.earth-policy.org/Books/PB2/PB2preface.htm.

[38] As above. China's economic growth is also confronting the nation with an enormous problem of pollution. See E.C. Economy, *The River Runs Black: The Environmental Challenge of China's Future*, (Cornell University Press, Ithaca and London, 2004).

[39] See: W.Youngquist, *GeoDestinies: The Inevitable Control of Earth Resources over Nations and Individuals*, (National Book Company, Portland OR, 1997); K.S. Deffeyes, *Hubberts Peak: The Impending World Oil Shortage*, (Princeton University Press, Princeton, 2001); J. Attarian, "Oil Depletion Revisited: Why the Peak is Probably Near", *The Social Contract*, Winter 2004-2005, pp. 129-146;

R.C. Duncan, "Big Jump in Ultimate Recovery Peak Would Ease, Not Reverse, Postproduction Decline" *Oil and Gas Journal*, July 19, 2004, pp.18-21; A.M. Samsam Bakhtiari, "World Oil Production Capacity Model Suggests Output Peak by 2006-2007", *Oil and Gas Journal*, April 26, 2004, pp.18-20; D. Goodstein, *Out of Gas: The End of the Age of Oil*, (W.W. Norton, New York and London, 2004); P.Roberts, *The End of Oil: On the Edge of a Perilous New World*, (Houghton Miflin, New York, 2004); M. Simmons, *Twilight in the Desert: The Coming Saudi Oil Shock and the World Economy*, (John Wiley and Sons, New York, 2005). Websites include: http://www.dieoff.com; http://www.hubbertpeak.com; http://www.technocracy.org; http://www.oilcrisis.com; http://www.peakoil.net/.

[40] "Oil Crisis by 2010: Expert" *The Advertiser*, (Adelaide), August 26 2006, p. 11, at www.news.com.au/adelaidenow.

[41] V. Darroch, "Not Enough New Fields to Offset Declines", at http://www.sundayherald.com/49065; "Oil Supply Shortage Likely After 2007" at http;//www.commondreams.org/news2004/012811.htm.

[42] Proponents of an abiogenic origin of oil, reject the conventional theory of a fossil origin of oil, and see petroleum formed from non-biological processes occurring deep in the Earth's crust and mantle. Petroleum is formed at high temperatures and pressures from inorganic carbon. This theory was defended by a number of Russian scientists, including the chemist Mendeleev, and also in by the astrophysicist Thomas Gold (1920-2004). See T. Gold, *The Deep Hot Biosphere*,(Copernicus Books, 1999). For an overview and criticism see G.P. Glasby, "Abiogenic Origin of Hydrocarbons: An Historical Overview", *Resource Geology*, vol.56, no.1, 2006, pp.83-96.

[43] V. Edwards, "Big Oil Says Reserves Are Plentiful", *The Australian*, September 12 2006, p.4.

[44] As above. The discovery of a petroleum pool in the Gulf of Mexico could be taken as support of Mr. Mark Nolan's view. This reserve would increase US reserves by more than 50 per cent as the site could hold between three and 15 billion barrels of oil and natural gas liquids. See "Oil Discovery Fuels US Hopes", *The Weekend Australian*, September 9-10 2006, p.11. However this discovery will not significantly reduce US dependence on foreign oil as the US consumes about 5.7 billion barrels of crude oil per annum.

---

[45] L.Grant, *The Collapsing Bubble: Growth an Fossil Fuel.* (Seven Locks Press, Santa Ana CA, 2002). p.25.

[46] M.R. Simmons, *Twilight in the Desert: The Coming Oil Shock and the World Economy* (Wiley, New York, 2005).

[47] J. Leggett, *The Empty Tank: Oil, Gas, Hot Air, and the Coming Global Financial Catastrophe,* (Random House, New York, 2005) and British Title: *Half Gone: Oil, Gas, Hot Air and the Global Energy Crisis,* (Portobello Books, London, 2005).

[48] J. Leggett, *The Empty Tank,* as above p.6.

[49] As above, p.22.

[50] As above, p.25.

[51] See B. Williams, "Debate Over Peak-Oil Issue Boiling Over, With Major Implications for Industry, Society," *Oil and Gas Journal*, vol.101, no.27, July 14 2003, pp.18-19, cited from Leggett, as above, p.27.

[52] M. Salameh, "How Realistic are OPEC's Proven Oil Reserves?" *Petroleum Review*, August 2004, cited from Leggett, as above p.27. See also I. Sandrea, "Deep Water Oil Discovery Rate May Have Peaked: Production Peak May Follow in 10 years', *Oil and Gas Journal*, vol. 102, no.28, July 26 2004, p.18.

[53] J. Leggett, *The Empty Tank,* cited note 47, p.1. See US Energy Information Administration at www.eia.doe.gov.

[54] As above, pp.141-142.

[55] As above, p. 157.

[56] As above. See also S. Leeb and G. Strathy, *The Coming Economic Collapse: How you can Thrive When Oil costs $200 a Barrel*, (Warner Business Books, New York, 2006).

[57] J.H. Kunstler, *The Long Emergency: Surviving the Converging Catastrophes of the Twenty-First Century* (Atlantic Books, London, 2005), pp.24-25.

---

[58] R.C. Duncan, "The Olduvai Theory: Energy, Population, and Industrial Civilization", *The Social Contract*, Winter 2005-2006, pp.134-144. Duncan sees an end to industrial civilization by 2030 due to limits to energy supply and a "complex matrix of delayed feedback interactions, including: depletion of non-renewable resources, lack of capital and operational investment funds, soil erosion, declining industrial and agricultural production, Peak Oil, global warming, pollution, deforestation, falling aquifers, unemployment, resource wars, pandemic diseases - to name just a few". (p.143).

[59] J.W. Smith, G. Lyons and E. Moore, *Global Anarchy in the Third Millennium: Race, Place and Power at the End of the Modern Age*, (Macmillan, London, 2000).

[60] See www.dieoff.org.

[61] J.W. Smith, (et. al.), *Global Meltdown: Immigration, Multiculturalism, and National Breakdown in the New World Disorder*, (Praeger, Westport, Connecticut and London, 1998).

[62] R.D. Kaplan, "The Coming Anarchy: How Scarcity, Crime, Overpopulation, Tribalism, and Disease are Rapidly Destroying the Social Fabric of Our Planet", *The Atlantic Monthly*, February 1994, pp.44-76.

[63] See for example W. Kininmonth, *Climate Change: A National Hazard,* (Multi-Science Publishing Co. Essex, 2004).

[64] See G.Sauer-Thompson and J.W.Smith, *The Unreasonable Silence of the World: Universal Reason and the Wreck of the Enlightenment Project*, (Ashgate, Aldershot, 1997).

[65] A.J. McMichael, *Planetary Overload: Global Environmental Change and the Health of the Human Species,* (Cambridge University Press, Cambridge, 1993).

[66] I.Lilley, "Whatever Happens, We'll Manage", *The Australian*, (Higher Education), May 17 2006, pp.44-45, and Hot Air - Now Nigh's The End?" *Griffith Review*, 12, at www.griffith.edu.au/griffithreview. W.Kininmonth, "Don't be Gored into Going Along", *The Australian* September 12 2006, p.12 . A recent attack on the theory of global warming and global climatic change, as well as environmentalism, is given by Christopher C. Horner in *The Politically Incorrect*

*Guide to Global Warming and Environmentalism*, (Regnery Publishing Inc, Washington DC, 2007). Horner sees environmentalism as anti-American, anti-capitalist, cosmopolitan (anti-nationalist) and anti-human - an attempt by neo-communists ("Green is the new red") "to gain more government control over the economy and individual activity". (p.3). Needless to say, even if this political critique was true of environmentalism, the scientific arguments given by the consensus view of climate change must still be systematically refuted by a scientific assessment of *all* the evidence, which today cannot be done by any one person.

[67] C.R. De Freitas, "Are Concentrations of Carbon Dioxide in the Atmosphere Really Dangerous?" *Bulletin of Canadian Petroleum Geology*, vol.50, no.2, June 2002, pp.297-327.

[68] I. Plimer, "All Hot and Bothered", *The Independent Weekly* (Adelaide), July 8-14 2006, p.12.

[69] As above.

[70] D.J. Karoly (*et al*), "Detection of a Human Influence on North American Climate", *Science*, vol.302, no 5648, November 14 2003, pp.1200-1203, doi: 10.1126/science. 1089159.

[71] See S. Bony (*et al*), "How Well Do We Understand and Evaluate Climate Change Feedback Processes?" *Journal of Climate*, vol. 19, no.15, August 2006, pp. 3445-3482; S. Menon and L.L. Rotstayn, "The Radiative Influence of Aerosol Effects on Liquid-Phase Cumulus and Stratiform Clouds Based on Sensitivity Studies with Two Climate Models", *Climate Dynamics*, vol.27, no.4, September 2006, pp.345-356.

[72] G. J. Chaitin, *Information, Randomness and Incompleteness: Papers on Algorithmic Information Theory*, (World Scientific, Singapore, 1987).

[73] C. Essex and R. McKitrick, *Taken by Storm: The Troubled Science, Policy and Politics of Global Warming*, (Key Porter Books Toronto, 2002). See generally, F. Pearce, "State of Denial", *New Scientist*, November 4 2006, pp.18-21.

[74] See C.R. De Freitas cited note 67.

[75] National Research Council, *Abrupt Climate Change: Inevitable Surprises*,

(National Academy Press, Washington DC, 2002).

[76] Plimer cited note 68.

[77] See W. Glen, *The Mass-Extinction Debates: How Science Works in a Crisis*,(Stanford University Press, Standford, 1994); V. Courtillo, *Evolutionary Catastrophes: The Science of Mass Extinction*, (Cambridge University Press, Cambridge, 1999); A.A. Hoffman and P.A. Parsons, *Extreme Environmental Change and Evolution*, (Cambridge University Press, Cambridge, 1997); A.Hallam and P.B.Wignall, *Mass Extinctions and Their Aftermath*,(Oxford University Press, Oxford, 1997); T.Hallam, *Catastrophes and Lesser Calamities: The Causes of Mass Extinctions,* (Oxford University Press, Oxford, 2004) and E. Kofbert, *Field Notes from a Catastrophe: Man, Nature and Climate Change*, (Bloomsbury Publishing, London, 2006).

[78] D.H. Erwin, *Extinction: How Life on Earth Nearly Ended 250 Million Years Ago*, (Princeton University Press, Princeton, 2006).

[79] Erwin, as above.

[80] As above.

[81] M. J. Benton, *When Life Nearly Died: The Greatest Mass Extinction of All Time*, (Thomas and Hudson, London, 2003). See also P.D. Ward, "Impact from the Deep", *Scientific American*, vol.295, no.4, October 2006, pp. 42-49 for an overview of the evidence that many past mass extinctions on Earth may have been used by a "killer greenhouse effect". Rapid global warming caused by the release of carbon dioxide and methane sets in motion a chain of events where hydrogen sulphide (H2S) gas is released into the atmosphere killing plants and animals and destroying the ozone shield in the troposphere. The influx of ultraviolet radiation kills off most of the remaining life on Earth. See further L.R. Kump (*et al*), "Massive Release of Hydrogen Sulfide to the Surface Ocean and Atmosphere During Intervals of Oceanic Anoxia", *Geology*, vol.33, no.5, May 2005, pp.397-400.

[82] Kininmonth, cited note 63.

[83] H. Svensmark and N. Calder, *The Chilling Stars: A New Theory of Climate Change*, (Icon Books, Cambridge, 2007).

---

[84] S. F. Singer and D.T. Avery, *Unstoppable Global Warming Every 1,500 years*, (Rowman and Littlefield, Lanham, 2007).

[85] W. Dansgaard (*et al*), "North Atlantic Climate Oscilliations Revealed by Deep Greenland Ice Core", in F.E.Hansen and T.Takahashi (eds.), *Climate Processes and Climate Sensitivity*,(American Geophysical Union, Washington D.C., 1984), Geophysical Monograph No. 29, pp.288-298.

[86] Singer and Avery, as above note 84, p.2.

[87] For one critical assessment of the cosmological/geological views see A. Thorpe, "A Fake Fight", *New Scientist*, March 17 2007, p. 24. L.K.Fenton (*et al.*), "Global Warming and Climate Forcing by Recent Albedo Changes on Mars", *Nature*, vol. 446, April 5, 2007, doi: 10.1038/nature05718, used a computer model, also used to study global warming on Earth and found that Mars warmed by a total of 0.65 C over the period of the 1970s to the 1990s, a temperature rise similar to the Earth's 0.6 C average rise over the 20th century. Some climate change skeptics have used the research of Fenton (*et al*) to attempt to discredit the theory of global warming. However, merely from that data alone, one cannot conclude that the temperature change on Earth is due to the same process. In fact, according to Fenton (*et al.*), it is not. They point out that the removal and deposition of bright Martian surface dust affects variations of the Martian surface albedo (albedo being the ratio of the intensity of light reflected from a thing, to that light received from the sun). Albedo alterations are involved in a positive feedback system, whereby albedo changes strengthen the winds, that in turn generate the changes. This process is unique to Mars.

[88] M.E. Mann (*et al*), "Global-Scale Temperature Patterns and Climate Forcing Over the Past Six Centuries," *Nature*, vol. 392, 1998, pp.779-787; M.E. Mann (*et al*), " Northern Hemisphere Temperatures During the Past Millennium: Inferences, Uncertainties, and Limitations", *Geophysical Research Letters*, vol.26, no 6, pp. 759-762; M.E. Mann (*et al*), "Testing the Fidelity of Methods Used in Proxy-Based Reconstructions of Past Climate", *Journal of Climate*, vol. 18, 2005, pp. 4097-4107.

[89] J.T. Houghton (*et al* eds.), *Climate Change 2001: The Scientific Basis*,(Cambridge University Press, Cambridge, 2001) p. 101.

[90] On this debate see R.S. Bradley and P.D. Jones, "Little Ice Age" Summer

Temperature Variations: Their Nature and Relevance to Recent Global Warming Trends," *Holocene*, vol.3, 1993, pp. 367-376; R.S. Bradley (*et al*), "Climate in Medieval Time", *Science*, doi: 10.1126/science. 1090372, 2003, pp. 404-405; T.J. Crowley and T.S.Lowery, "How Warm was the Medieval Warm Period? A Comment on 'Man-made Versus Natural Climate Change'", *Ambio*, vol.29, 2000, pp. 51-54.

[91] S. McIntyre and R. McKitrick, "Corrections to Mann *et al* (1998) Proxy Data Base and Northern Hemisphere Average Temperature Series", *Energy and Environment*, vol.14, 2003, pp. 751-771, "The M&M Critique of MBH98 Northern Hemisphere Climate Index: Update and Implications", *Energy and Environment*, vol.16, 2005, pp. 69-100, "Hockey Sticks, Principal Components, and Spurious Significance", *Geophysical Research Letters*, vol.32, 2005, L03710, doi: 10. 1029/2004GLO21750.

[92]G.Bürger and U. Cubasch, "Are Multiproxy Climate Reconstructions Robust?" *Geophysical Research Letters*, vol.32, 2005, L23711, doi: 10. 1029/2005GLO24155; S.Rutherford (*et al*), "Proxy- Based Northern Hemisphere Surface Reconstructions: Sensitivity to Method, Predictor Network, Target Season, and Target Domain", *Journal of Climate*, vol.18, 2005, pp. 2308-2329; H.E. von Storch and F.Zorita, "Comment on 'Hockey Sticks, Principal Components, and Spurious Significance, by S.M. McIntrye and R.L. McKitrick", *Geophysical Research Letters*, vol.32, 2005, L20701, doi: 10. 1029/2005GLO22753.

[93] E.J. Wegman (*et al*), *Ad Hoc Committee Report on the 'Hockey Stick' Global Climate Reconstruction* at http://energy.commerce.house.gov/l08/home/07142006_Wegman_Report.pdf.

[94] As above, pp. 4-5.

[95] National Research Council of the National Academies (NRC), Committee on Surface Temperature Reconstructions for the Last 2,000 Years, Board on Atmospheric Sciences and Climate Division on Earth and Life Studies, *Surface Temperature Reconstructions for the Last 2,000 years* (2006) at http://newton.nap.edu/pdf/0309102251/pdf_image.

[96] As above p. 17.

[97] R.A. Kerr, "Politicians Attack, But Evidence For Global Warming Doesn't

Wilt", Science, vol.313, no. 5786, July 28 2006, p. 421,doi: 10.1126/science.313.5786.421.

[98] Wegman (*et al*) cited note 93, p. 65.

[99] As above p. 50. See also C.Wunsch, "Abrupt Climate Change: An Alternative View," *Quarterly Research*, vol. 65, 2006, pp. 191-203.

[100] G.C. Hegerl (*et al*), "Climate Sensitivity Constrained by Temperature Reconstructions over the Past Seven Centuries, "*Nature*, vol. 440, April 20 2006, pp. 1029-1032, doi: 10.1038/nature04679.

[101] F.Pearce, "Grudge Match", *New Scientist,* March 18 2006, pp. 40-43.

[102] R. Black, "Global Warming Risk 'Much Higher'", BBC News, May 23 2006 at http://news.bbc.co.uk/2/hi/science/nature/5006970.stm.

[103] As above.

[104] C.Turley (*et al*), "Reviewing the Impact of Increased Atmospheric CO2 on Ocean pH and the Marine Ecosystem", in H.J.Schellnhuber (*et al*), *Avoiding Dangerous Climate Change*, (Cambridge University Press, Cambridge, 2006), pp. 65-70.

[105] As above p. 65.

[106] C. Henderson, "Paradise Lost", *New Scientist*, August 5 2001, pp. 28-33; S.C. Doney,"The Dangers of Ocean Acidification", *Scientific American*, vol.294, no.3, March 2006, pp. 38-45; K.Calderia and M.E. Wickett, "Anthropogenic Carbon and ocean pH", *Nature*, vol.425, September 25 2003, p.365; J.C.Orr, (*et al*), "Anthropogenic Ocean Acidification Over the Twenty-First Century and Its Impact on Calcifying Organisms", *Nature*, vol.437, September 29 2005, pp. 681-686; K. Lee, "Global Relationships of Total Alkalinity with Salinity and Temperature in Surface Waters of the World's Oceans," *Geophysical Research Letters*, vol. 33, 2006, L. 19605, doi:10.1029/2006GLO27207.

[107] Turley (*et al*), cited note 104, p. 67; U.Reibesell (*et al*), "Reduced Calcification of Marine Plankton in Response to Increased Atmospheric CO", *Nature,* vol.407, 2000, pp. 364-367.

[108] Turley (*et al*), cited note 104, p. 67; R.J.Charlson (*et al*) "Oceanic Phytoplankton, Atmospheric Sulphur, Cloud Albedo and Climate", *Nature*, vol.326, 1987, pp. 655-661.

[109] Turley, as above.

[110] K. Emanuel, "Increasing Destructiveness of Tropical Cyclones over the Past 30 Years" *Nature*, vol.436, 2005, pp. 686-688; R.A. Pielke, Jr., "Are There Trends in Hurricane Destruction?" *Nature*, vol.438, December 22/29, 2005, pp. E11, doi: 10. 1038/nature04426; C.W. Landsea, "Hurricanes and Global Warming, *Nature*, vol.438, 22/29 December, 2005, pp. E11-E13, DOI: 10. 1038/nature04477; K. Emanuel, " Emanuel Replies",*Nature*, vol.438, December 22/29, 2005, pp. E13, doi: 10. 1038/nature04427.

[111] B.O'Keefe, "Warming ' Can't be Blamed for Storms'", *The Australian*, February 21, 2006, p.6.

[112] J.B. Elsner, "Evidence in Support of the Climate Change - Atlantic Hurricane Hypothesis", *Geophysical Research Letters*, vol.33, 2006, L167705, DOI: 10. 1029/2006GLO26869.

[113] Q. Schiermeier, "Trouble Brews Over Contested Trend in Hurricanes", *Nature*, vol. 435, June 23 2005, pp.1008-1009, doi: 10.1038/4351008b, "Hurricane Link to Climate Change is Hazy", *Nature*, vol. 437, September 22 2005,doi: 10.1038/437461a; "Storms Get Fewer But Fiercer", *Nature*, September 15 2005, doi: 10.1038/news/050912-11; A.Witze, "Tempers Flare at Hurricane Meeting," *Nature*, vol. 441, May 4 2006, doi: 10.1038/441011a; J.C.L. Chan, "Comment on "Changes in Tropical Cyclone Number, Duration, and Intensity in a Warming Environment"," Science, vol. 311, no.5768, March 24 2006, doi:10. 1126/science. 1121522.

[114] Cited from O'Keefe, cited note 111.

[115] C.D. Hoyos (*et al*), "Deconvolution of the Factors Contributing to the Increase in Global Hurricane Intensity", *Science*, vol. 321, no. 5770, April 7 2006, pp. 94-97, doi: 10. 1126/science.1123560.

[116] B.D. Santer (et. al.), "Forced and Unforced Ocean Temperature Changes in Atlantic and Pacific Tropical Cyclogenesis Regions", *Proceedings of the National*

*Academy of Sciences*, vol.103, no.38, September 19 2006, pp. 13905-13910, doi:10. 1073/pnas.0602861103.

[117] M.McCarthy, "Sea Levels are Rising Faster Than Predicted, Warns Antarctic Survey," *The Independent* (UK) at http://news.independent.co.uk/environment/article1621770.ece; J.C. Fyfe, "Southern Ocean Warming Due to Human Influence", *Geophysical Research Letters*, vol. 33, 2006, L19701, doi:10.1029/2006GLO27247.

[118] J.A. Church and J.M. Gregory, "Changes in Sea Level", in J.T. Houghton (*et al* eds.), *Climate Change 2001: The Scientific Basis*, (Cambridge University Press, Cambridge, 2001), pp. 639-693.

[119] R.J. Nicholls and J.A. Lowe, "Climate Stabilisation and Impacts of Sea-Level Rise", in H.J. Schellnhuber (*et al* eds.), *Avoiding Dangerous Climate Change*, (Cambridge University Press, Cambridge, 2006), pp. 195-202, cited p.197. Apart from rising sea levels, global climatic change also results in accelerated effects of coastal erosion. Rising sea levels may have a differential impact on shorelines. A computer model by Slott (*et al*) found that shoreline retreat rates were much higher than expected from sea level rise alone with the sea in some region encroaching up to 10 times further than would otherwise be expected. See J.M. Slott, "Coastline Responses to Changing Storm Patterns", *Geophysical Research Letters*, vol.33, 2006, L18404. doi:10.1029/2006/GLO27445.

[120] R. Black, "'Major Melt' for Alpine Glaciers", BBC News 4 April 2006 at http://news.bbc.co.uk/1/hi/sci/tech/4874224.stm.

[121] As above.

[122] L.G. Thompson (*et al*), "Abrupt Tropical Climate Change: Past and Present", *Proceedings of the National Academy of Sciences*, vol. 103, no.28, July 11 2006, pp. 10536-10543, doi: 10.1073/pnas.0603900103.

[123] S. Hainzl (*et al*), "Evidence for Rainfall-Triggered Earthquake Activity", *Geophysical Research Letters*, vol.33, 2006, L19303. doi:10.1029/2006/GLO27642.

[124] B. McGuire, "Earth: Fire and Fury", *New Scientist*, May 27 2006, pp. 32-36.

[125] As above.

---

[126] As above. See also B.McGuire, *A Guide to the End of the World: Everything You Never Wanted To Know*, (Oxford University Press, Oxford, 2003) and *Surviving Armageddon: Solutions for a Threatened Planet*, (Oxford University Press, Oxford, 2005).

[127] S.J. Hassol and R.W.Corell, "Arctic Climate Impact Assessment ",in H.J. Schellnhuber (*et al* eds.), *Avoiding Dangerous Climate Change*, (Cambridge University Press, Cambridge, 2006), pp. 205-213, cited p.205.

[128] As above p.206.

[129] As above.

[130] As above.

[131] R.A. Kerr, "A Worrying Trend of Less Ice, Higher Seas", *Science*, vol.311, no 5768, March 24 2006, pp. 1698-1701, doi: 10.1126/science.311.5768.1698.

[132] " Scientists See Arctic Passing "Tipping Point": Meltdown Fears as Arctic Ice Cover Falls to Record Winter Low", *The Guardian* (UK), May 15 2006 at htttp://www.heatisonline.org.

[133] J.C. Comiso, "Abrupt Decline in Arctic Winter Sea Ice Cover", *Geophysical Research Letters*, vol. 33, 2006, L18504, doi:10.1029/2006GLO27341.

[134] T. Folkestad (*et al*), "Evidence and Implications of Dangerous Climate Change in the Arctic", in H.J. Schellnhuber (*et al* eds.), *Avoiding Dangerous Climate Change*, (Cambridge University Press, Cambridge, 2006), pp. 215-218, cited p.217.

[135] See note 132.

[136] Folkestad, cited note 134.

[137] M.D. Walker (*et al*), "Plant Community Responses to Experimental Warming Across the Tundra Biome", *Proceedings of the National Academy of Sciences*, vol. 103, no.5, January 31 2006, pp. 1342-1346, doi: 10.1073/pnas.0503198103; F. Pearce, "Dark Future Looms for Arctic Tundra", *New Scientist*, January 21 2006, p.15.

---

[138] J. Hansen (*et al*), "Global Temperature Change",*Proceedings of the National Academy of Sciences*, vol. 103, no.39, September 26 2006, pp. 14288-14293, doi: 10.1073/pnas.0606291103.

[139] F. Pearce, "One Degree and We're Done For", *New Scientist*, September 30 2006, p.8.

[140] C. Rapley, "The Antarctic Ice Sheet and Sea Level Rise", in H.J. Schellnhuber (*et al* eds.), *Avoiding Dangerous Climate Change*, (Cambridge University Press, Cambridge, 2006), pp. 25-27.

[141] D. Kennedy and B. Hudson, "Ice and History", *Science*, vol.311, March 24 2006, pp. 1673, doi: 10.1126/science.1127485.

[142] G. Ekström (*et al*), "Seasonality and Increasing Frequency of Greenland Glacier Earthquakes", *Science*, vol.311, March 24 2006, pp. 1756-1758, doi: 10.1126/science.1122112.

[143] I. Joughlin, "Greenland Rumbles Louder as Glaciers Accelerate," *Science*, vol.311, March 24 2006, pp.1709, doi: 10.1126/science.1124496.

[144] R. Bindshadler, "Hitting the Ice Sheets Where It Hurts," *Science*, vol.311, no 5768, March 24 2006, pp. 1720-1721, doi: 10.1126/science.1125226.

[145] As above.

[146] " Antarctic Warming: Hot Air", *New Scientist*, April 8 2006, p.7; J. Turner, "Significant Warming of the Antarctic Winter Troposphere", *Science*, vol.311, no 5769, March 31 2006, pp. 1914-1917, doi: 10.1126/science.1121652; D.P.Schneider (*et al*), " Antarctic Temperatures Over The Past Two Centuries from Ice Cores", *Geophysical Research Letters*, vol.33, 2006, L16707. doi:10.1029/2006/GLO27057.

[147] J. Turner cited note 146.

[148] As above.

[149] I. Velicona and J. Wahr, "Measurements of Time-Variable Gravity Show Mass Loss in Antartica", *Science*, vol.311, no 5768, March 24 2006, pp. 1754-

1756, doi: 10.1126/science.1123785.

[150] As above.

[151] "Icy Warning", *New Scientist*, February 4 2006, p.7.

[152] J.M. Gregory (*et al*), "Climatology: Threatened Loss of the Greenland Ice-Sheet", *Nature*, vol. 428, April 8 2004, doi:10.1038/428616a.

[153] T.M. Lenton, "Millennial Timescale Carbon Cycle and Climate Change in an Efficient Earth System Model", *Climate Dynamics*, vol.26, 2006, pp. 687-711, DOI:10.1007/s00382-0060-0109-9; F. Pearce, "If We Don't Stop Burning Oil...", *New Scientist*, February 18 2006, p.10.

[154] "Greenland's Water Loss has Doubled in a Decade", *New Scientist*, February 25 2006, p.20.

[155] E. Rignot and P. Kanagaratnam, "Changes in Velocity Structure of the Greenland Ice Sheet", *Science*, vol.311, no 5763, February 17 2006, pp. 986-990, doi: 10.1126/science.1121381.

[156] As above.

[157] As above.

[158] J.A. Lowe (*et al*), "The Role of Sea-Level Rise and the Greenland Ice Sheet in Dangerous Climate Change: Implications for the Stabilisation of Climate", in H.J. Schellnhuber (*et al* eds.), *Avoiding Dangerous Climate Change*, (Cambridge University Press, Cambridge, 2006), pp. 29-36.

[159] A. Nesje (*et al*), "Were Abrupt Late Glacial and Early-Holocene Climate Changes in Northwest Europe Linked to Freshwater Outbursts to the North Atlantic and Arctic Oceans?" *The Holocene*, vol.14, 2004, pp. 299-310; J.F. McManus (*et al*), "Collapse and Rapid Resumption of Atlantic Meridional Circulation Linked to Deglacial Climate Change", *Nature*, vol. 428, 2004, pp. 834-837; L. Tarasov and W.R. Peltier, "Arctic Freshwater Forcing of the Younger Dryas Cdd Reversal", *Nature*, vol. 435, 2005, pp. 662-665.

[160] S. Battersby, "Deep Trouble", *New Scientist,* April 15 2006 at http://global.factiva.com/ha/default.aspx.

---

[161] M.E. Schlesinger (*et al*), "Assessing the Risk of a Collapse of the Atlantic Thermohaline Circulation", in H.J. Schellnhuber (*et al* eds.), *Avoiding Dangerous Climate Change*, (Cambridge University Press, Cambridge, 2006), pp. 37-47.

[162] As above at p. 45.

[163] M.S. Torn and J. Harte, "Missing Feedbacks, Asymmetric Uncertainties, and the Underestimation of Future Warming", *Geophysical Research Letters*, vol.33, 2006, L10703, doi:10.1029/2005/GLO225540.

[164] As above.

[165] As above.

[166] M. Scheffer (*et al*), "Positive Feedback Between Global Warming andAtmospheric CO2 Concentrations Inferred from Past Climate Change", *Geophysical Research Letters*, vol.33, 2006, L10702, doi:10.1029/2005/GLO25044.

[167] J. Harte (*et al*) "Global Warming and Soil Microclimate: Results from a Meadow-Warming Experiment", *Ecological Applications*, vol. 5, no.1, 1995, pp. 132-150; D.A. Lashof (*et al*), "Terrestrial Ecosystem Feedbacks to Global Climate Change", *Annual Review of Energy and the Environment*, vol. 22, 1997, pp. 75-118.

[168] See Lashof as above.

[169] P.M. Cox (*et al*), "Conditions for Sink-to Source Transitions and Runaway Feedbacks from Land Carbon Cycle", in H.J. Schellnhuber (*et al* eds.), *Avoiding Dangerous Climate Change*, (Cambridge University Press, Cambridge, 2006), pp. 155-161; C.D. Jones (*et al.*), "Impact of Climate-Carbon Cycle Feedbacks on Emissions Scenarios to Achieve Stabilisation", in H.J. Schellnhuber, as cited, pp.323-331.

[170] As above.

[171] R.S. Lindzen (*et al*), "Does the Earth have an Adaptive Infrared Iris?" *Bulletin of the American Meteorological Society*, vol.82, 2001, pp.417-432.

---

[172] B.J. Soden (*et al.*), "The Radiative Signature of Upper Tropospheric Moistening", *Science*, vol. 310, 2005, pp. 841-844.

[173] M. Wild (*et al*), "From Dimming to Brightening: Decadal Changes in Solar Radiation at Earth's Surface", *Science*, vol. 308, May 6 2005, pp. 847-850.

[174] " Brighter Sun Adds to Fears of Climate Change", *The Sunday Times*, 26 March 2006 at <http://www.timesonline.co.uk/article/0,,2087-2104022, 00.html>.

[175] Q. Schiermeier, "Cleaner Skies Leave Global Warming Forecasts Uncertain", *Nature*, vol. 435, May 12 2005, p. 135.

[176] As above.

[177] S. Clark, "Saved by the Sun", *New Scientist*, September 16 2006, pp. 32-36.

[178] R. Hooper, "Something in the Air", *New Scientist*, January 21 2006, pp. 40-43.

[179] As above.

[180] As above.

[181] As above. Another issue that requires further scientific investigation is the significance of the finding that *living* plants emit methane. See F. Keppler and T. Röckmann, "Methane, Plants and Climate Change", *Scientific American* 00368733, vol. 296, no.2, February 2007.

[182] See M.Meinshausen, "What Does a 2° C Target Mean for Greenhouse Gas Concentrations? A Brief Analysis Based on Multi-Gas Emission Pathways and Several Climate Sensitivity Uncertainty Estimates", in H.J. Schellnhuber (*et al* eds.), *Avoiding Dangerous Climate Change*, (Cambridge University Press, Cambridge, 2006), pp. 265-279; M.D. Mastrandrea and S.H. Schneider, "Probabilistic Assessment of 'Dangerous' Climate Change and Emission Scenarios: Stakeholder Metrics and Overshoot Pathways", in H.J. Schellnhuber (*et al* eds.), as cited, pp. 253-264; D. Stainforth (*et al.*) "Risks Associated with Stabilisation Scenarios and Uncertainty in Regional and Global Climate Change Impacts", in H.J. Schellnhuber (*et al* eds.), pp. 317-321; P.G. Challenar (*et al*), "Towards the Probability of Rapid Climate Change", in H.J. Schellnhuber (*et al*

eds.), pp. 55-63.

[183] The Stern Review, *The Economics of Climate Change*, October 31 2006 at www.sternreview.org.uk.

[184] T.M.L.Wigley and M.E. Schlesinger, "Analytical Solution for the Effect of Increasing CO2 on Global Mean Temperature", *Nature*, vol. 315, 1985, pp. 649-652.

[185]M. Allen (*et al*), "Observational Constraints on Climate Sensitivity", in H.J. Schellnhuber (*et al* eds.), *Avoiding Dangerous Climate Change*, (Cambridge University Press, Cambridge, 2006), pp. 281-289.

[186]As above p. 281.

[187]As above p. 282.

[188]As above.

[189]As above p. 288.

[190]For a survey see R. Warren, "Impacts of Global Climate Change at Different Annual Mean Global Temperature Increases", in H.J. Schellnhuber (*et al* eds.), *Avoiding Dangerous Climate Change*, (Cambridge University Press, Cambridge, 2006), pp. 93-131. For a more "popular" account see F. Linden, *The Winds of Change: Climate, Weather, and the Destruction of Civilizations*, (Simon and Schuster, New York, 2006).

[191]K. Hennessy (*et al*), *Climate Change Impacts on Fire-Weather in South East Australia*,(CSIRO, Canberra, 2006); S.W. Running, "Is Global Warming Causing More, Larger Wild Fires?" *Science*, vol.311, no. 5789, August 18 2006, pp. 927-928, doi: 10.1126/science.1130370; A.L. Westerling (*et al*), "Warming and Early Spring Increase Western US Forest Wildfire Activity", *Science*, vol.311, no. 5789, August 18 2006, pp. 940-943, doi: 10.1126/science.1128834.

[192] Warren cited note 190, p. 95.

[193] As above. See further B.Hare, "Relationship Between Increases in Global Mean Temperature and Impacts on Ecosystems, Food Production, Water and Socio-Economic Systems", in H.J. Schellnhuber (*et al* eds.), *Avoiding Dangerous*

*Climate Change*, (Cambridge University Press, Cambridge, 2006), pp. 177-185.

[194] T. E. Lovejoy and L.L. Hannah (eds.), *Climate Change and Biodiversity*, (Yale University Press, New Haven and London, 2005).

[195] A. Nyong and I. Niang-Diop, "Impacts of Climate Change in the Tropics: The African Experience",in H.J. Schellnhuber (*et al* eds.), *Avoiding Dangerous Climate Change*, (Cambridge University Press, Cambridge, 2006), pp. 235-241.

[196] Warren cited note 190, p. 95.

[197] The Stern Review, *The Economics of Climate Change*, October 31 2006 at www.sternreview.org.uk, p. 57.

[198] As above p. vi.

[199] " Fix Global Warming Now, or Pay Later", *The Australian*, October 31 2006, p. 8.

[200] The Stern Review, cited note 197, p. vi. Some critics of the Stern Review have argued that in the case of an energy-based economy such as Australia, the impact of Stern's recommendations would be between AUS $ 15 billion and AUS $ 66 a year, reducing Australian wages by 20 per cent. See D. Shanahan and M.Warren, "Howard Defiant in Face of Global Warming Warning", *The Australian*, October 31 2006, p. 1. On the other hand Sir Nicholas Stern has said that Australia's present water crisis could be seven times worse on the business as usual scenario. S. Briggs, "Worse to Come in Water Crisis", *The Advertiser* (Adelaide), November 4 2006, p. 5.

[201] B. Carter, "British Report the Last Hurrah of Warmaholics", *The Australian*, November 3 2006, p. 14. M.Warren "The Clean Green Dream:", *The Weekend Australian*, November 4-5 2006, p. 19; Also according to Sir Nicholas Stern Australia's Great Barrier Reef is likely to be destroyed by global warming on the business as usual scenario. See P.Kent, "Great Barrier Reef ' Already Doomed'", *The Advertiser,* November 6 2006, p. 8; B.Lomberg, "Stern Scare Blunted by the Figures", *The Australian*, November 6 2006, p. 16.

[202] B.L. Preston (*et al*), *Climate Change in the Asia/Pacific Region*, October 9 2006 at http://www.csiro.au/files/files/p9xj.pdf.

[203]As above p. 2.

[204]As above.

[205]As above p. 3.

[206]As above.

[207]As above p. 4.

[208]As above. See also pp. 60-61.

[209] The Stern Review, cited note 197, pp. 112-114; "Climate Change, Disasters, Desertification Can Affect Security, Survival of Developing Countries, Second Committee Told", UN Press Release GA/EF/13046, October 17 2003 at http://www.un.org/News/Press/docs/2003/gaef3046.doc.htm.

[210] A. Dupont and G. Pearman, *Heating Up the Planet: Climate Change and Security*, Lowy Institute Paper no.12, 2006 at http://www.lowyinstitute.org/, p. viii.

[211] P.Schwartz and D. Randall, "An Abrupt Climate Change Scenario and Its Implications for United States National Security", October 2003 at http://www.environmentaldefense.org/documents/3566_AbruptClimateChange.pdf.

[212] Dupont and Pearman, p. 82.

[213] See for example World Health Organization (WHO), *Heat Waves: Risks and Responses,* (Health and Global Environmental Change Series No2, WHO Regional Office for Europe, Denmark, 2004); A.J.Michael and A. Githeko (*et al*), "Human Health", Working Group II, Third Assessment Report. Intergovernmental Panel on Climate Change, *Climate Change 2001: Impacts, Adaptation, and Vulnerability,*(Cambridge University Press, Cambridge, 2001), pp. 453-485; A.J. Mc Michael (*et al.* eds.), *Climate Change and Human Health: Risks and Responses*, (World Health Organization, Geneva, 2003); R.T. Watson and A.J. Mc Michael, "Global Climate Change- The Latest Assessment: Does Global Warming Warrant a Health Warning?" *Global Change and Human Health*, vol. 2, 2001, pp. 64-75; A.J. McMichael, "Population, Environment,

Disease and Survival: Past Patterns, Uncertain Futures", *Lancet*, vol.359, 2002, pp. 1145-1148; A.J. McMichael, "The Biosophere, Health and Sustainable Development", *Science* , vol.297, 2002, p. 1093; R.R. Colwell, "Global Warming and Infectious Diseases, *Science* , vol.274, 1996, pp. 2025-2031; E. Lindgren and R. Gustafson, "Tick-borne Encephalitis in Sweden and Climate Change", *Lancet*, vol.358, 2001, pp. 16-87; R.R. Colwell and J.A. Patz, *Climate, Infectious Disease and Health*, (American Academy of Microbiology, Washington DC, 1998); P. Martens and A.J. McMichael, *Environmental Change, Climate and Health*, (Cambridge University Press, Cambridge, 2002); P.R. Epstein, "Climate and Health", *Science* , vol.285, 1999, pp. 347-348; J.A. Patz (*et al.*), "Global Climate Change and Emerging Infectious Diseases", *Journal of the American Medical Association*, vol.275, 1996, pp. 217-223; W.J.M. Martens, *Health and Climate Change: Modelling the Impacts of Global Warming and Ozone Depletion*, (Earthscan, London, 1998); W.J.M. Martens (*et al*), "Climate Change and Future Populations at Risk of Malaria", *Global Environmental Change*, vol. 9 (Supplement), 1999, S89-S107; S.Lindsay and W.J.M. Martens, "Malaria in the African Highlands: Past, Present and Future", *Bulletin of the World Health Organization*, vol. 78, 2000, pp. 33-45; P.R. Epstein, "Is Global Warming Harmful to Health?", *Scientific American*, vol. 283, no.2, 2000, pp. 50-57; D.J. Rogers and S.F. Randolph, "The Global Spread of Malaria in a Future, Warmer World", *Science*, vol. 289, 2000, pp. 1763-1765.

[214] C.McMurray and R. Smith, *Diseases of Globalization: Socioeconomic Transitions and Health*, (Earthscan, London, 2001); K.Lee (*et al.*), "Global Change and Health- The Good, the Bad and the Evidence", *Global Change and Human Health*, vol.3, no.1, 2002, pp. 16-19.

[215] R.Ornstein and P. Ehrlich, *New World New Mind: Moving Toward Conscious Evolution*, (Doubleday, New York, 1989), p. 127.

[216] D. MacKenzie, "Bird Flu Outruns the Vaccines", *New Scientist*, November 4 2006, pp. 8-9.

[217] A.S. Monto, "The Threat of an Avian Influenza Pandemic", *New England Journal of Medicine*, vol. 352, no 4, 2005, pp. 323-325.

[218] S. Kennedy (*et al.*), *A Primer on the Macroeconomic Effects of an Influzena Pandemic*, Australian Treasury Working Paper February 2006.

[219] W.J. Mc Kibben and A.A. Sidarenko, *Global Macroeconomic Consequences*

*of Pandemic Influenza*, Lowy Institute for International Policy, February 2006 at http://www.lowyinstitute.org/Publication.asp?pid=345.

][220] R.T. Levins (*et al*), "The Emergence of New Diseases", *American Scientist*, vol. 82, 1994, pp. 52-60; P.Daszak (*et al.*), "Emerging Infectious Diseases of Wildlife - Threats to Biodiversity and Human Health", *Science*, vol. 287, 2000, pp. 443-449.

[221] A.J. Mc Michael (*et al.*), *Climate Change and Human Health*, (World Health Organization, Geneva, 1996); A. Kruess and T. Tschamtke, "Habit Fragmentation, Species Loss and Biological Control", *Science*, vol. 264, 1994, pp. 1581-1584; C.D. Harwell (*et al.*), "Diseases in the Ocean: Emerging Pathogens, Climate Links, and Anthropogenic Factors", *Science*, vol. 285, 1999, pp. 1505-1510.

[222] Millennium Ecosystem Assessment, *Ecosystems and Human Well Being:Health Synthesis*, (World Health Organization, Geneva, 2005).

[223] As above p.12,

[224] M. Enserink, "During a Hot Summer, Bluetongue Virus Invades Northern Europe", *Science*, vol. 313, 2006, pp. 1218-1219.

[225] A.J. McMichael (*et al.*), "Climate Change and Human Health: Present and Future Risks", *The Lancet*, Februrary 9 2006, doi:10.1016/S0140-6736(06) 68079-3; T.Hampton, "Researchers Study Health Effects of Environmental Change", *Journal of the American Medical Association*, vol. 296, no.8, 2006, pp. 913-920.

[226] S. Vandentorren (*et al*), "August 2003 Heat Wave in France: Risk Factors for Death of Elderly People Living at Home", *European Journal of Public Health*, vol.16, no. 6, 2006, pp. 583-591.

[227] B.L. Preston (*et al*), *Climate Change in the Asia/Pacific Region*, (CSIRO, Melbourne, 2006) at http://www.csiro.au/resources/pfkd.html.

[228] As above p. 36.

[229] As above p. 1.

---

[230] A.Dupont and G. Pearman, *Heating Up the Planet: Climate Change and Security*, (Lowy Institute Paper no 12, 2006) at http://www.lowyinstitute.org.

[231] As above p. 28.

[232] As above p. 29. Consultative Group on International Agricultural Research. Inter-Center Working Group on Climate Change, *The Challenge of Climate Change: Research to Overcome its Impact on Food, Scarcity, Poverty, and National Resource Degradation in the Developing World*, (Consultative Group on International Agricultural Research, Inter-Center Working Group on Climate Change ,2002), at http://www.cgiar.org/pdf/climatechange.pdf , pp. 1-2.

[233] Dupont and Pearman, as above, pp.29-30.

[234] As above p. 30.

[235] As above p. 33.

[236] As above. See also A.Dupont, *East Asia Imperilled: Transnational Challenges to Security*, (Cambridge University Press, Cambridge, 2001), p. 17.

[237] M. Mimura (*et al*), "Small Island States", IPCC Working Group II, Fourth Assessment Report, *Climate Change 2007: Climate Change Impacts, Adaptation and Vulnerability*, at http://www.ipcc.ch.

[238] J. Hay (*et al.*), *Climate Variability and Change and Sea-level Rise in the Pacific Islands Region*, (South Pacific Regional Environment Programme, Samoa, 2003); J. Barnett and W.N. Adger, "Climate Dangers and Atoll Countries", *Climate Change*, vol. 61, 2003, pp. 321-337; J. Barnett, "Adapting to Climate Change in Pacific Island Countries: The Problem of Uncertainty", *World Development*, vol. 29, 2001, pp. 977-993.

[239] Mimura (*et al*), cited note 237; K.L. Ebi (*et al.*), *Climate Variability and Change and their Health Effects on Small Island States*, (Report on Regional Workshops and Conference Convened by WHO, WMO and UNEP); Hay (*et al.*) cited note 238.

[240] As above.

[241] Center for Health and the Global Environment, Harvard Medical School, *Climate Change Futures: Health, Ecological and Economic Diemensions*, (Center for Health and Global Environment, Harvard Medical School, 2005).

[242] P. Macumbi, "Plague of My People", *Nature*, vol. 430. August 19 2004, p.925.

[243] B. Greenwood, "Between Hope and a Hard Place", *Nature*, vol. 430, August 19 2004, pp. 926-927; *Africa Malaria Report 2003*, (WHO/UNICEF, Geneva, 2003); R. Snow, "The Invisible Victims", *Nature*, vol.430, August 19 2004, pp. 934-935.

[244] Macumbi, cited note 242.

[245] As above .

[246] J.A. Patz, "A Human Disease Indicator for the Effects of Recent Global Climate Change", *Proceedings of the National Academy of Sciences*, vol. 99, no. 20, 2002, pp. 12506-12508; S.I. Hay (*et al.*), "Urbanization, Malaria Transmission and Disease Burden in Africa", *Nature Reviews Microbiology*, vol. 3 no.1, 2005, pp. 81-90; S.I. Hay (*et al.*), "Climate Variability and Malaria Epidemics in the Highlands of East Africa", *Trends in Parasitology*, vol. 21, no.2, 2005, pp. 52-53.

[247] J.A. Patz, (*et al.*), "Global Climate Change and Emerging Infectious Diseases", *Journal of the American Medical Association*, vol. 275, 1996, pp. 217-233; M. van Lieshout (*et al.*), "Climate Change and Malaria: Analysis of the SRES Climate Change and Socio-Economic Scenarios", *Global Environmental Change*, vol. 14, 2004, pp. 87-99.

[248] P.Reiter, "Climate Change and Mosquito-Borne Diseases", *Environmental Health Perspectives*, vol. 109, March 2001, pp. 141-161.

[249] L.J. Bruce-Chwatt and J de Zulueta, *The Rise and Fall of Malaria in Europe: A Historico-Epidemiological Study*, (Oxford University Press, Oxford, 1980).

[250] Reiter cited note 248.

[251] D.L. Hartmann (*et al*), "Can Ozone Depletion and Global Warming Interact to Produce Rapid Climate Change?" *Proceedings of the National Academy of*

*Sciences,* vol.97, no.4, February 15 2000, pp. 1412-1417.

[252] R. Stott, "Contraction and Convergence: Healthy Response to Climate Change", *British Medical Journal,* vol.332, June 10 2006, pp. 1385-1387, cited p. 1385.

**Chapter 3**

[1] J. Diamond, *Collapse: How Societies Choose to Fail or Survive,* (Allan Lane, London, 2005), pp. 504-505.

[2] Joseph Wayne Smith , quoted from R. Dahl, "A Changing Climate of Litigation", *Environmental Health Perspectives,* vol. 115, no.4, April 2007, pp. A 204-A 207, cited p. A 207.

[3] On the legal aspects of the Kyoto Protocol see D. Freestone and C. Streck (eds.), *Legal Aspects of Implementing the Kyoto Protocol Mechanisms: Making Kyoto Work,* (Oxford University Press, Oxford, 2005). For material on climate change litigation see E.M. Peñalver, "Acts of God or Toxic Torts? Applying Tort Principles to the Problem of Climate Change", *Natural Resources Journal,* vol. 38, Fall 1998, pp. 563-601; D.A. Grossman, "Warming up to a Not-So-Radical Idea: Tort-Based Climate Change Litigation", *Columbia Journal of Environmental Law,* vol. 28, 2003, pp.1-61, A.L. Strauss, "The Legal Option: Suing the United States in International Forums For Global Warming Emissions", *Environmental Law Institute,* vol. 33, 2003, pp.10185-10191 at http://www.eli.org; B.C. Mank, "Standing and Global Warming: Is Injury to All Injury to None?" *Environmental Law,* vol. 35, 2005, pp. 1-84; R.Verheyen, *Climate Change Damage and International Law: Prevention Duties and State Responsibility,* (Martinus Nijhoff Publishers, Leiden, 2005); W.C.G. Burns, "Potential Causes of Action for Climate Change Damages in International Fora: The Law of the Sea Convention", *McGill International Journal of Sustainable Development Law and Policy,* vol. 2, no. 1, March 2006, pp. 27-51, and H. Osofsky and W.C.G. Burns, *Adjudicating Climate Change: Sub-National, National and Supra-National Approaches,* (Cambridge University Press,

Cambridge, 2007).

[4] J.Smith and D. Shearman, *Climate Change Litigation: Analysing the Law, Scientific Evidence and Impacts on the Environment, Health and Property,* (Presidian Legal Publications, Adelaide, 2006).

[5] *Massachusetts v Environmental Protection Agency* 549 US_ (2007) at http://laws.findlaw.com/us/000/05-1120.html>.

[6] *State of California v General Motors Corporation et al*, Case No C06-05755, US District Court, Northern District of California.

[7] *Gray v The Minister For Planning and Ors* [2006] NSWLEC 720

[8] As above at [134]-[135]. On the precautionary and intergenerational equity principles see generally Sharon Beder, *Environmental Principles and Policies: An Interdisciplinary Approach*, (University of New South Wales Press, Sydney, 2006).

[9] As above at [152].

[10] Editorial, "Pie-In-The-Sky On Coal", *The Australian*, November 30 2006, p. 11.

[11] *Re Xstrata Coal Queensland Pty Ltd and Ors* [2007] QLRT 33

[12] As above at [8].

[13] As above at [16].

[14] As above at [18].

[15] Even a rise of only a fraction of a degree C can have substantial environmental effects as we have seen from the evidence reviewed in chapter 2 of this book. The scientific basis upon which the learned presiding member concluded that the temperature rise over the 20th century, as stated by the IPCC, was "surprising low", is not stated in the judgment.

[16] As above at [19]-[20]. As is noted at [20], the Tribunal "is empowered by statute to "inform itself of anything in the way it considers appropriate"": *Land*

*and Resources Tribunal Act 1999*, section 49(2)(b).

[17] As above at [21].

[18] As above at [21]-[22].

[19] As above at [23].

[20] Xstrata Coal states its commitment to sustainable development on its website: http://www.xstrata.com/sustainability.

[21] *Massachusetts v Environmental Protection Agency*, cited note 5.

[22] As above at p.19.

[23] "Environmentalists Hail US Emissions Ruling" at http://www.abc.net.au/news/newsitems/200704/s1888809.htm. Catherine Fitzpatrick of Greenpeace Australia is quoted as saying: "When you get the world's largest polluter of greenhouse gas emissions having their Supreme Court rule that greenhouse gases need to be taken into account and possibly regulated, then it's a loud signal to the rest of the world, not just the US but to many countries around the world". The case may return to a lower courts for further proceedings consistent with the US Supreme Court ruling. At the time of writing Friends of the Earth is suing the Canadian government for breaches of the Kyoto Protocol: *Friends of the Earth V Her Majesty the Queen, the Minister of the Environment and the Minister of Health*, (Application, Federal Court of Canada - Trial Division, May 28 2007). Canada's greenhouse gas emissions are 34 per cent above its six per cent reduction target. The application for judicial review is made pursuant to sections 18 and 18.1 of the *Federal Court Act* (Canada) for an alleged breach of section 166(1) of the *Canadian Environmental Protection Act, 1999* which relates to matters of international air pollution.

[24] *The State of California v General Motors Corporation et al*, cited note 6.

[25] "California Sues Automobile Companies for Climate Change Damage", at http://www.climatelaw.org/media/CA%20auto%20companies.

[26] P.Wilson, "Blair Puts Carbon Targets Into Law", *The Australian*, March 15 2007, p. 8.

[27] "Binding Carbon Targets Proposed", BBC News, March 13 2007 at http://news.bbc.co.uk/2/hi/uk_news/politics/6444145.shm.

[28] See www.dtistats.net/energystats/et_mar07.pdf.

[29] As above p. 22; "UK Carbon Emissions Highest Since Labour Came to Power", Friends of the Earth Press Release, March 29 2007 at http://www.foe.co.uk/resource/press_releases/uk_carbon_emissions_highes_2903 200....

[30] J.W. Smith (ed.), *Immigration, Population and Sustainable Environments: The Limits to Australia's Growth*, (Flinders Press, Bedford Park, 1991).

[31] See generally K. Betts and M. Gilding, "The Growth Lobby and Australia's Immigration Policy", *People and Place*, vol. 14, no. 4, 2006, pp. 40-52.

[32] See A. Bradbrook (*et al.*,eds.), *The Law of Energy for Sustainable Development*, (Cambridge University Press, New York, 2005) and R.Lyster and A. Bradbrook, *Energy Law and the Environment*, (Cambridge University Press, Melbourne, 2006).

[33] M.Hillman, "Personal Carbon Allowances", *British Medical Journal*, vol. 332, 2006, pp. 1387-1388, at p. 1387.

[34] M.Hillman with T.Fawcett, *How We Can Save the Planet*, (Penguin Books, London, 2004).

[35] Hillman, cited note 33, p.1387.

[36] As above.

[37] As above.

[38] R. Starkey and K. Anderson, *Domestic Tradable Quotas: A Policy Instrument of Reducing Greenhouse Gas Emissions from Energy Use*, Tyndall Centre Technical Report, No. 39 at www.tyndall.ac.uk/publications/tech_reports/tech_reports.shtml.

[39] Hillman with Fawcett, cited note 34.

[40] A. Meyer, *Contraction and Convergence: The Global Solution to Climate Change*, (Green Books, Devon, 2000).

[41] "The Ideas and Algorithms behind Contraction and Convergence and CC Options", at http://www.gci.org.uk/model/ideas_behind_cc.html.

[42] Tiempo Climate Newswatch, "Contraction and Convergence", at http://www.cru.uea.ac.uk/tiempo/newswatch/comment060704.htm.

[43] As above.

[44] A.Meyer, "The United States has it Right on Climate Change in Theory", at http://www.opendemocracy.net/debates/article-6-129-2462.jsp; "Greenhouse Gas 'Plan B' Gaining Support". *New Scientist,* December 10 2003 at http://www.newscientist.com/article.ns?id=dn4467.

[45] "Greenhouse Gas 'Plan B' Gaining Support", as above.

[46] www.climatecare.org, cited from *New Internationalist*, July 2006, p. 14

[47] I. Roberts, Review of M.Hillman with T.Fawcett, *How We Can Save the Planet*", *British Medical Journal*, vol. 332, June 10 2006, p. 1398 at p. 1398.

[48] D. Freestone and C. Streck (eds.), cited note 3.

[49] P.F.Smith, "Contraction and Convergence: Myth and Reality". *British Medical Journal*, vol. 332, June 24 2006, p. 1509.

[50] M.Lynas, "Shares in the Sky", *New Internationalist*, no. 357, June 2003 at http://www.newint.org/issue357/shares.htm, at p.2.

[51] As above. See also T. Athanasiou and P. Baer, *Dead Heat: Global Justice and Global Warming*, (Seven Stories Press, New York, 2002); E. Neumayer, "In Defence of Historical Accountability for Greenhouse Gas Emissions", *Ecological Economics*, vol. 33, 2000, pp. 185-192.

[52] J-F., Revel, *Anti-Americanism*, (Encounter Books, New York, 2004); P. Driessen, *Eco-Imperialism: Green Power, Black Death*, (Merril Press, Bellevue, WA, 2003).

[53] J.W. Smith, *Reductionism and Cultural Being*, (Marinus Nijhoff, The Hague, 1984).

[54] C. Flavin and G. Gardner, "China, India, and the New World Order", in L.Starke (ed.), *State of the World 2006*, (Earthscan, London, 2006), pp. 3-23.

[55] J. Frior and J.E. Jacobsen, *The Crowded Greenhouse: Population, Climate Change, and Creating a Sustainable World*, (Yale University Press, New Haven, 2002), pp. 187-188.

[56] M. McCarthy and C. Coonan,"The Great Pall of China", *The Independent*, April 26 2007 at http://news.independent.co.uk./environment/climate_change/article2483839.ece.

[57] As above.

[58] C. Murray, *Human Accomplishment: The Pursuit of Excellence in the Arts and Sciences, 800 BC to 1950*, (HarperCollins, New York, 2003).

[59] L.P. King and S. Szelenyi, *Theories of the New Class: Intellectuals and Power*, (University of Minnesota Press, Minneapolis, 2004).

[60] F.E. Trainer, *The Conserver Society: Alternatives for Sustainability*, (Zed Books, London, 1995).

[61] T.E. Trainer, *Abandon Affluence!* (Zed Books, London, 1985).

[62] M.Hillman with T.Fawcett, cited note 34, p.38.

[63] As above p. 163. George Monbiot has said on this point: "If the biosphere is wrecked, it will not be done by those who couldn't give a damn about it, as they now belong to a diminishing minority. It will be destroyed by nice, well-meaning, cosmopolitan people who accept the case for cutting emissions, but who won't change by one iota the way they live. I know people who profess to care deeply about global warming, but who would sooner drink Toilet Duck than get rid of their agas, patio heaters and plasma TVs, all of which are staggeringly wasteful." G. Monbiot, "How Much Reality Can You Take?" at http://www.monbiot.com/archives/2006109/21/how-much-reality-can-you-take/.

[64] I. Lowe, *A Big Fix: Radical Solutions for Australia's Environmental Crisis*, (Black Inc., Melbourne, 2005).

[65] L.Brown, *Plan B 2.0: Rescuing a Planet Under Stress and a Civilization in Trouble*, (W.W. Norton, New York, 2006).

[66] Lowe, cited note 64, p.20.

[67] As above.

[68] B. Lomborg, *The Skeptical Environmentalist*, (Cambridge University Press, Cambridge, 2001).

[69] Lowe, cited note 64, p.15.

[70] As above, pp. 74-75.

[71] As above.

[72] As above at p. 81.

[73] See also F.Pearce, "Ecopolis Now", *New Scientist*, June 17 2006, pp. 36-42.

[74] A. Simms, "To the Rescue", *New Scientist*, February 2006, p. 50.

[75] Brown, cited note 65, p. 254.

[76] As above p. 259.

[77] G. Monbiot, *Heat: How to Stop the Planet Burning*, (Allen Lane, London, 2006).

[78] C.D. Jones (*et al.*), "Strong Carbon Cycle Feedbacks in a Climate Model with Interactive CO2 and Sulphate Aerosols", *Geophysical Research Letters*, vol. 30, May 9 2003, doi:10.1029/2003GLO16867, p. 1479.

[79] Monbiot, cited note 77, pp. 16-17.

[80] D. Fleming "Energy and the Common Purpose: Descending the Energy

Staircase with Tradeable Energy Quotas (TEQs)", at http://www.teqs.net/book/teqs.pdf.

[81] Monbiot, cited note 77, p. 46.

[82] As above p. 60.

[83] J.D. Khazzoom, "Economic Implications of Mandated Efficiency Standards for Household Appliances", *Energy Journal,* vol. 1, 1980, pp. 21-39.

[84] Monbiot, cited note 77, p. 61.

[85] As above.

[86] As above.

[87] As above, p. 63 .

[88] As above, p. 63-72.

[89] As above, p. 74.

[90] As above, p. 203.

[91] As above, p. 48.

[92] R.H. Nelson, *Economics as Religion: From Samuelson to Chicago and Beyond,* (Pennsylvania State University Press, University Park, Pennsylvania, 2001).

[93] As above p. xv.

[94] J.W. Smith (*et al.*), *The Bankruptcy of Economics: Ecology, Economics and the Sustainability of the Earth*, (Macmillan, London, 1999).

[95] R.D. North, *Life on a Modern Planet: A Manifesto for Progress*, (Manchester University Press, Manchester, 1995), p. 254.

[96] R.H. Coase, "The Problem of Social Cost", *Journal of Law and Economics",* vol. 3, 1960, pp. 1-44.

[97] As above, p.15.

[98] As above, p. 2.

[99] R.H. Coase, *The Firm, the Market and the Law*, (University of Chicago Press, Chicago and London, 1988).

[100] K. Hanly, "The Problem of Social Cost: Coase's Economics versus Ethics", *Journal of Applied Philosophy*, vol. 9, 1992, pp. 77-83; M.L. Cropper and W.F. Oates, "Environmental Economics: A Survery", *Journal of Economic Literature*, vol. 30, 1992, pp. 674-740; E.J. Mishan, "Pangloss on Pollution", *Swedish Journal of Economics*, vol.73, no. 1, 1971, pp. 113-120.

[101] T.Smith, "The Case Against Free Market Environmentalism", *Journal of Agricultural and Environmental Ethics*, vol. 8, no. 2, 1995, pp. 126-144; T.Smith, "Response to Narveson", *Journal of Agricultural and Environmental Ethics*, vol. 8, no. 2, 1995, pp. 157-158; M. Sagoff, "Some Problems with Environmental Economics", *Environmental Ethics*, vol. 10, 1988, pp. 55-74; J.M. Gowdy and P.G. Olsen, "Further Problems with Neoclassical Environmental Economics", *Environmental Ethics*, vol. 16, 1994, pp. 161-171.

[102] P.Hawken, A.Lovins and L.H.Lovins, *Natural Capitalism: Creating the Next Industrial Revolution*, (Little, Brown and Company, Boston, 1999). See further J.Kavel, *The Enemy of Nature: The End of Capitalism or the End of the World?* (Zed Books, London, 2006), and for an opposing view to Kavel, J. Porritt, *Capitalism as if the World Matters*, (Earthscan, London, 2006).

[103] See J.W. Smith (*et al.*) cited note 94.

[104] Stern Review, *The Economics of Climate Change* (2006) at www.sternreview.org.uk.

[105] The scientific foundations of the Stern Review are based on the consensus view of climate change, but some researchers, critical of this position have subjected the science of the Stern Review to a critique. See R.M. Carter (*et al.*), "The Stern Review: A Dual Critique, Part I: The Science", *World Economics*, vol. 7, no. 4, October-December 2006, pp. 167-198.

[106] W. Nordhaus, "The *Stern Review* on the Economics of Climate Change"

(2006) at http://nordhaus.econ.yale.edu/SternReviewDz.pdf.

[107] I. Byatt (*et al.*), "The Stern Review: A Dual Critique, Part II: Economic Aspects", *World Economics*, vol. 7, no. 4, October-December 2006, pp. 199-229.

[108] D.Pearce, *Economic Values and the Natural World*, (Earthscan, London, 1993), p. 54.

[109] See generally M. Sagoff, *The Economy of the Earth*, (Cambridge University Press, Cambridge, 1988); J. Broome, *Counting the Cost of Global Warming*, (White Horse Press, Isle of Harris, UK, 1992); J. Broome, *Ethics Out of Economics*, (Cambridge University Press, Cambridge, 1999); J. Broome, *Weighing Lives*, (Oxford University Press, Oxford, 2004); T.Cowan and D. Parfit, "Against the Social Discount Rate", in P.Laslett and J. Fishkin (eds.), *Justice Between Age Groups and Generations*, (Yale University Press, New Haven, 1992), pp.144-161; S.M. Gardiner, *Why Do Future Generations Need Protection?* E.D.F-Ecole Polytechnic, July 2006.

[110] It is argued by economists that $X is not equivalent to $X in 5 years time. If this money is saved today at a positive interest rate, one will get more money in the future, i.e. $X + i. Subjectively the rational economic subject considers $X now to be equivalent to $X + i in the future, so a discount rate is needed to give an accurate present valuation of the value of money in the future. See I. Fisher, *The Theory of Interest*, (Macmillan, New York, 1930); L. van Liedekerke, "John Rawls and Derek Parfit's Critique of the Discount Rate", *Ethical Perspectives*, vol. 11, 2004, pp. 72-83.

[111] See J. Broome, "Discounting the Future", in J. Broome, *Ethics Out of Economics*, (Cambridge University Press, Cambridge, 1999), at p. 60; J. Broome, "Should We Value Population?" *Journal of Political Philosophy*, vol. 13, no.4, 2005, pp. 339-413, at pp. 411-413; J.Rawls, *A Theory of Justice*, (Harvard University Press, Cambridge, 1971), pp. 297-298; D.Parfit, *Reasons and Persons*, (Clarendon Press, Oxford, 1984), p. 486; J.V. Krutilla and A.C. Fisher, *The Economics of Natural Environments*, (John Hopkins University Press, Baltimore MD, 1985).

[112] E.Padilla, "Climate Change, Economic Analysis and Sustainable Development", *Environmental Values*, vol. 13, 2004, pp. 523-544; M.H.Prager and K.W.Shertzer, "Remembering the Future": A Commentary on "Intergenerational Discounting: A New Intuitive Approach"", *Ecological*

*Economics*, vol. 60, 2006, pp. 24-26, at p.24.

[113] Padilla, cited note 112, p. 533. For further discussion see C.Price, *Time, Discounting and Value*, (Blackwell, Oxford, 1993); J.Adams, "Cost-Benefit Analysis: The Problem Not the Solution", *The Ecologist*, vol. 26, no.1, 1996, pp. 2-4; D.Jamieson, "Ethics, Public Policy, and Global Warming", in A. Light and H. Rolston III (eds.), *Environmental Ethics: An Anthology*, (Blackwell Publishing, Oxford, 2003), pp. 371-379.

[114] *After the Stern Review: Reflections and Responses; Paper B: Value Judgements, Welfare Weights and Discounting: Issues and Evidence*, February 12 2007 at www.sternreview.org.uk.

[115] Nordhaus, cited note 106, p.12.

[116] As above p.12.

[117] Stern Review, cited note 104, p.163.

[118] As above p.197.

[119] As above p. 201.

[120] T. Trainer, "The Stern Review: Critical Notes on its Abatement Optimism", Unpublished manuscript, November 2006. Contact: T.Trainer@unsw.edu.au.

[121] B.Lomborg, "Stern Scare Blunted by the Figures", *The Australian*, November 6 2006, p.16.

[122] As above.

[123] National Emissions Trading Taskforce, *Possible Design for a National Greenhouse Gas Emission Trading Scheme* (2006) at www.emissionstrading.net.au; S. Kverndokk, "Tradeable CO2 Emissions Permits: Initial Distribution as a Justice Problem", *Environmental Values*, vol.4, 1995, pp. 129-148.

[124] A. Hodge, "Householders Cut Greenhouse Gases", *The Australian*, June 2 2006, p. 9; M.Warren, "Carbon Trading Market to Open 'As Early as 2011'", *The Australian*, May 29 2007, p.1. See also www.australiancarbontraders.com.

---

[125] F. Zakaria, "The Case for a Global Carbon Tax", Newsweek April 16, vol.149, no. 16 2007.

[126] S.Morris, "Mini Tax on Carbon to Prepare for Future", *The Australian*, April 5, 2007, p.1.

[127] J.R.Kahn and D.Franceschi, "Beyond Kyoto:A Tax-Based System for the Global Reduction of Greenhouse Gas Emissions", *Ecological Economics*, vol. 58, 2006, pp. 778-787.

[128] D. Montgomery and A.E.Smith, "Price, Quantity and Technology Strategies for Climate Policy", in M.Schlesinger (*et al.* eds.), *Human-Induced Climate Change:An Interdisciplinary Assessment*, (Cambridge University Press, Cambridge, Forthcoming 2007).

[129] N. Mabey (*et al.*), *Argument in the Greenhouse: The International Economics of Controlling Global Warming*, (Routledge, London and New York, 1997), p. 317.

[130] R. Gerlagh and E.Papyrakis, "Are the Economic Costs of (Non-) Stabilizing the Atmosphere Prohibitive? A Comment", *Ecological Economics*, vol. 46, 2003, pp. 325-327; D.C. Hall and R.J. Behl, "Integrating Economic Analysis and the Science of Climate Instability", *Ecological Economics*, vol. 57, 2006, pp. 442-465.

[131] R.H.Socolow and S.W. Pacala, "A Plan to Keep Carbon in Check", *Scientific American,* September 2006, pp. 28-35, cited pp. 29-30.

[132] As above.

[133] See A.Gore, *An Inconvenient Truth* (2006) DVD and *An Inconvenient Truth,*

(Rodale Emmaus, PA, 2006).

[134] B.Metz and D. van Vuuren, "How, and at What Costs, Can Low-Level Stabilization be Achieved?- An Overview", in H.J. Schellnhuber (*et al.* eds.), *Avoiding Dangerous Climate Change*, (Cambridge University Press, Cambridge, 2006), pp. 337-345, cited p.339.

[135] W.R. Moonaw (*et al.*), "Technological and Economic Potential of Greenhouse Gas Emissions Reduction", in B.Metz (*et al.*eds.), *Climate Change 2001: Mitigation. Contribution of Working Group III to the Third Assessment Report of the IPCC*, (Cambridge University Press, Cambridge, 2001), cited from Metz, note 134.

[136] R.T.Watson (*et al.*), *Climate Change 2001: Synthesis Report. A Contribution of Working Groups I, II, and III to the Third Assessment Report of the IPCC*, (Cambridge University Press, Cambridge, 2001), at http://www.ipcc.ch/pub/online.htm, pp 106-122; T. Banuri, *Summary for Policymakers. Climate Change 2001: Mitigation. A Report of Working Group III of the Intergovernmental Panel on Climate Change*, (Cambridge University Press, Cambridge, 2001), http://www.ipcc.ch/pub/online.htm.

[137] J.W. Smith, *The High Tech Fix: Sustainable Ecology or Technocratic Megaprojects for the 21st Century*, (Avebury, Aldershot, 1991).

[138] B.Crystall, "The Big Clean-Up", *New Scientist*, September 3 2005, pp. 30-31 at p.30.

[139] As above.

[140] As above.

[141] R.Baker, 'Burying the Problem', *The Age* (Melbourne), July 30 2005, p.3.

[142] As above. See also J.Gibbins (*et al.*), "Scope for future CO2 Emission Reductions from Electricity Generation through the Deployment of Carbon Capture and Storage Technologies", in H.J.Schellnhuber (*et al.* eds.), *Avoiding Dangerous Climate Change*, (Cambridge University Press, Cambridge, 2006), pp. 379-383.

[143] B.Davis, "A Greener Shade of Black", *New Scientist*, September 3 2005, pp. 38-40; D.G.Hawkins (*et al.*), "What to Do About Coal", *Scientific American*, September, 2006, pp. 44-51. There are plans to build the world's largest clean coal project at Kwinana, South of Perth, Australia, by BP and Rio Tinto. The plan would bury 90 per cent of carbon emissions, but would cost three times as much as a conventional coal-fired power station to build and 50 per cent more to run. The project requires a government subsidy. See N.Wilson, "BP, Rio in Clean

Coal Plan", *The Australian*, May 22 2007, pp. 1-4.

[144] A.Hodge, "Climate Pact has its Critics Fuming", *The Australian*, July 28 2005, p. 6.

[145] As above.

[146] As above.

[147] R. Baker, "Revealed: How Big Energy Won the Battle on Climate Change", *The Age*, July 30 2005, pp 1,2.

[148] As above at p.1.

[149] As above.

[150] "Scant Protection", *New Scientist*, January 21 2006, p.7; B.S. Fisher (*et al.*), *Technological Development and Economic Growth.* (Australian Bureau of Agricultural and Resource Economics, Canberra, 2006).

[151] D.M. Kammen, "The Rise of Renewable Energy", *Scientific American*, September 2006, pp. 60-69, cited p.61.

[152] T. Trainer, *Renewable Energy Cannot Sustain Consumer Society*, (Springer, Dordrecht, 2007).

[153] H.C.Hayden, *The Solar Fraud* (2nd edition), (Vales Lake, Pueblo West, 2004).

[154] H. Grabl (*et al.* eds.), *World in Transition:Towards Sustainable Energy Systems*, (Earthscan, London, 2004).

[155] As above p. 10.

[156] S.F. Singer and D.T.Avery, *Unstoppable Global Warming Every 1,500 years*,(Rowman and Littlefield, Lanham, 2007), pp. 13-14. Emphasis added. See also D.T.Avery, *Saving the Planet with Pesticides and Plastic*, (Hudson Institute, Indianapolis, 1995).

[157] M.J.Hoffert (*et al.*), "Advanced Technology Paths to Global Climate Stability: Energy for a Greenhouse Planet", *Science*, vol. 298, 2002, pp. 981-987.

[158] As above, p. 981.

[159] As above, p. 984.

[160] J.Leake and R.Booth, "Iceland's Hot Rock to Power Europe", *The Australian*, May 14 2007, p. 13.

[161] K. Orchison, "450 years' Electricity in Hot Rocks", *The Weekend Australian*, September 9-10 2006, (Power Generation Supplement), p.2.

[162] A.O'Brien, "Carpenter Warms to Geothermal Energy", *The Weekend Australian*, November 18-19 2006, p.2.

[163] "Clean Power Under Our Feet", *New Scientist*, January 27 2007, p.4; J.W.Tester (*et al.*), *The Future of Geothermal Energy: Impact of Enhanced Geothermal Systems (EGS) on the United States in the 21st Century* (2007) at http://geothermal.inel.gov/publications/future_of_geothermal_energy.pdf.

[164] "Clean Power Under Our Feet", as above.

[165] Trainer cited note 152, chapter 8.

[166] S.C. Hunt (*et al.*), "Cultivating Renewable Alternatives to Oil", in L.Starke (ed.), *State of the World 2006*, (Earthscan, London, 2006), pp. 61-77.

[167] J.Hill (*et al.*), "Environment, Economic, and Energetic Costs and Benefits of Biodiesel and Ethanol Biofuels", *Proceedings of the National Academy of Sciences*, vol. 103, no.30, 2006, pp.11206-11210.

[168] C.F. Runge and B.Senauer, "How Biofuels Could Starve the Poor", *Foreign Affairs*, May/June 2007 at http://www.foreignaffairs.org.

[169] J.Ogden, "High Hopes for Hydrogen", *Scientific American*, September 2006, pp. 70-77.

[170] S.F.Lincoln, *Challenged Earth*, (Imperial College Press, London, 2006), p. 389.

[171] J.J.Romm, *The Hype About Hydrogen: Fact and Fiction in the Race to Save*

*the Climate*, (Island Press, Washington DC, 2005), p.9.

[172] Trainer, cited note 152.

[173] W.W. Gibbs, "Plan B for Energy" *Scientific American*, September 2006, pp. 78-87.

[174] Hoffert (*et al.*), cited note 157, p.985.

[175] See L.Bengtsson, "Geo-Engineering To Confine Climate Change: Is It At All Feasible?" *Climatic Change*, vol. 77, 2006, pp. 229-234; D.Jamieson, "Ethics and Intentional Climate Change", *Climatic Change*, vol. 33, 1996, pp. 323-336.

[176] R.Douthwaite, *The Growth Illusion,* (Resurgence, Bideford, Devon, 1992); H.E. Daly, *Steady-State Economics*, (Earthscan, London, 1992).

[177] J.Lovelock, *The Revenge of Gaia: Why the Earth is Fighting Back - and How We Can Still Save Humanity,* (Allen Lane, London, 2006).

[178] As above p.91.

[179] As above.

[180] H.Caldicott, *Nuclear Power is Not the Answer to Global Warming or Anything Else*,

(New Press, New York, 2006).

[181] R. Edwards, "Who Will Pay for a Nuclear Future?" *New Scientist*, June 10 2006, p.8; A.Makhijani (*et al.* eds.), *Nuclear Wastelands: A Global Guide to Nuclear Weapons Production and Its Health and Environmental Effects*, (MIT Press, Cambridge, Massachusetts, 2000).

[182] D. Williams and K.Baverstock, "Chernobyl and the Future: Too Soon for a Final Diagnosis", *Nature*, vol. 440, 2006, pp. 993-994.

[183] Z.Merali, "Return of the Atom", *New Scientist*, September 16, 2006, pp. 6-7; J.M.Deutch and G.J. Moniz, "The Nuclear Option", *Scientific American*, September 2006, pp. 57-59.

[184] See Lovelock, cited note 177, p.89.

[185] J.W.S. Leeuwin and P.Smith, "Can Nuclear Power Provide Energy for the Future; Would it Solve the CO2 Emission Problem? (2003) at www.oprit.rug.nl/deenen/Technical.html.

[186] Hoffert (*et al.*), cited note 157.

[187] See D.Shearman and C.Butler, Submission by Doctors for the Environment Australia, *Uranium Mining, Processing and Nuclear Energy Review* (July 2006) at www.dea.org.au. The following reproduces, with revision, material written by D.Shearman in the section "Environmental Issues: Nuclear Power, Climate Change and Sustainability".

[188] S.White, "The Nuclear Power Option - Expensive, Ineffective and Unnecessary", *The Sydney Morning Herald,* July 13 2005, at <http://www.smh.com.au/news/Opinion/The_nuclear_power_option_expensive_ineffective_and_unneccessary/2005/06/12/1118514925517.html>.

[189] Nuclear reactors have a lifetime of 30-40 years; to make efficient use of these reactors, many of which would need to built in the next few years, it is clear that 300TW of energy would have to be produced over a considerably longer period than two decades. That is, nuclear energy could only complement, rather than replace other ways of generating electrical energy in the next few decades.

[190] Total recoverable supply may be less than currently thought, for the same reasons that global oil stocks appear to have been exaggerated.

[191] Editorial, "Recycling the Past", *Nature*, vol. 439, 2006, pp. 509-510.

[192] As above.

[193] An all-out push to promote nuclear energy would change the shape of the CO2 emission curve, complicating the calculation of peak CO2. Assume under scenario A that CO2 emissions peak in 2020. Atmospheric CO2 would then peak in about 2080. However, before 2080 (in this scenario) CO2 emissions would again start to rise (because of the depletion of high-grade U), and would at least approach the amount released in 2020. Thus CO2 levels could again rise sometime after 2080 (though they might fall slightly and temporarily after 2080).

[194] J.Vidal, "Nuclear Plants Bloom. Is the Reviled N-Power the Answer to Global Warming?" *The Guardian*, August 12 2004 at http://www.guardian.co.uk/life/feature/story/0,,1280884,00.html.

[195] R.Stone, "Nuclear Trafficking: 'A Real and Dangerous Threat'" *Science*, vol. 292, 2001, pp. 1632-1636.

[196] See Leeuwin and Smith, cited note 185.

[197] M.Meacher, "Why Plan for a Nuclear Future When World Uranium Supplies are Running Out?" *Guardian Weekly* July 21-27 2006, p. 17.

[198] J.Leggett, *The Empty Tank: Oil, Gas, Hot Air, and the Coming Global Financial Catastrophe*, (Random House, New York, 2005).

[199] As above p.91.

[200] T.Trainer, *The Conserver Society: Alternatives for Sustainability*, (Zed Books, London, 1995) and "The Simper Way" website: http://www.socialwork.arts.unsw.edu.au/tsw/.

**Chapter 4**

[1] J. Gray, *Straw Dogs: Thoughts on Humans and Other Animals*, (Granta Books, London, 2002), pp. 16-17.

[2] T.Trainer, "Social Responsibility: The Most Important, and Neglected, Problem of All?" at http://socialwork.arts.unsw.edu.au/tsw/ D98.SocialResponsibility.html.

[3] As above p.1.

[4] As above p.2.

---

[5] R.H. Rossi, *Down and Out in America: The Origins of Homelessness*, (University of Chicago Press, Chicago,1991); J. Blau, *The Visible Poor: Homelessness in the United States*, (Oxford University Press, New York, 1993); R. Fincher and P. Saunders, *Creating Unequal Futures? Rethinking Poverty, Inequality and Disadvantage*, (Allen and Unwin, Crows Nest NSW, 2001); P. Dierterlen, *Poverty: A Philosophical Approach*, (Rodopi Philosophical Studies/Amsterdam, New York, 2005); C.Jencks, *The Homeless*, (Harvard University Press, Cambridge, Massachusetts, 2005).

[6] R. Paehlke, *Environmentalism and the Future of Progressive Politics* (Yale University Press, New Haven, 1989); D.Leonard-Barton, "Living Lightly Can Mean Greater Independence, Richer Lives", *The Christian Science Monitor*, October 21 1980, p. 20; J.B. Schor, *The Overworked American: The Unexpected Decline of Leisure*, (Basic Books, New York,1991).

[7] A. de Botton, *Status Anxiety*, (Pantheon Books, New York 2004).

[8] D.Elgin, *Voluntary Simplicity*, (William Morrow, New York, 1981); A.Etzioni,

" Voluntary Simplicity: Characterization, Select Psychological Implications and Societal Consequences", in T. Jackson (ed.), *The Earthscan Reader in Sustainable Consumption*, (Earthscan, London, 2006), pp. 159-177.

[9] C.Hamilton and R. Denniss, *Affluenza: When Too Much is Never Enough*, (Allen and Unwin, Crows Nest, New South Wales, 2005).

[10] D.W. Kidner, "Why Psychology is Mute About the Environmental Crisis", *Environmental Ethics*, vol. 16, 1994, pp. 359-376.

[11] As above, p.359.

[12] B.F. Skinner, "Why We Are Not Acting to Save the World", in *Upon Further Reflection*, (Prentice-Hall, Englewood Cliffs, New Jersey, 1987).

[13] M.E.Soule, "What is Conservation Biology?" *BioScience*, vol.35, 1985, pp. 727-734 and *Conservation Biology: The Science of Scarcity and Diversity*, (Sinauer Associates, Sunderland, MA,1986). Conservation medicine has been

discussed in chapter 2 of this book.

[14] On Carol D. Saunders see "Growing Green Kids", *Chicago Wilderness Magazine*, fall 2004, at http://chicagowildernessmag.org/issues/fall2004/greenkids.html. See also http://www.conservationpsychology.org/.

[15] C. D. Saunders "The Emerging Field of Conservation Psychology," *Human Ecology Review*, vol.10, no.2, 2003, pp.137-149, at p.138.

[16] C. D. Saunders (*et al.*), "Using Psychology to Save Biodiversity and Human Well Being", *Conservation Biology*, vol. 20, no. 3, 2006, pp.702-705.

[17] J.L.Anderson, "Stone Age Minds at Work in the 21st Century Science: How Cognitive Psychology Can Inform Conservation Biology", *Conservation Biology in Practice*, vol. 2, 2001, pp. 18-25; A.T. Brook, "What is Conservation Psychology"? *Population and Environmental Psychology Bulletin*, vol. 27, no.2, 2001, pp.1-2; S.Clayton and A.Brook, "Can Psychology Help Save the World? A Model for Conservation Psychology", *Analyses of Social Issues and Public Policy*, vol.5, 2005, pp. 87-102; E.S. Reed, *Encountering the World: Toward an Ecological Psychology*, (Oxford University Press, New York, 1996); D.D. Winter and S.M. Koger, *The Psychology of Environmental Problems*, 2nd edition, (Lawrence Erlbaum Assoc., Mahwah, New Jersey, 2004).

[18] M.A.Schroll, "Remembering Ecopsychology's Origins: A Chronicle of Meetings, Conversations, and Significant Publications", at http://www.ecopsychology.org/journal/ezine/ep_origins.html.

[19] M. Sabin (ed.) *The Earth has a Soul: The Nature Writings of C.G.Jung*, (North Atlantic Books, Berkley, California, 2002).

[20] J.Kuhn, "Toward an Ecological Humanistic Psychology", *Journal of Humanistic Psychology*, vol. 41, 2001, pp. 9-24.

[21] T. Rozak (*et al* eds.), *Ecopsychology: Restoring the Earth, Healing the Mind*, (Sierra Club Books, San Francisco, 1995).

[22] See T. Jackson (ed.), *The Earth Scan Reader in Sustainable Consumption*, (Earthscan, London, 2006).

[23] T. Jackson "Readings in Sustainable Consumption", as above, pp. 1-23, cited p.1.

[24] *Agenda 21* (1972), report on the United Nations Conference on Environment and Development, Rio de Janeiro, June 3-14, at www.un.org/esa/sustdev/documents/agenda21/english/Agenda 21.pdf, sections 4.3-4.6.

[25] DTI, *Changing Patterns- UK Government Framework for Sustainable Consumption and Production*, (Department of Trade and Industry, London, 2003).

[26] Defra (Department for Environment, Food and Rural Affairs), *Securing the Future - Developing UK Sustainable Development Strategy*, (The Stationery Office, London, 2005).

[27] Tony Blair, cited from Jackson, cited note 23, at p.7.

[28] See for example A.Kollmus and J.Agyeman, "Mind the Gap: Why Do People Act Environmentally and What are the Barriers to Pro-Environmental Behavior?" *Environmental Education Research*, vol. 8, no.3, 2002, pp.239-260; P.C.Stern, "Psychology and the Science of Human-Environment Interactions", *American Psychologist*, vol. 55, 2000, pp. 523-530; J. Vining and A. Ebreo, "Emerging Theoretical and Methodological Perspectives on Conservation Behavior", in R.Bechtel and A.Churchman (eds.), *Handbook of Environmental Psychology*, (John Wiley, New York, 2002), pp. 541-558.

[29] For a concise summary see "Behavior Change Theories and Models" at http://www.csupomona.edu/~jvgrizzell/best_practices/bctheory.html.

[30] N.K. Janz and M.H. Becker, "The Health Belief Model: A Decade Later", *Health Education Quarterly*, vol. 11, 1984, pp. 1-47.

[31] A. Bandura, "Self-Efficacy: Toward a Unifying Theory of Behavior Change", *Psychological Review*, vol. 84, 1977, pp. 191-215; A Bandura "Self-Efficacy Mechanism in Human Agency", *American Psychologist*, vol. 37, 1982, pp. 122-147; C.C. DiClemente, "Self Efficacy and Smoking Cessation Maintenance: A

Preliminary Report", *Cognitive Therapy and Research*, vol.5, 1981, pp.175-187; C.C. DiClemente, "Self Efficacy and the Addictive Behaviors", *Journal of Social and Clinical Psychology*, vol. 4, 1986, pp. 302-315; C.C. DiClemente (*et al.*), "The Process of Smoking Cessation: An Analysis of Precontemplation, Contemplation and Contemplation/Action", *Journal of Consulting and Clinical Psychology*, vol. 59, 1991, pp. 295-304; J.O. Prochaska, "Strong and Weak Principles for Progressing from Precontemplation to Action on the Basis of Twelve Problem Behaviors", *Health Psychology*, vol.13, 1994, pp. 47-51; J.O. Prochaska and C.C. DiClemente, "Stages and Processes of Self-Change of Smoking: Toward an Integrative Model of Change", *Journal of Consulting and Clinical Psychology*, vol. 51, 1983, pp. 390-395.

[32] O. Hernàndez and M.C. Monroe, "Thinking About Behavior", in B.A. Day and M.C. Monroe (eds.), *Environmental Education and Communication for a Sustainable World: A Handbook for International Practitioners*, (Academy for Educational Development, Washington DC, 2000), pp. 7-15.

[33] G.L.Zimmerman (*et al.*), "A 'Stages of Change' Approach to Helping Patients Change Behavior", *American Family Physician*, March 1 2000 at http://www.aafp.org/afp/20000301/1409.html.

[34] R.M. Muth and J.C.Hendee, "Technology Transfer and Human Behavior", *Journal of Forestry*, 1980, pp. 141-144.

[35] See Hernàndez and Monroe, cited note 32, p.12.

[36] S. Clayton and A. Brook, "Can Psychology Help Save the World? A Model for Conservation Psychology", *Analyses of Social Issues and Public Policy*, vol. 5, 2005, pp.87-102.

[37] As above, p.2.

[38] As above, p.3. See also M.Deutsch and H.G. Gerard, "A Study of Normative and Informational Social Influence Upon Individual Judgment", *Journal of Abnormal Social Psychology*, vol.51, 1955, pp. 629-636; L.D. Ross, "The Intuitive Psychologist and his Shortcomings: Distortions in the Attribution Process", in L. Berkowitz (ed.), *Advances in Experimental Social Psychology*, vol. 10, (Academic Press, New York, 1977); L.D. Ross and R.F. Nisbett, *The Person and the Situation*, (McGraw-Hill, New York, 1991). (References from Clayton and Brook, cited note 36).

---

[39] M.B. Brewer, "The Social Self: On Being the Same and Different at the Same Time", *Personality and Social Psychology Bulletin*, vol.17, 1991, pp. 475-482; A.T. Brook, *Effects of Contingencies of Self-Worth on Self-Regulation of Behavior*, (Unpublished doctoral dissertation, University of Michigan, Ann Arbor, 2005).

[40] S. Clayton and S. Opotow (eds.), *Identity and the Natural Environment*, (MIT Press, Cambridge MA, 2003), cited from Clayton and Brook, note 36, p.4.

[41] Clayton and Brook, cited note, 36, p.4; M.T. Klare, *Resource Wars: The New Landscape of Global Conflict*, (Metropolitan Books, New York, 2001).

[42] Clayton and Brook, as above, p.5.

[43] D.Winter and S.Koger, *The Psychology of Environmental Problems*, 2nd edition, (Lawrence Erlbaum Associates, Publishers, Mahwah, New Jersey, 2004).

[44] As above p.215; P.O. Hallin, "Environmental Concern and Environmental Behavior in Foley, a Small Town in Minnesota", *Environment and Behavior*, vol. 27, no. 4, 1995, pp. 558-578.

[45] As above p.216.

[46] As above p.227.

[47] See in general http://www.sustainable-development.gov.uk.

[48] See in general, UK Secretary of State for Environment, Food and Rural Affairs, *The UK Government Sustainable Development Strategy, Securing the Future: Delivering UK Sustainable Development Strategy*, (March, 2005) at http://www.sustainable-development.gov.uk/publications/uk-strategy/index.htm.

[49] As above p.26.

[50] As above p.29.

[51] As above p.30.

[52] See Prime Minister Strategy Unit, *Personal Responsibility and Changing*

*Behavior: The State of Knowledge and Its Implications for Public Policy*, (February, 2004) at http://www.number10.gov.uk/files/pdf/pr.pdf.

[53] T.Jackson, *Motivating Sustainable Consumption - A Review of Evidence on Consumer Behaviour and Behavioural Change* (2005) at http://www.sd-research.org.uk/documents/MotivatingSCfinal.pdf.

[54] The UK Government Sustainable Development Strategy, cited note 48, p.36.

[55] UK Sustainable Consumption Roundtable, *I Will If You Will: Towards Sustainable Consumption*, (Sustainable Development Commission and National Consumer Council, May 2006), at http://www.sd-commission.org.uk/publications/downloads/I-Will-If-You-Will. pdf.

[56] As above p.39.

[57] As above.

[58] As above p.57.

[59] As above.

[60] UK Department for Environment, Food and Rural Affairs (Defra), *Triggering Widespread Adoption of Sustainable Behaviour*, (Behaviour Change: A Series of Practical Guides for Policy Makers and Practitioners, Number 4, Summer 2006) at http//www.defra.gov.uk/scienceproject_data/Documentlibrary/SD_14006/SD14006_3804_INF.pdf.

[61] M. Gladwell, *The Tipping Point: How Little Things Can Make a Big Difference*, (Back Bay Books, Los Angeles, 2002); P.Ball, *Critical Mass: How One Thing Leads to Another*, (Farrar, Straus and Giroux, New York, 2004); M. Buchanan, *Ubiquity: Why Catastrophes Happen*, (Three Rivers Press, New York, 2002).

[62] Defra, cited note 60, p.7.

[63] As above, p.8.

[64] D. Halpern and C.Bates (with G. Beales and A.Heathfield), Prime Minister's

Strategy Unit, Cabinet Office, *Personal Responsibility and Changing Behaviour: The State of Knowledge and Its Implications for Public Policy*, (February, 2004) at http://www.number10.gov.uk/files/pdf/pr.pdf.

[65] T.Jackson, *Motivating Sustainable Consumption - A Review of Evidence on Consumer Behaviour and Behavioural Change* (2005) at http://www.sd-research.org.uk/documents/MotivatingSCfinal.pdf.

[66] D.Miller (ed.), *Acknowledging Consumption: A Review of New Studies*, (Routledge, London and New York, 1995).

[67] A.Appadurai, *The Social Life of Things: Commodities in Cultural Perspective*, (Cambridge University Press, Cambridge, 1986); R.Bagozzi (*et al.*), *The Social Psychology of Consumer Behaviour*, (Open University Press, Buckingham, 2002); C.Campbell, *The Romantic Ethic and the Spirit of Modern Consumerism*, (Basil Blackwell, Oxford, 1987); M.Douglas, *Natural Symbols: Explorations in Cosmology*, (Barrie and Rockliff, London,1970); M.Douglas and B.Isherwood, *The World of Goods: Towards an Anthropology of Consumption*, (Routledge, London and New York, 1996); T.Edwards, *Contradictions of Consumption: Concepts, Practices and Politics in Consumer Society*, (Open University Press, Milton Keynes, 2000); M.Featherstone, *Consumer Culture and Postmodernism*, (Sage, London,1991); Y.Gabriel and T.Lang, *The Unmanageable Consumer: Contemporary Consumption and Its Fragmentation*, (Sage, London,1995); G.McCracken, *Culture and Consumption*, (Indiana University Press, Bloomington and Indianapolis, 1990).

[68] G.Mead, *Mind, Self and Society*, (University of Chicago Press, Chicago, 1934); P.Stringer (ed.), *Confronting Social Issues: Applications of Social Psychology*, (Academic Press, London,1982).

[69] Jackson, cited note 65, p.121; E.Geller, "Evaluating Energy Conservation Programs: Is Verbal Report Enough?" *Journal of Consumer Research*, vol.8, 1981, pp. 331-335.

[70] Jackson, cited note 65, p.121; P.Stern and T.Dietz, "The Value Basis of Environmental Concern", *Journal of Social Issues*, vol. 50, 1994, pp. 65-84.

[71] As above, p.122.

[72] As above.

[73] As above, p.123-124.

[74] As above, p.127; G. Gardner and P.Stern, *Environmental Problems and Human Behavior*, 2nd edition, (Pearson, Boston MA, 2002), p. 31.

[75] As above, p.134.

[76] As above, p.130.

[77] A. Darnton (*et al.*), *Promoting Pro-Environmental Behaviour: Existing Evidence to Inform Better Policy Making*, (October 2006), at http://www.defra.gov.uk/science/project_data/DocumentLibrary/SD14002/SD14002_3712_FRP.pdf.

[78] As above p.4.

[79] As above p.5.

[80] As above.

[81] As above.

[82] As above, p.6.

[83] As above.

[84] For a review see P.Stern, "Toward a Coherent Theory of Environmentally Significant Behavior", *Journal of Social Issues*, vol. 56, no.3, 2000, pp.407-424.

[85] D.Ballard, "Using Learning Processes to Promote Change for Sustainable Development", *Action Research*, vol. 3, no.2, 2005; J.Chapman, *System Failure*, 2nd edition, (Demos. London, 2004), cited from Darnton (*et al.*), cited note 77, p.21.

[86] Darnton (*et al.*), cited note 77, p.28.

[87] As above, pp.54-60.

[88] As above, pp.54.

[89] As above, pp.69.

[90] A.Darnton, "Strategic Thinking: Impact of Sustainable Development on Public Behaviour", (May 2004), at http://www.comminit.com/strategic thinking/st2006/thinking-1693.html.

[91] D.Shearman and J.W.Smith, *The Climate Change Challenge and the Failure of Democracy*, (Praeger, Westport, 2007).

**Chapter 5**

[1] Walter Lippman, *Public Opinion*, (Harcourt Brace and Company, New York, 1922), p. 248 cited in A. Gore, *The Assault on Reason*, (Bloomsbury, London, 2007), p. 10.

[2] See J.W. Smith (*et al*), *The Bankruptcy of Economics: Ecology, Economics and the Sustainability of the Earth*, (Macmillan, London, 1999).

[3] R. A. Posner, *Public Intellectuals: A Study of Decline*, (Harvard University Press, Cambridge MA, 2001), and H. Small (ed.), *The Public Intellectual*, (Blackwell Publishing, Oxford, 2002).

[4] R. Jacoby, *The Last Intellectuals: American Culture in the Age of Academe*, (Basic Books, New York, 2000).

[5] Peter Bockley, *Pre-Budget Talk, Ockum's Razor* Australian Broadcasting Corporation (ABC) Radio National, 9 May 1999, and J. Richardson, "Brand Aid", *The Australian* April 17 2002, p. 20.

[6] David Kirp, *Shakespeare, Einstein and the Bottom Line: The Marketing of Higher Education*, (Harvard University Press, Cambridge, MA, 2003).

[7] T. Coady, (ed.), *Why Universities Matter*. (Allen and Unwin, Sydney 2000), and

F.Crowley, *Degrees Galore: Australia's Academic Teller Machines*, (Published by Frank Crowley, 48 Clifton Drive, Port Macquarie, NSW 2444, Australia, 1998), chapter 3.

[8] See Education Watch International at http://edwatch.blogspot.com.

[9] See on this debate, for example: J.K.Wilson, *The Myth of Political Correctness: The Conservative Attack on Higher Education*, (Duke University Press, Durham, 1995); G.A. Tobin (*et al*), *The Uncivil University: Politics and Propaganda in American Education*, (Institute for Jewish and Community Research, San Francisco, 2005).

[10] P. Johnson, "Universities? We'd be Better Off Without Them", *The Australian*, 18 September 1991, p. 21.

[11] P.P. McGuinness, "Editorial: The Decline of Universities", *Quadrant*, April 1999, pp. 2-5; Cardinal J.H. Newman, *The Idea of a University*, introduction to G.N. Shuster, (Image Books, Garden City, New York, 1959).

[12] A. Irvine, "Latest Find from the Frege Archives", *Quadrant*, April 1999, pp. 53-55. On Frege's work see *Conceptual Notation and Related Articles*, translated and edited by T.W. Bynum, (Clarendon Press, Oxford, 1972) and *The Foundations of Arithmetic*, translated by J.L. Austin, (Blackwell, Oxford, 1950).

[13] Irvine, as above note 12, p. 53.

[14] C. Hamilton and S.Maddison (eds.), *Silencing Dissent*, (Allen and Unwin, Crows Nest New South Wales, 2007).

[15] As above, p. 49.

[16] D.Horowitz, *The Professors: The 101 Most Dangerous Academics in America*, (Regnery Publishing, Washington DC, 2006).

[17] Free Exchange on Campus, at http://www.freeexchangeoncampus.org.

[18] M. Levine, "America's "Most Dangerous" Professors". *Mother Jones* March 6 2006 at http://www.motherjones.com/commentary/columns/2006/03/the_professors.html.

[19] Campus Watch "Keeping an Eye on Professors Who Teach About the Middle East", at History News Network, http://hnn.us/articles/986.html.

[20] As above.

[21] C. Birch, *On Purpose*, (University of New South Wales University Press Kensington, New South Wales, 1990).

[22] F. Crowley, *Degrees Galore: Australia's Academic Teller Machines,* cited note 7.

[23] J. McCulmon, "How Learned Books Nearly Perished", *The Australian*, May 6 1998, p. 40.

[24] As above.

[25] B.Jones, *A Thinking Reed*, (Allen and Unwin, St. Leonards, New South Wales, 2006).

[26] G. Richards, "Academic Loses E-Mail After Criticising Uni", *The Age*, April 21, 1999.

[27] P. Craven, "Corporate Criticism Finds Popular Voice", *The Australian*, September 30 1998; T. Coady (ed.), *Why Universities Matter*, cited note 7.

[28] D. Illing, "Hounded for Green Counsel", *The Australian*, June 2 1999.

[29] B. Martin (eds.*et al.*), *Intellectual Suppression*, (Angus and Robertson, North Ryde, New South Wales, 1986).

[30] As above p. 72.

[31] As above p.192-193.

[32] As above p. 187.

[33] T. Veblen, *The Higher Learning in America: A Memorandum on the Conduct*

*of Universities by Business Men*, (Sagamore Press, New York, 1957).

[34] C. Wright Mills, *Power, Politics, and People: The Collected Essays of C.Wright Mills*, edited by I.L. Horowitz, (Oxford University Press, New York, 1963), p. 297.

[35] Reviewed in *Background Briefing*, "Academic Freedom", Australian Broadcasting Corporation, Radio National, 6 December 1998.

[36] P.A. Mitchell, *The Making of the Modern Law of Defamation*, (Hart Publishing, Oxford, 2005).

[37] See note 35.

[38] As above.

[39] As above.

[40] Universal Declaration of Human Rights. Adopted and Proclaimed by the General Assembly, resolution 217 A (III) of December 10 1948.

[41] Martin (ed. *et al*) cited note 29.

[42] See. A. Sajó (ed.) *Human Rights with Modesty: The Problem of Universalism*, (Martinus Nijhoff, Leiden and Boston, 2004), and A. Sajó (ed.), *Abuse: The Dark Side of Fundamental Rights*, (Eleven International Publishing Co, Amsterdam, 2006).

[43] " Santos School of Petroleum Engineering a "Far-sighted Investment" in Industry Future". *The Adelaidian* (University of Adelaide), no. 8, August 13 1999.

[44] As above.

[45] See D.Shearman and J.W.Smith, *The Climate Change Challenge and the Failure of Democracy*, (Praeger, Westport, 2007), chapter 2.

[46] "University Takes Tobacco "Blood Money" ", *Guardian Weekly*, December 7 2000.

[47] As above.

[48] "Professor Quits Over Tobacco Company's £3.8m Donation". *Guardian Weekly* May 24 2001.

[49] "Tobacco Firm Finance New University Link-up", *Guardian Weekly* December 14 2000.

[50] R. Barnett, *The Idea of Higher Education*, (SRHE and Open University Press, Buckingham, 1990).

[51] Cardinal J.H. Newman cited note 11.

[52] K. Jaspers, *The Idea of a University*, edited by K. Deutsch, (Owen, London, 1960).

[53] See F.Crowley, *Degrees Galore*, cited note 7.

[54] See generally Cardinal J.H. Newman, cited note 11 and Jaspers, cited note 52.

[55] J.Niland, "The Fate of University Science: The Future of Australian Universities", Address to the National Press Club, Canberra February 25, 1998.

[56] See note 40.

[57] As above.

[58] As above.

[59] *World Declaration on Higher Education for the Twenty-First Century: Vision and Action*, UNESCO, October 9 1998.

[60] As above.

[61] As above.

[62] J.Kovel, *The Enemy of Nature*, (Zed Books, London, 2002).

[63] WHO. *Preamble to the Constitution of the World Health Organization* as

adopted by the International Health Conference, New York, June 19-22 1946, and entered into force on April 7 1948.

[64] Report of the United Nations Conference on Environment and Development, Rio de Janeiro, June 3-14 1992 at http://www.un.org/documents/ga/confl51/aconfi15126-1annex1.html.

[65] D. Shearman with G. Sauer-Thompson, *Green or Gone: Health, Ecology, Plagues, Greed and Our Future*, (Wakefield Press, Adelaide, 1997).

[66] As above.

[67] D.Botting, *Humboldt and the Cosmos*, (Joseph, London, 1973).

[68] W. Fox. *Towards a Transpersonal Ecology*, (State University of New York Press, New York, 1995) pp. 31-32.

[69] See note 40.

[70] See T. Campbell (et. al.), *Protecting Human Rights: Instruments and Institutions*, (Oxford University Press, Oxford, 2003); C. Gearty, *Can Human Rights Survive?* (Cambridge University Press, Cambridge, 2006); X Li, *Ethics, Human Rights and Culture: Beyond Relativism and Universalism*, (Palgrave Macmillan, New York, 2006).

[71] See. A. Sajó (ed.) cited note 42.

[72] Universal Declaration of Human Responsibilities, UNESCO, March 25-28 1997, at www.interactioncouncil.org/udhr/declaration/udhr.pdf.

[73] As above.

[74] P.K. Feyerabend, *Science in a Free Society*, (New Left Books, London, 1978).

[75] A.Gore, *The Assault on Reason*, (Bloomsbury, London, 2007).

[76] J.W.Smith, *Reason, Science and Paradox*, (Croom Helm, London, 1985).

[77] W.H. Newton-Smith, *The Rationality of Science*, (Routledge and Kegan Paul, Boston, 1981).

[78] K.Lehrer and C. Wagner, *Rational Consensus in Science and Society*, (D. Reidel, Dordrecht, 1981).

[79] Opinion Interview, "Soul Man: An Interview with Satsh Kumar", *New Scientist*, June 17 2000. p.46.

[80] J.H. Kunstler, *The Long Emergency*, (Atlantic Books, London, 2005).

[81] T. L. Friedman, "The Power of Green". *International Herald Tribune*, April 5 2007.

[82] D.Shearman and J.W.Smith, *The Climate Change Challenge and the Failure of Democracy*, (Praeger, Westport, 2007).

[83] J.W.Smith, *Reason, Science and Paradox*, cited note 76.

[84] F. Crick, *Of Molecules and Men*, (University of Washington Press, Seattle, 1966) and *The Astonishing Hypothesis: The Scientific Search for the Soul*, (Scribner, New York, 1994).

[85] Crick as above.

[86] D.R. Griffin (ed.), *The Reenchantment of Science*, (State University of New York Press, New York, 1988).

[87] As above.

[88] A.Gore, cited note 75.

[89] See D.Shearman and J.W.Smith, cited note 82.

[90] E.O. Wilson, *Consilience: The Unity of Knowledge*, (Knopf, New York, 1998).

[91] As above.

[92] D.Shearman and J.W.Smith, cited note 92.

[93] Feyerabend, cited note 74.

[94] J. Gray, *Enlightenment's Wake: Politics and Culture at the Close of the Modern Age*, (Routledge, London and New York, 1995); B.Appleyard, *Understanding the Present: Science and the Soul of Modern Man*, (Picador/Pan Books, London, 1993); J.W.Smith, *Reductionism and Cultural Being*, (Martinus Nijhoff, The Hague, 1984).

[95] As above.

[96] V.I. Vernadsky, *The Biosphere*, (Springer-Verlag, New York, 1998).

[97] J.Lovelock, *Gaia: A New Look at Life on Earth*, (Oxford University Press, Oxford, 1979).

[98] M.J. Tyler (*et al*), "Inhibition of Gastric Acid Secretion in the Gastric Brooding Frog *Rheobatrachus Silus*", *Science*, vol. 220, 1983, pp. 609-610.

[99] As above.

[100] J.C. Fanning (*et al*), "Converting a Stomach to a Uterus: The Microscopic Structure of the Gastric Brooding Frog *Rheobatrachus Silus* ", *Gastroenterology*, vol. 82, 1982, pp. 62-70.

[101] G.Monbiet, *Heat*, (Allen Lane, London, 2006).

[102] S. Stick, (Editorial), "Environmental Tobacco Smoke-Physicians Must Avoid Fanning the Flames", *Australian and New Zealand Journal of Medicine*, vol. 30, 2000, pp. 436-439.

[103] Monbiet, cited note 101.

[104] P.D.Thacker, "The Many Travails of Ben Santer", *Environmental Science and Technology* (On-line), August 9 2006 at http://pubs.acs.org/subscribe/journals/esthag-w/2006/aug/policy/pt-santer.html.

[105] Quoted from Geraldine Doogue and Clive Hamilton, ABC Radio Saturday Extra April 21 2007 at http://www.abc.net.au/rn/saturdayextra/stories/2007/1901158.htm; Lavoisier Group, *Submission to JSCOT Inquiry into the Kyoto Protocol* at http://www.lavoisier.com.au/papers/submissions/JSCOT.pdf. For a discussion see

C.Hamilton, *Scorcher: The Dirty Politics of Climate Change*, (Black Inc, Melbourne, 2007).

[106] Editorial, "Control is All. Who Needs Crude Censorship When Corporate Bodies Call the Shots in Research?" *New Scientist*, June 5, 1999 at http://www.newscientist.com/contents/issue/2189.html.

[107] As above.

[108] A. J. Sutton (et. al.), *Methods for Meta-Analysis in Medical Research*, (Wiley, Chichester, 2000).

[109] D. Sackett (*et al*), *Evidence-based Medicine*, (Churchill Livingstone, New York, 1997).

[110] See http://www.austmus.gov.au/consensus/02.htm.

[111] See http://www.promedmail.org/pls/promed/f?p=2400:1000.

[112] *Nuclear power – Truth and Ideology, An Energy Policy for Australia: Doctors for the Environment Australia* at http://dea.org.au/docs/DEA_e_p_App2NuclearPowerTruthandIdeology.pdf

[113] As above.

[114] As above.

[115] As above.

[116] C. Busby, "Poisoning in the Name of Progress", *The Ecologist*, vol. 29, no.7, 1999, pp. 395-401.

[117] As above.

[118] As above.

[119] See D. Shearman with G.Sauer-Thompson, *Green or Gone: Health, Ecology, Plagues, Greed and our Future*, (Wakefield Press, Adelaide, 1997).

[120] Beryl Lieff Benderly, "A Hippocratic Oath" for Scientists?", *Science Careers*

*Forum*, March 2 2007 at http://sciencecareers.sciencemag.org/career_development/previous_issues/articles/2007_03_02/caredit_a0700030.

[121] T. Flaherty, "Pilchards: Will We Ever Know?" *South Australian Regional Ripples*, vol. 6 no. 1, Autumn, 1999.

[122] As above.

[123] M.J.Tyler, *Australian Frogs*, (Viking O'Neil, Penguin Books, Melbourne, 1989).

[124] J. Lovelock, *The Revenge of Gaia*, (Allen Lane, London, 2006).

[125] B.Martin, (ed. *et al.*), *Intellectual Suppression*, cited note 29, p. 138.

[126] P.Cohen, "Your Mission Is…", *New Scientist*, July 3 1999, pp. 38-41.

[127] G. Vogel, "Picking Up the Pieces After Hwang", *Science*, vol. 312, April 28 2006, pp. 516-517.

[128] Justice Louis Brandells, quoted by A.G. Wallace, "Educating Tomorrow's Doctors: The Thing that Really Matters is that We Care", *Academic Medicine*, vol. 72, no 4, April 1997, pp. 253-258, at p. 253.

[129] See note 72.

[130] As above.

[131] R. Tytler, *Re-Imagining Science Education*, (Australian Council for Education Research, Camberwell, Victoria, 2007).

[132] T. Homer-Dixon, *The Upside of Down: Catastrophe, Creativity and the Renewal of Civilization*, (Island Press, Washington DC, 2006).

[133] See Chapter 1 of D.Shearman and J.W.Smith, *The Climate Change Challenge and the Failure of Democracy*, cited note 82.

[134] N. Harvey (*et al.*), "Geography and Environmental Studies in Australia: Symbiosis for Survival in the 21st Century", *Australian Geographical Studies*,

vol. 40, no.1, March 2002, pp. 21-32.

[135] R.Carson, *Silent Spring*, (Penguin Book, London,1962).

[136] Harvey (*et al.*), cited note 134.

[137] As above.

[138] D. Smith, "How Will it All End?" in B. Cartledge (ed), *Health and the Environment*, (Oxford University Press, Oxford, 1994).

[139] Harvey (*et al.*), cited note 134

[140] As above.

[141] J.L. Simon, *The Ultimate Resource*, (Princeton University Press, Princeton, 1981).

[142] S.Kuznets, *Economic Growth of Nations: Total Output and Production Structure*, (Belknap Press of Harvard University Press, Cambridge, Massachusetts, 1971).

[143] As above.

[144] R. Bailey, (ed.), *The True State of the Planet*, (Free Press, New York, 1995).

[145] D.W. Orr, "What is Education For", *Trumpeter*, vol. 8, no. 3, Summer 1991, pp. 99-102.

[146] D.W. Orr, "Modernization and the Ecological Perspective", in D.W. Orr and M.S. Soroos (eds.), *The Global Predicament*, (University of North Carolina Press, Chapel Hill, 1979), pp. 75-89, at p.76.

[147] D.W. Orr, "Technological Fundamentalism", *Conservation Biology*, vol. 8, no.2, June 1994, pp. 335-337.

[148] D.W. Orr, "Speed", *Conservation Biology*, vol. 12, no.1, February 1998, p. 4-7, cite p. 7.

[149] D. H. Meadows (*et al*), *The Limits to Growth*, (Potomac Association, London,

1972).

[150] A. Lovins, *Non-Nuclear Futures*, (Friends of the Earth, San Francisco, 1975).

[151] W. Berry, *The Unsettling of America*, (Sierra Club books, San Francisco, 1986).

[152] J.W.Smith (*et al.*), Healing a Wounded World, (Praeger, Westport, 1997).

[153] D.W. Orr, "Slow Knowledge", *Conservation Biology*, vol. 10, no.3, June 1996, pp. 669-702, cited pp. 700-701.

[154] T.Bethell, *The Politically Incorrect Guide to Science*, (Regnery Publishing, Washington DC, 2005).

[155] E. O. Wilson, *The Future of Life Little*, (Knopf, New York, 2002), p.134.

[156] E. O. Wilson, *Biophilia*, (Harvard University Press, Cambridge, MA, 1984).

[157] S.S. Gould, quoted from D.W.Orr, *Earth in Mind: On Education, Environment and the Human Prospect*, (Island Press, Washington DC, 1994), p. 43.

[158] G.Hardin, *The Immigration Dilemma: Avoiding the Tragedy of the Commons*, (Federation for American Immigration Reform, Washington DC, 1995).

[159] D.W. Orr, cited note 157, p. 5.

[160] As above.

[161] D.W. Orr, cited note 153.

[162] N.Maxwell, *From Knowledge to Wisdom*, (Basil Blackwell, Oxford, 1984)

[163] D.W. Orr, cited note 157, p. 14.

[164] D. Yencken (*et al*), *Environment, Education and Society in the Asia-Pacific: Local Traditions and Global Discourses*, (Routledge, London, 2000).

[165] United Nations Environment Programme, *GEO 2000*, Press Release, Nairobi,

September 15 1999.

[166] A. Gore, *Earth in the Balance: Ecology and the Human Spirit*, (Houghton Mifflin, Boston, 1992).

[167] As above, p. 269.

[168] D. Shearman and J.W.Smith, cited note 82.

[169] UNEP, *Global Environmental Outlook - 2000*, at http://www.unep.org./geo2000/.

[170] D.W. Orr, cited note 157.

[171] As above.

[172] As above

[173] D. Hutton and L. Connors, *A History of the Australian Environment Movement*, (Cambridge University Press, Melbourne, 1999).

[174] D.Shearman and J.W.Smith, cited note 82.

**Chapter 6**

[1] Ludwig Wittgenstein, remarks made in a letter to the philosopher Norman Malcolm, cited in N. Malcolm, *Ludwig Wittgenstein: A Memoir*, (Oxford University Press, Oxford, 1958), p.39.

[2] D. Hume, *A Treatise of Human Nature*, edited by L.A. Selby-Bigge,(Oxford

University Press, Oxford, 1978), book I, part iv, sec. Vii, p.272.

[3] A. C. Grayling, *Among the Dead Cities: The History and Moral Legacy of the WWII Bombing of Civilians in Germany and Japan*, (Walker and Company, New York, 2006), pp. 209-210.

[4] See F. Ferre, *Basic Modern Philosophy of Religion*, (George Allen and Unwin, London, 1968); G.G. Grisez, *Beyond the New Theism: A Philosophy of Religion*, (University of Notre Dame Press, Notre Dame, 1975); M. Martin, *The Case Against Christianity*, (Temple University Press, Philadelphia, 1991).

[5] S.C. Pepper, *World Hypotheses: A Study in Evidence*, (University of California Press, Berkeley, 1957); J.W.Smith, *Reductionism and Cultural Being: A Philosophical Critique of Sociobiological Reductionism and Physicalist Scientific Unificationism*, (Martinus Nijhoff, The Hague, 1984).

[6] J.W.Smith, "Formal Logic: A Degenerating Research Programme in Crisis", *Cogito*, vol. 2, no.3, September 1984, pp.1-8.

[7] E. Gellner, "Reflections on Philosophy, Especially in America", in *The Devil in Contemporary Philosophy*, (Routledge, London, 1974), pp.37-38. Cited from J.Kekes, *The Nature of Philosophy*, (Basil Blackwell, Oxford, 1980), p.13.

[8] R.M. Pirsig, *Zen and the Art of Motorcycle Maintenance: An Inquiry into Values*, (Perennial Classics, New York, 1999).

[9] J. Gaarder, *Sophie's World: A Novel About the History of Philosophy*, (Phoenix, London, 1996).

[10] A. Camus, *The Myth of Sisyphus*, (Penguin, London, 2000).

[11] A.A. Long, *The Cambridge Companion to Early Greek Philosophy*, (Cambridge University Press, Cambridge, 1999).

[12] N. Wolterstorff, *On Universals: An Essay on Ontology*, (University of Chicago Press, Chicago, 1970); K. Campbell, *Abstract Particulars*, (Blackwell, Oxford, 1990); R. Teichmann, *Abstract Entities*, (Macmillan, Houndmills,1992).

[13] A. E. Taylor, *Socrates*, (Hyperion Press, Westport, 1979); M.G. Levin, *The Socratic Method: A Novel*, (Simon and Schuster, New York, 1987).

[14] See for example, Plato, *Meno*, edited and translated by R.W. Sharples, (Bolchazy-Carducci, Chicago, 1985); J.M. Day (ed.), *Plato's Meno in Focus*, (Routledge, London, 1994).

[15] R. Rucker, *Infinity and the Mind: The Science and Philosophy of the Infinite*, (Harvester Press, Sussex, 1982).

[16] S. Körner, *The Philosophy of Mathematics: An Introductory Essay*, (Hutchinson, London, 1960); P. Maddy, *Realism in Mathematics*, (Clarendon Press, Oxford, 1990); M.Tiles, "Philosophy of Mathematics", in N. Bunnin and E.P. Tsui-James (eds.), *The Blackwell Companion to Philosophy*, (Blackwell Publishing, Oxford, 2003), 2nd edition, pp.345-374.

[17] K. Lehrer, *Knowledge*, (Clarendon Press, Oxford,1974); P.K. Moser, *Empirical Justification*, (D. Reidel, Dordrecht, 1985).

[18] See P.J. Zwart, *About Time: A Philosophical Inquiry into the Origin of Nature and Time*, (North Holland Publishing Co, Amsterdam, 1976).

[19] C. Howson, *Hume's Problem: Induction and the Justification of Belief*, (Clarendon Press, Oxford, 2000).

[20] E.M. Curley, *Descartes Against the Skeptics*, (Basil Blackwell, Oxford, 1978).

[21] S.Gaukroger, *Descartes: An Intellectual Biography*, (Clarendon Press, Oxford,1995).

[22] J.W.Smith, *The Progress and Rationality of Philosophy as a Cognitive Enterprise*, (Avebury, Aldershot, 1988).

[23] A. Kenny, *The Heritage of Wisdom: Essays in the History of Philosophy*, (Basil Blackwell, Oxford, 1987).

[24] W.H.Newton-Smith, *The Rationality of Science*, (Routledge and Kegan Paul, London, 1981); P.K. Feyerabend, *Against Method*, (New Left Books, London, 1975) and *Science in Free Society*, (New Left Books, London, 1978).

[25] M. Jammer, *The Philosophy of Quantum Mechanics: The Interpretation of Quantum Mechanics in Historical Perspective*, (Wiley, New York, 1974).

[26] See J.W.Smith (*et al*), "Fingernails on the Mind's Blackboard", in *The Bankruptcy of Economics: Ecology, Economics and the Sustainability of the Earth,* (Macmillan, London, 1999), pp. 55-88.

[27] J. Horton, *The End of Science: Facing the Limits of Knowledge in the Twilight of the Scientific Age*, (Addison-Wesley, Reading MA, 1996).

[28] A. Cohen and M. Dascal (eds.), *The Institution of Philosophy: A Discipline in Crisis?* (Open Court, La Salle, 1989).

[29] J-F. Lyotard, *The Postmodern Condition: A Report on Knowledge,* (University of Minnesota Press, Minneapolis, 1984); J-F. Lyotard and J-L Thébaud, *Just Gaming*, (University of Minnesota Press, Minneapolis, 1985).

[30] See D.Papineau, "Is Epistemology Dead?" *Proceedings of the Aristotelian Society*, vol. 82, 1982, pp. 129-142; E.Sosa, "Nature Unmirrored, Epistemology Naturalized",*Synthese,* vol. 55, 1983, pp. 49-72; D. Nails (ed.), *Naturalistic Epistemology*, (D.Reidel, Dordrecht, 1987); S.Haack, "Recent Obituaries of Epistemology", *American Philosophical Quarterly*, vol. 27, no.3, July 1990, pp. 199-212.

[31] F.Nietzsche, *Schopenhauer as Educator*, translated by J.W. Hillesheim and M.R. Simpson, (South Bend, Indiana, Gateway, 1965),p.36.

[32] J.Gray, *Enlightenment's Wake: Politics and Culture at the Close of the Modern Age*, (Routledge, London and New York, 1995); J. Topolski (ed.), *Historography: Between Modernism and Postmodernism*, (Rodopi, Amsterdam, 1994).

[33] N. Fraser and L. Nicholson, "Social Criticism Without Philosophy: An Encounter Between Feminism and Postmodernism", in Cohen and Dascal cited note 28, pp. 283-302.

[34] See J. Stick, "Can Nihilism be Pragmatic?" *Harvard Law Review*, vol.100, 1986, pp. 332-401; I. Levison, "Law as Literature", *Texas Law Review*, vol.60, 1982, pp. 371-403; D.M. Trubek, "Where the Action is: Critical Legal Studies and Empiricism", *Stanford Law Review*, vol. 36, 1984, pp. 575-622; G.E. White, "The Inevitability of Critical Legal Studies", *Stanford Law Review*, vol. 36, 1984, pp. 649-672.

[35] M.Midgley, *Science as Salvation: A Modern Myth and its Meaning*, (Routledge, London and New York, 1992).

[36] D. Shearman and G. Sauer-Thompson, *Green or Gone: Health, Ecology, Plagues, Greed and Our Future*, (Wakefield Press, Adelaide, 1997).

[37] J. Derrida, *Margins of Philosophy,* (University of Chicago Press, Chicago, 1982); C. Romano, "The Illegality of Philosophy, in Cohen and Dascal cited note 28, pp. 199-211. Romano says on this point (p. 200): Disagreement about what constitutes a justification thus keeps the philosophical engines grinding as much as any other cause.... For if a crisis attends contemporary philosophy, it's not something new that can be blamed directly on the work of Derrida or Richard Rorty, but something old that these two thinkers among others, have successfully returned to the spotlight- the embarrassing fact that philosophy as a discipline, refuses to take 'Yes' for an answer. It remains a field that declines to agree on a set of ground rules, widely enough accepted to be authoritative, that would allow genuine answers to its dilemmas to emerge and be passed along to the rest of the culture.

[38] G. Lakoff and M. Johnson, *Metaphors We Live By*, (University of Chicago Press, Chicago, 1980).

[39] L.V. Cheney, *Telling the Truth: Why Our Culture and Our Country Have Stopped Making Sense- And What We Can Do About It*, (Simon and Schuster, New York, 1995); P. Smith, *Killing the Spirit: Higher Education in America*, (Viking, New York, 1990); R. Kimball, *Tenured Radicals: How Politics has Corrupted Our Higher Education*, (Harper and Row, New York, 1990).

[40] A. Bloom, *The Closing of the American Mind: How Higher Education has Failed Democracy and Impoverished the Souls of Today's Students*, (Simon and Schuster, New York, 1987), p. 346.

[41] B. Readings, *The University in Ruins*, (Harvard University Press, Cambridge, Massachusetts, 1996), p. 55.

[42] R. Kimball, *Tenured Radicals*, cited note 39.

[43] L.W. Levine, *The Opening of the American Mind: Canons, Culture, and History*, (Beacon Press, Boston, 1996), p. 36.

[44] J.W.Smith, *The Progress and Rationality of Philosophy as a Cognitive Enterprise*, (Avebury, Aldershot, 1988); J.F. Harris, *Against Relativism: A Philosophical Defense of Method*, (Open Court, La Salle, 1993).

[45] G. Sauer-Thompson and J.W. Smith, *The Unreasonable Silence of the World: Universal Reason and the Wreck of the Enlightenment Project*, (Ashgate, Aldershot, 1997).

[46] J.R. Saul, *The Unconscious Civilisation*, (Penguin Books, Ringwood, Victoria, 1997).

[47] J. Franklin, "The Sydney Philosophy Disturbance", *Quadrant*, April, 1999, pp. 16-21.

[48] P. Ryckmans, *The View from the Bridge: The 1996 Boyer lectures*, (ABC Books, Sydney, 1996), p. 14.

[49] R. Downie, and J. MacNaughton, "Should Medical Students Read Plato?" *Medical Journal of Australia*, vol. 170, 1999, pp. 125-127.

[50] A. Tatum, "Benefits of Humanities Highlighted", *Australian Doctor*, March 19 1999.

[51] K.M. Sayre, "An Alternative View of Environmental Ethics", *Environmental Ethics*, vol. 13, 1991, pp. 195-213.

[52] As above.

[53] T. Trainer, *The Nature of Morality*, (Avesbury, Aldershot, 1991).

[54] For example, T. Trainer, *The Conserver Society: Alternatives for Sustainability*, (Zed Books, London and New Jersey, 1995); W. Ophuls, *Requiem for Modern Politics: The Tragedy of the Enlightenment and the Challenge of the New Millennium*, (Westview Press, Boulder, CO, 1998).

[55] E.Hargrave, "What's Wrong? Who is to Blame?" *Environmental Ethics*, vol.25, 2003, pp. 3-4.

[56] D. Jones, "The Ir/ Relevance of Environmental Ethics", *Environmental Ethics*,

vol.75, 2003, pp. 223-224.

[57] D.A.Brown, "Environmental Ethics and Public Policy", *Environmental Ethics*, vol.26, 2004, pp. 111-112.

[58] For example, "googling" "postmodern feminism" on December 22 2006 at 2:30 (EST) gave 33,100 results. A few seconds later googling "ethics and global climate change" and "philosophy and global climate change" yielded "search did not match any documents". A *Philosopher's Index* search a few minutes later for "feminism" gave 3426 hits, but "climate change" only 32.

[59] Ernest Gellner, cited note 7.

[60] J.Gray, *Straw Dogs: Thoughts on Humans and Other Animals*, (Granta Books, London, 2002).

[61] As above p. 4.

[62] See J.W.Smith, *Reductionism and Cultural Being*, cited note 5.

[63] Gray, cited note 60, p.17.

[64] As above p. 28.

[65] Gray refers to the work of Benjamin Libet which Libet takes to show that the electrical activity in the brain that causes action occurs before the conscious decision to act is made. See B. Libet, *Neurophysiology of Consciousness: Selected Papers and New Essays*, (Birkhauser, Boston, 1993), *Mind Time: The Temporal Factor in Consciousness*, (Harvard University Press, Cambridge, 2004) and B. Libet (*et al* eds.), *The Volitional Brain: Towards a Neuroscience of Free Will*, (Imprint Academic, Thorverton, 1999). For criticism see A.R. Mele, *Free Will and Luck*, (Oxford University Press, Oxford, 2006).

[66] Gray cited note 60 p. 82.

[67] As above p. 82.

[68] Cited from D.B. Fink, "Judaism and Ecology: A Theology of Creation", at http://environment.harvard.edu/religion/religion/judaism/index.html.

[69] H. Cox, *The Secular City*, (Pelican, London, 1968).

[70] D. Christie-Murray, *A History of Heresy*, (Oxford University Press, Oxford and New York, 1989); D.F. Ford, *The Modern Theologians: A Introduction to Christian Theology in the Twentieth Century*, (Volume 1), (Basil Blackwell, Oxford, 1989).

[71] J.M. Robertson, *The Historical Jesus: A Survey of Positions*, (Watts and Co, London, 1916); H.B. Bonner, *Christianity and Conduct,* (Watts and Co, London, 1919); E. Greenly, *The Historical Reality of Jesus,* (Watts and Co, London, 1927).

[72] R. Morgan and J. Barton, *Biblical Interpretation*, (Oxford University Press, Oxford, 1988), p. 17.

[73] Quoted from D.F.Strauss, *The Life of Jesus Critically Examined,* (SCM Press, London, 1973).

[74] D.F.Strauss, *The Old Faith and the New: A Confession*, (Asher and Co, London, 1973).

[75] A. Schweitzer, *Quest for the Historical Jesus: A Critical Study of its Progress from Reimarus to Wrede*, (3rd edition), (Adam and Charles Black, London, 1956).

[76] V.A Harvey, "New Testament Scholarship and Christian Belief", in R.J.Hoffman and G.A.Larue (eds.), *Jesus in History and Myth*, (Prometheus Books, New York, 1986), pp. 193-200, cited p. 193.

[77] Hoffman and Larue, as above, p. 8.

[78] G.Higgins, *Anacalypsis: An Inquiry into the Origin of Languages, Nations and Religions* (1833), (Kessinger Publishing, Kila MT, 1997); G. Massey, *Ancient Egypt, The Light of the World: A Work of Reclamation and Restitution in Twelve Volumes*,(Kessinger Publishing, Kila MT, 2001); A.B. Kuhn, *The Lost Light: An Interpretation of Ancient Scriptures* (1940), (Kessinger Publishing, Kila MT, 1997). See also J. Assman, *Moses the Egyptian: The Memory of Egyptian Western Monotheism*, (Harvard University Press, Cambridge MA, 1998); E. Carpenter, *Pagan and Christian Creed: Their Origin and Meaning, (Kessinger Publishing, Kila MT,* 1997); T.W. Doana, *Bible Myths and Their Parellels in Other Religions,* (Kessinger Publishing, Kila MT, 1997); D. Fideler, *Jesus Christ, Sun of God:*

*Ancient Cosmology and Early Christian Symbolism*, (Quest Books, Wheaton IL, 1993); M. Gadalla, *Historical Deception: The Untold Story of Ancient Egypt*, (Bastet Publishing, Eire, PA, 1996); D.Rushkoff, *Nothing Sacred: The Truth About Judaism*, (Crown Publishing, New York, 2003); J. Wheless, *Forgery in Christianity: A Documented Record of the Foundations of the Christian Religion* (1930), (Kessinger Publishing, Kila MT, 1997).

[79] T. Harpur, The *Pagan Christ: Is Blind Faith Killing Christianity?* (Allen and Unwin, Sydney, 2004), p. 34. We are indebted to Harpur's book for the references to the work of Higgins, Massey, Kuhn and Graves.

[80] K. Graves, *The World's Sixteen Crucified Saviours: Christianity before Christ*, (Kessinger Publishing, Kila MT, 1999).

[81] Harpur, cited note 79, p. 37; T. Freke and P. Gandy, *The Jesus Mysteries: Was the "Original Jesus" a Pagan God?* (Random House, New York, 1999); E. Carpenter, *Pagan and Christian Creed: Their Origin and Meaning*, (Kessinger Publishing, Kila MT, 1997).

[82] Harpur, cited note 79, pp. 68-69. See also W.E.A. Budge, *Gods of the Egyptians*, (Dover Publications, New York, 1969) and *Egyptian Religion*, (Citadel Press, New York, 1997); E. Hornung, *Conceptions of God in Ancient Egypt*, (Cornell University Press, Ithaca, 1982).

[83] Harpur, cited note 79, p. 11.See also T. Flynn, "Matthew vs. Luke", *Free Inquiry*, vol. 25, no.1, 2004, pp.34-35, T.W. Doana, *Bible Myths and Their Parellels in Other Religions,* (Kessinger Publishing, Kila MT, 1997).

[84] I. Finkelstein and N.A. Silberman, *The Bible Unearthed: Archaeology's New Vision of Ancient Israel and the Origin of its Sacred Texts*, (Free Press, New York, 2001).

[85] As above pp. 139-140. Harpur, cited note 79, is of the opinion from a consideration of the archaeological evidence that not even one gold goblet or brick has been found to indicate the historical reality of King Solomon and that "the country of a people so rich that King David in his poverty, could collect millions of pounds to build a temple is found to have been without art, sculptures, mosaics, bronzes, pottery or precious stones," cited p.119.

[86] Finkelstein and Silberman, cited note 84, p.118. See further: D. Lazare, "False

Testament: Archaeology Refutes the Bible's Claim to History", *Harper's Magazine*, March 2002, pp. 39-47.

[87] See K. Keating, *What Catholics Really Believe - Setting the Record Straight*, (Servant Publications, Ann Arbor, 1992), pp. 34-38; M.J. Harris, *Three Crucial Questions About Jesus*, (OM Publishing, Carlisle, 1994); J. S. Spong, *Rescuing the Bible from Fundamentalism*, (HarperSanFrancisco, New York, 1991), *Born of a Woman: A Bishop Rethinks the Birth of Jesus*, (HarperSanFrancisco, New York, 1992), *Why Christianity Must Change or Die*, (HarperSanFrancisco, New York, 1998). There are literally hundreds of problematical passages. Some involve conflict with generally accepted moral norms and principles of rationality - Ex. 21,7; Deut 21, 10-14; Ex 21: 20, 21; Ex 22: 18; Lev 20; 27; Lev 20:6; Ex 22:20; Deut 13:6-10; Num 15:30; Deut 17:12; Ex 31: 14-15; Ex 35: 2-3; Num 15: 32-36; Lev 7, 22-25; Gen 17:14; Num 4:15; Num 4:20; Lev 22: 8-9. Some passages seem to be mutually inconsistent, for example (Rom 15:33 & Ex 15:3); (James 1:13 & Gen 22:1); (Num 23:19 & Gen 6:6); (Prov. 12:22 & 1Kings 22:23); (Mark 10: 27 & Judges 1:19); (1 Cor 14:33 & Is. 45:17); (Ex 33:20 & Gen 33:30); (John 1:18 & Amos 14:1); (1 Peter 2:13 & Acts 5:29); (Ps. 104: 5&2 Peter 3:10); (Rom 4:5,6& James 2: 14, 17, 24, 25); (John 20:17 & Matt 28: 1, 9); (Mark 16:8 & Luke 24: 8,9); (Luke 24: 50, 51 & Acts 1:9, 12 & Mark 16:14, 19 & Matt 28, 16-20); (Acts 1:18 & Matt 27: 5-7); (Acts 26:23: John 11:43, 44); (Mark 15:25 & John 19:14, 15); (Acts 2:30 & Matt 1:18); (2 Sam 10:18& 1 Chron 19:18), to name but some. See further, E. Doherty, *The Jesus Puzzle: Did Christianity Begin with a Mythical Christ?* (Canadian Humanist Publications, Ottawa, 1999); H.E. Barnes, *The Twilight of Christianity*, (Richard R. Smith Inc., New York, 1931); H.Leidner, *The Fabrication of the Christ Myth*, (Surrey Books, Tampa Fl, 1999).

[88] R.Bultmann, *Jesus Christ and Mythology*, (SCM Press, London, 1960), "New Testament and Mythology", in H.W. Bartsch (ed.), *Kerygma and Myth*, (Harper Torchbooks, New York, 1961), pp. 1-44, *Faith and Understanding I*, (SCM Press, London, 1969).

[89] M. Heidegger, *Being and Time*, (Basil Blackwell, Oxford, 1967).

[90] See H.W. Bartsch (ed.), *Kerygma and Myth*, cited note 88; S. Barton, "Problems in Preaching a Demythologized Christ I", *Interchange*, vol. 24, 1978, pp. 223-225 and "Problems in Preaching a Demythologized Christ II", *Interchange*, vol. 25, 1979, pp. 61-67; R.F. Brown, *The Birth of the Messiah*, (Doubleday, Gordon City, New York, 1977); E.Käsemann, *Essays in New Testament Themes*, (SCM Press, London, 1981); J.G. Gager, "The Gospels and

Jesus: Some Doubts about Method", *The Journal of Religion*, vol.54, 1974, pp. 244-272; N. Frye, *The Great Code: The Bible and Literature*, (Harcourt Brace Jovanovich Publishers, New York and London, 1981).

[91] J.A.T.Robinson, *Honest to God*, (SCM Press, London, 1963) and *Christian Morals Today*, (SCM Press, London, 1964).

[92] I.M. Russell (ed.), *Feminist Interpretation of the Bible*, (Basil Blackwell, Oxford, 1985); E.S. Fiorenza, *But She Said: Feminist Practices of Biblical Interpretation*, (Beacon Press, Boston, 1992); R. Nash (ed.), *Liberation Theology*, (Mott Media, Milford, Michigan, 1984); J. Macquarrie, *In Search of Deity: An Essay in Dialectical Theism*, (SCM Press, London, 1984); E. Pagels, *The Origin of Satan*, (Vintage Books, New York, 1996); D. Groothuis, *Unmasking the New Age*, (InteVarsity, Downers Grove, 1986); T.J.J. Altizer and W. Hamilton, *Radical Theology and the Death of God*, (Penguin Books, Harmondsworth, 1968); J. Allegro, *The Sacred Mushroom and the Cross*, (Hodder and Stoughton, London, 1970); R.Chandler, *Understanding the New Age*, (Word, Waco,1988); T. Peters, *The Comic Self*, (Harper, San Francisco, 1991); M.Fox, *The Coming of the Cosmic Christ*, (Collins Dove, Melbourne, 1990); M. Smith, *Jesus the Magician*, (Victor Gollancz, London, 1978); B. Thiering, *Jesus and the Riddle of the Dead Sea Scrolls*, (Harper, San Francisco, 1992); A.E. McGrath, *A Passion for Truth*, (Apollos, Leicester, 1996); H. Frei, *The Identity of Jesus Christ*, (Fortress, Philadelphia, 1975).

[93] L.White, "The Roots of Our Ecological Crisis", *Science*, vol. 155, 1967, pp. 1203-1207.

[94] E. Whitney, "Lyn White, Ecotheology, and History", *Environmental Ethics*, vol.15, 1993, pp. 151-169; M. Marangudakis, "The Medieval Roots of Our Ecological Crisis", *Environmental Ethics*, vol.23, 2001, pp. 243-260; J.B. Callicott, "Genesis Revisited: Murian Musings on the Lynn White Jr Debate", *Environmental History Review*, vol. 14, 1990, pp. 65-92.

[95] D.L. Eckberg and T.J. Blocker, "Varieties of Religious Involvement and Environmental Concern", *Journal for the Scientific Study of Religion*, vol.28, 1989, pp. 509-517; A. Greeley, "Religion and Attitudes toward the Environment", *Journal for the Scientific Study of Religion*, vol.32, no. 1, March 1993, pp. 19-28.

[96] C.K. Chapple and M.E. Tucker (eds.), *Hinduism and Ecology: The Intersection*

*of Earth, Sky and Water*, (Harvard Divinity School, Cambridge, 2000). On Judaism and the environment, see M.D. Yaffe, (ed.), *Judaism and Environmental Ethics: A Reader*,(Lexington Books, Lanham, 2001); M. Gerztenfeld, *Judaism, Environmentalism and the Environment: Mapping and Analysis*,(Jerusalem Institute for Israel Studies, Jerusalem,1998).

[97] R. Goodman, "Taoism and Ecology", *Environmental Ethics*, vol.2, 1980, pp. 73-80; I. Po-keung, "Taoism and the Foundations of Environmental Ethics", *Environmental Ethics*, vol.5, 1983, p. 335; J. Baird Callicott, "Conceptual Resources for Environmental Ethics in Asian Traditions of Thought: A Propaedeutic", *Philosophy East and West*, vol. 37, 1987, pp. 115-131; H.Rolston III, "Can the East Help the West to Value Nature?" *Philosophy East and West*, vol. 37, 1987, pp. 172-190; R.P. Peerenboom, "Beyond Naturalism: A Reconstruction of Daoist Environmental Ethics", *Environmental Ethics*, vol.13, 1991, pp. 3-22; K.L. Lai, "Conceptual Foundations for an Environmental Ethics: A Daoist Perspective", *Environmental Ethics*, vol.25, 2003, pp. 247-266.

[98] See Richard Folz, "Islam and Ecology Bibliography", at http://environment.harvard.edu/religion/religion/islam/bibliography.html.

[99] D.R. Griffin (ed.), *Spirituality and Society*, (State University of New York Press, Albany, 1988); D.R. Griffin (ed.), *The Reenchantment of Science,* (State University of New York Press, Albany, 1988);); D.R. Griffin (ed.), *Sacred Interconnections: Postmodern Spirituality, Political Economy and Art,* (State University of New York Press, Albany, 1990); C. Birch and J.B. Cobb Jr., *The Liberation of Life: From Cell to the Community*, (Environmental Ethics Books, Denton, 1990); J.B.Cobb Jr, *Is it Too Late? A Theology of Ecology*, (Environmental Ethics Books, Denton, 1995); T. Hartmann, *The Last Hours of Ancient Sunlight: The Fate of the World and What We Can Do Before It's Too Late*, (Three Rivers Press, New York, 2004); A. Primavesi, *Gaia's Gift*, (Routledge, London, 2003) and *Sacred Gaia*, (Routledge, London, 2000).

[100] J.B.Martin-Schramm and R.L. Stivers, *Christian Environmental Ethics: A Case Method Approach*, (Orbis Books, Maryknoll, New York, 2003); R.S. Gottlieb, *A Greener Faith: Religious Environmentalism and Our Planet's Future*, (Oxford University Press, Oxford, 2006).

[101] P. Bristow, "The Roots of Our Ecological Crisis", *Answers in Genesis (TJ Archive)*, vol. 15, no.1, April 2001, pp. 76-79 at http://www.answersingenesis.org/tj/v15/il/ecology.asp.

---

[102] See for example J.K. Guelke, "Looking for Jesus in Christian Environmental Ethics", *Environmental Ethics*, vol.26, 2004, pp. 115-134.

[103] A.Dent, *Ecology and Faith: The Writings of Pope John Paul II*, (Arthur James, Berkhamsted, 1997); H. Paul Santmire, *Nature Reborn: The Ecological and Cosmic Promise of Christian Theology*, (Fortress Press, Minneapolis, 2000).

[104] R. Dawkins, *The God Delusion*, (Bantam, London, 2006).

[105] S. Harris, *The End of Faith: Religion, Terror and the Future of Reason*, (W.W. Norton, New York, 2005).

[106] As above, p.12.

[107] B. York, "Mayers Proves that the Left can be Blinded by Zeal", http://www.hillnews.com/thehill/export/TheHill/Comment/ByronYork/021005.html; B. Carnell, "Did James Watt Really Say This?", http://brian.carnell.com/archives/years/2005/01/000047.html. Watt, according to Carnell did say to the US Congress in 1981: "I do not know how many future generations we can count on before the Lord returns. Whatever it is, we have to manage with a skill to leave the resources needed for future generations".

[108] T. F. LaHaye and J.B. Jenkins, *Left Behind: A Novel of the Earth's Last Days (Left Behind No.1)* (Tyndale House Publishers, Carol Stream IL, 1996).

[109] There is a "Rapture Index" to measure how close we are to end times: www.raptureready1com/rapz.html. See J.Carroll, "Fasten Your Seatbelts: The Rapture Index", *The San Francisco Chronicle*, February 11 2005 at http://www.commondreams.org/views05/0211_22.htm.

[110] Carroll, as above.

[111] B. McKibben, "The Christian Paradox: How a Faithful Nation Gets Jesus Wrong", *Harper's Magazine*, August 2005, pp. 31-37.

[112] R. Koch and C. Smith, *Suicide of the West*, (Continuum, London, 2006), p. 44.

[113] I. Semeniuk, "Can E.O. Wilson Really Save the World?" *New Scientist*, September 30 2006, pp. 54-56.

[114] E.O. Wilson, *The Creation: An Appeal to Save Life on Earth*, (WW Norton, New York, 2006).

[115] As above p. 5.

[116] As above p168.

[117] Statement of the Evangelical Climate Initiative, "Climate Change: An Evangelical Call to Action", at http://www.christiansandclimate.org/statement.

[118] Semeniuk, cited note 113, p. 54.

[119] " Pope's Environmentalist Appeal - Save World from Degradation", at http://www.corriere.it/english/articoli/2006/08_Agosto/28/pope.shtml.

[120] "Address of His Holiness Ecumenical Patriarch Bartholomew", at the Environmental Symposium, Santa Barbara California, November 8 1997, quoted from R.S. Gottlieb, *A Greener Faith: Religious Environmentalism and Our Planet's Future*, (Oxford University Press, Oxford, 2006), pp. 83-84.

[121] "Philosophers have hitherto only interpreted the world in various ways; the point is to change it". K. Marx, *Theses on Feuerbach* (1845) at http://www.marxists.org/archive/marx/works/1845/theses/theses.htm.

[122] See G. Sauer-Thompson and J.W. Smith, *The Unreasonable Silence of the World*, cited note 45.

[123] See J.W.Smith, *Reductionism and Cultural Being*, cited note 5.

[124] As above.

[125] As above.

[126] See G. Sauer-Thompson and J.W. Smith, *The Unreasonable Silence of the World*, cited note 45.

[127] J.W. Smith (*et al*), *Global Meltdown*, (Praeger, Westport, 1998).

[128] N. Maxwell, *From Knowledge to Wisdom: A Revolution in the Aims and*

*Methods of Science*, (Blackwell, Oxford, 1987).

**Chapter 7**

[1] A.O'Hear, *After progress: Finding the Old Way Forward*, (Bloomsbury, London, 1999), p. 237.

[2] H. Shue, "Avoidable Necessity: Global Warming, International Fairness, and Alternative Energy," in I.Shapiro and J.W. DeCew (eds.), *Theory and Practice, Nomas XXXVII*, (New York University Press, New York and London, 1995), pp. 239-265, cited pp. 257-258.

[3] W.Ophuls, *Requiem for Modern Politics: The Tragedy of the Enlightenment and the Challenge of the New Millennium*, (Westview Press, Boulder, CO, 1997), p. 269.

[4] O. Meuvrel, "Cattle a Cause of Global Warming," *The Advertiser* (Adelaide), October 1 2005, p.69.

[5] D.C. Korten, *When Corporations Rule the World*, (Earthscan, London, 1995); J.Mander and E.Goldsmith (eds.), *The Case Against the Global Economy and a Turn Toward the Local*, (Sierra Club Books, San Francisco,1996); J.R.Saul, *The Collapse of Globalism and the Reinvention of the World*, (Viking/Penguin, London, 2005).

[6] For an overview see. J.W. Smith (*et al*), *Healing a Wounded World: Economics, Ecology and Health for a Sustainable Life*, (Praeger, Westport and London, 1997), *Global Meltdown: Immigration, Multiculturalism and National Breakdown in the New World Disorder*, (Praeger, Westport and London, 1998), *The Bankruptcy of Economics: Ecology, Economics and the Sustainability of the Earth*, (Macmillan, London, 1999), *Global Anarchy in the Third Millennium?* (Macmillan, London, 2000).

[7] N. Georgescu-Roegen, *The Entropy Law and the Economic Process*, (Harvard

University Press, Cambridge MA, 1971); C.E.Smith and J.W. Smith, "Economics, Ecology, and Entropy: The Second Law of Thermodynamics and the Limits to Growth", *Population and Environment*, vol.17, no.4, March 1996, pp.309-321; J.W. Smith and G. Sauer-Thompson, "Civilization's Wake: Ecology, Economics and the Roots of Environmental Destruction and Neglect", *Population and Environment*, vol.19, no.6, July 1998, pp.541-575.

[8] W.R. Catton, *Overshoot: The Ecological Basis of Revolutionary Change*, (University of Illinois Press, Urbana, 1982).

[9] H.Daly, *Steady-State Economics*, 2nd edition, (Earthscan, London, 1992).

[10] T. Trainer, *The Conserver Society: Alternatives for Sustainability*, (Zed Books, London and New Jersey, 1995); C.Hamilton and R.Denniss, *Affluenza: When Too Much is Never Enough*, (Allen and Unwin, Crows Nest, 2005).

[11] See the references on voluntary simplicity in chapter 4 of this book.

[12] T. Banuri (*et al*), *Climate Change 2001: Mitigation: A Report of Working Group III of the Intergovernmental Panel on Climate Change, Summary for Policymakers*, (Cambridge University Press, Cambridge, 2001), at http://www.ipcc.ch/pub/online.htm, p.137.

[13] W.H. Newton-Smith, *The Rationality of Science*, (Routledge and Kegan Paul, Boston, 1981).

[14] S.F. Haller, *Apocalypse Soon? Wagering on Warnings of Global Catastrophe*, (McGill-Queen's University Press, Montreal and Kingston, 2002), p.xii.

[15] J.M. Wallace, "What Science Can and Cannot Tell Us About Greenhouse Warming, " *Bridges*, vol. 7, nos. 1-2, 2000, pp.1-15.

[16] See E.Soule, "Assessing the Precautionary Principle", *Public Affairs Quarterly*, vol. 14, no.4, October 2000, pp. 309-328; N.A. Manson, "Formulating the Precautionary Principle", *Environmental Ethics*, vol.24, 2002, pp. 263-274; D. Turner and L.Hartzell, " The Lack of Clarity in the Precautionary Principle", *Environmental Values*, vol.13, 2004, pp.449-460; P.Sandin, "The Precautionary Principle and the Concept of Precaution", *Environmental Values*, vol.13, 2004, pp.461-475; S. Vanderheiden, "Knowledge, Uncertainty, and Responsibility: Responding to Climate Change", *Public Affairs Quarterly*, vol. 18, no.2, 2004,

pp. 141-158; S.M.Gardiner, "A Core Precautionary Principle", *Journal of Political Philosophy*, vol.14. no.1, 2006, pp. 33-60 and S.Beder, *Environmental Principles and Policies*, (University of New South Wales Press, Sydney, 2006).

[17] A.Simms, *Ecological Debt: The Health of the Planet and the Wealth of Nations*, (Pluto Press, London, 2005).

[18] As above, p. 103

[19] As above, p. 27

[20] *Summary for Policy Makers*, cited note 12, p. 638.

[21] H. Dittmar, *The Social Psychology of Material Possessions: To Have Is To Be*, (Harvester Wheatsheaf, Hemel, Hempstead, 1992).

[22] *Summary for Policy Makers*, cited note 12, p. 638.

[23] As above, pp. 639-640.

[24] S.M. Gardiner, "Ethics and Global Climate Change", *Ethics*, vol.114, April 2004, pp. 555-600. Gardiner begins his review essay by saying: "Very few moral philosophers have written on climate change". (p. 555). This is puzzling for a number of reasons, for among other things, many working in the climate change field have seen the problem as ultimately an ethical issue. Gardiner says that the neglect of this problem by professional philosophers is due to the necessary interdisciplinary nature of the field, which not only makes the area intellectually demanding but gives the impression that the issues are for others to deal with. However, these types of factors have not prevented philosophers from examining other interdisciplinary fields which are also intellectually demanding such as bioethical issues in the genetics revolution. For some excellent overview papers of various aspects of the debate on ethics and global climatic change, see. J. Lemons, "Atmospheric Carbon Dioxide: Environmental Ethics and Environmental Facts", *Environmental Ethics*, vol.5, 1983, pp. 21-32; D. Jamieson, "Ethics, Public Policy, and Global Warming", *Science, Technology, and Human Values*, vol.17, no.2, 1992, pp. 139-153; T.Williamson and A. Radford, "Building, Global Warming and Ethics", in W. Fox (ed.), *Ethics and the Built Environment*, (Routledge, London and New York, 2000), pp. 57-72; M.L. Menestrel (*et al*), "Processes and Consequences in Business Ethical Dilemmas: The Oil Industry and Climate Change", *Journal of Business Ethics*, vol. 41, 2002, pp. 251-266; D.

Jamieson, *Morality's Progress: Essays on Humans, Other Animals, and the Rest of Nature*, (Clarendon Press, Oxford, 2002): R.Hood, "Global Warming", in R.G. Frey and C.H. Wellman, (eds.), *A Companion to Applied Ethics: Blackwell Companions to Philosophy*, (Blackwell Publishing, Oxford, 2003), pp. 674-684; D.G. Arnold and K. Bustos, "Business Ethics and Global Climate Change", *Business and Professional Ethics Journal*, vol.24, nos. 1-2, 2005, pp. 103-130.

[25] R. Kirkman, *Skeptical Environmentalism: The Limits of Philosophy and Science*, (Indiana University Press, Bloomington, 2002); L.Pinguelli-Rosa and M. Munasinghe (eds.), *Ethics, Equity and International Negotiations on Climate Change*, (Edward Elgar, Cheltenham, 2002); P.G. Harris, *International Equity and Global Environmental Politics: Power and Principles in U.S. Foreign Policy*, (Ashgate, Aldershot, 2001).

[26] F.O. Hampson and J. Reppy (eds.), *Earthly Goods: Environmental Change and Social Justice*, (Cornel University Press, Ithaca and London, 1996); M. Paterson, *Global Warming and Global Politics*, (Routledge, London and New York, 1996).

[27] H. Shue, "Climate" in D. Jamieson (ed.), *A Companion to Environmental Philosophy: Blackwell Companions to Philosophy*, (Blackwell Publishers, Oxford, 2001), pp. 449-459; H. Coward and T. Hurka (eds.), *Ethics and Climate Change: The Greenhouse Effect*, (Wilfred Laurier University Press, Waterloo, Ontario, 1993).

[28] A. de-Shalit, *Why Posterity Matters: Environmental Policies and Future Generations*, (Routledge, London and New York, 1995), p. 2.

[29] E.Page, "Intergenerational Justice and Climate Change", *Political Studies*, vol. 47, 1999, pp. 53-66; W.Beckerman and J. Pasek, *Justice, Posterity, and the Environment*,(Oxford University Press, Oxford, 2001); F.Arler, "Global Partnership, Climate Change and Complex Equality", *Environmental Values*, vol. 10, no.3, 2001, pp. 301-329; C.Wolf, "Intergenerational Justice", in R.G. Frey and C.H. Wellman, (eds.), *A Companion to Applied Ethics: Blackwell Companions to Philosophy*, (Blackwell Publishing, Oxford, 2003), pp. 279-294; H. Shue, "Legacy of Danger: The *Kyoto Protocol* and Future Generations", in K. Horton and H. Patapan (eds.), *Globalisation and Equality*, (Routledge, London and New York, 2004), pp. 164-178.

[30] P.Singer, "Famine, Affluence and Morality", *Philosophy and Public Affairs*, vol. 1, 1972, pp. 229-244; R.E. Goodin, "What is So Special About Our Fellow

Countrymen?" *Ethics*, vol. 98, 1988, pp. 663-686 and *Utilitarianism as a Public Philosophy*, (Cambridge University Press, Cambridge, 1995); G. Elfstrom, *Ethics for a Shrinking World,* (MacMillan, London, 1990); A.Ellis, "Utilitarianism and International Ethics", in T.Nardin and D.R.Mapels (eds.), *Traditions of International Ethics*, (Cambridge University Press, Cambridge, 1992), pp. 158-179.

[31] D.Gauthier, *Morals by Agreement*, (Clarendon Press, Oxford, 1986); J.Rawls, *A Theory of Justice*, Oxford University Press, Oxford, 1973).

[32] R. Dworkin, *Taking Rights Seriously*, (Duckworth, London, 1977); H.Shue, *Basic Rights*, (Princeton University Press, Princeton, New Jersey, 1980); J.Feinberg, *Rights, Justice and the Bounds of Liberty*, (Princeton University Press, Princeton, New Jersey, 1980); L.W. Sumner, *The Moral Foundation of Rights*, (Clarendon Press, Oxford, 1987); J.W. Nickel, *Making Sense of Human Rights*, (University of California Press, London, 1987).

[33] I.Kant, *Foundations of the Metaphysics of Morals*, (Macmillan, London,1990); O.O'Neill, *Faces of Hunger: An Essay on Poverty, Justice and Development*, (Allen and Unwin, London, 1986) and *Towards Justice and Virtue: A Constructive Account of Practical Reasoning*, (Cambridge University Press, Cambridge, 1996).

[34] A. MacIntyre, *After Virtue*, 2nd edition, (Duckworth, London, 1985); G.P.Fletcher, *Loyalty: An Essay on the Morality of Relationships*, (Oxford University Press, Oxford, 1993); D. Miller, *On Nationality*, (Clarendon Press, Oxford, 1995).

[35] See A. de-Shalit, *Why Posterity Matters*, cited note 28 for telling criticisms of all non-communitarian positions such as utilitarianism:. De-Shalit advances his own communitarian response to the problem of obligations to future generations.

[36] D. Parfit, *Reasons and Persons*, (Oxford University Press, Oxford, 1984).

[37] As above p. 445.

[38] Mac Intyre cited note 34, p. 69.

[39] B. Williams, *Ethics and the Limits of Philosophy*, (Harvard University Press, Cambridge, 1985), p. 174.

[40] R.B. Brandt, "Morality and Its Critics", *American Philosophical Quarterly*, vol. 26, no.2, April 1989, pp. 89-100, at p. 89; T. Trainer, *The Nature of Morality*,

(Avebury, Aldershot, 1991). For a defense of moral particularism, the thesis that morality and moral thought does not require general moral principles, see J.Dancy, *Ethics Without Principles*, (Clarendon Press, Oxford, 2004).

[41] For technical arguments along these lines see S.P. Schwartz, "Why It Is Impossible to be Moral", *American Philosophical Quarterly*, vol. 36, no.4, October 1999, pp. 351-360.

[42] G. Cullity, "Asking Too Much", *Monist*, vol. 86, 2003, pp. 402-418 and *The Moral Demands of Affluence*, (Clarendon Press, Oxford, 2004). Cullity rejects the "severe demand" of morality and after deliberation reaches this conclusion (p. 415):

I should continue to contribute to aid agencies increments of time and money each of which is large enough to allow those agencies to save a life, until either:

> (a) there are no longer any lives to be easily saved by those agencies, or
>
> (b) my overall sacrifice, with respect to those non-altruistically-focused life-enhancing goods that can impartially be defended as permissible, is significant enough to make it justifiable for me to refuse to save a life directly at that cost.
>
> where a non-altruistically-focused life-enhancing good can be impartially defended as permissible when either:
>
> (i) there is no cheaper non-altruistically-focused alternative that it would be no worse for me to pursue, or
>
> (ii) a preparedness to abandon this good in favour of a cheaper alternative that is no worse would itself be life-impoverishing.

The demand is sufficiently "severe" though to mean that an academic philosopher should perhaps stop writing articles about moral philosophy and join an aid agency to save lives. With all due respect, few philosophers who believe that people have extensive moral obligations to give aid have made the break and resigned from their chairs.

[43] S. Kagan, *The Limits of Morality*, (Clarendon Press, Oxford, 1989).

[44] As above p.1.

[45] A. Gini, *Why It's Hard to be Good*, (Routledge, New York and London, 2006).

[46] D. Moellendorf, *Cosmopolitan Justice*, (Westview Press, Boulder, Colorado, 2002); R.Attfield, *Environmental Ethics: An Overview for the Twenty-First Century*, (Polity Press, Cambridge, 2003).

[47] L.P. Francis, "Global Systemic Problems and Interconnected Duties", *Environmental Ethics*, vol. 25, 2003, pp. 115-128.

[48] J. Carens, "Aliens and Citizens: The Case for Open Borders", in W. Kymlicka (ed.), *The Rights of Minority Cultures*, (Oxford University Press, Oxford, 1995), pp. 331-349; B. Barry, *The Liberal Theory of Justice*, (Clarendon Press, Oxford, 1973); C. Beitz, *Political Theory and International Relations*, (Princeton University Press, Princeton, New Jersey, 1979).

[49] J. Thompson, *Justice and World Order: A Philosophical Inquiry* , (Routledge, London, 1992).

[50] C. Jones, *Global Justice: Defending Cosmopolitanism*, (Oxford University Press, Oxford, 1999), p. 16; J. Gray, *Liberalism*, (Open University Press, Milton Keynes, 1986), p. x.

[51] T. Athanasiou and P. Baer, *Dead Heat: Global Justice and Global Warming*, (Seven Stories Press, New York, 2002), p. 37.

[52] P. Singer, *One World: The Ethics of Globalization*, (Yale University Press, New Haven and London, 2002), p.31. See also L. Pinguelli-Rosa and M. Munasinghe (eds.), *Ethics, Equity and International Negotiations on Climate Change*, (Edward Elgar, Northampton MA, 2003)

[53] P.G. Harris, "Fairness, Responsibility, and Climate Change", *Ethics and International Affairs*, vol. 17, no.1, 2003, pp. 149-156, cited p. 150.

[54] H. Shue, "Global Environment and International Inequality", *International*

*Affairs*, vol. 75, 1999, pp. 531-545, cited p. 534.

[55] S. McBride and J.Wiseman, *Globalization and Its Discontents*, (Macmillan, London, 2000).

[56] See the references in note 5.

[57] D. Parfit, *Reasons and Persons*, cited note 36.

[58] G. Curran, "Environment, Equality and Globalisation", in K. Horton and H. Patapan (eds.), *Globalisation and Equality*, (Routledge, London and New York, 2004), pp. 146-163, cited pp. 146-147.

[59] M. MacCracken, "Climate Change Discussions in Washington: A Matter of Contending Perspectives", *Environmental Values*, vol. 15, 2006, pp. 381-395, cited p. 389.

[60] See for example P. Singer, *One World*, cited note 52, p. 35.

[61] See the reference in note 6.

[62] The cosmopolitan approach to distributive justice is also challenged by communitarians. Michael Walzer, for example, argues that there are no global accounts of distributive justice, only local accounts. See M. Walzer, *Spheres of Justice: A Defence of Pluralism and Equality*, (Basil Blackwell, Oxford, 1983), p.314.

[63] T. Trainer, "Development: The Radical Alternative View", at http://socialwork.arts.unsw.edu.au/tsw/D99.Development.RadView.html.

[64] C. Brown, *International Relations Theory: New Normative Approaches*, (Harvester-Wheatsheaf, Hemel Hempstead,1992), p. 75; J.Gupta, "The Climate Change Regime: De Jure and De Facto Commitment to Sustainable Development", in J.Paavola and I.Lowe (eds.), *Environmental Values in a Globalising World: Nature, Justice and Governance*, (Routledge, London and New York, 2005), pp. 159-178.

[65] T. Trainer, cited note 63. For further documentation see http://www.socialwork.arts.unsw.edu.au/tsw/DocsGLOBALISATION.html and F.E. Trainer, *Developed to Death: Rethinking Third World Development*, (Green

Print, London, 1989).

[66] G.Hardin, *Living Within Limits: Ecology, Economics, and Population Taboos,*

(Oxford University Press, and Oxford, 1993)

[67] T. Erkins (ed.), *The Living Economy: A New Economics in the Making*, (Routledge and Kegan Paul, London, 1986), and "Trade and Self-Reliance", *The Ecologist*, vol.19. September/October 1989, pp. 186-190; R. Batra, *The Myth of Free Trade: A Plan for America's Economic Revival*, (Charles Scribner's Sons, New York, 1993).

[68] J. Galtung, *Development, Environment and Technology: Towards a Technology for Self-Reliance*, (United Nations, New York, 1979), *Environment, Development and Military Activity*, (Norwegian Universities Press, Oslo,1982), "Towards a New Economics: On the Theory and Practice of Self-Reliance", in P.Ekins (ed.), *The Living Economy*, (Routledge and Kegan Paul, London, 1986), pp. 97-109; J. Galtung (*et al*), *Self-Reliance: A Strategy for Development*, (Bogle-L'Ouverture Publications Ltd, London, 1980).

[69] H. Elliot, *Ethics for a Finite World: An Essay Concerning a Sustainable Future*,

(Fulcrum Publishing, Golden, Colorado, 2005).

[70] As above p.14.

[71] G.Monbiot, *Heat: How to Stop the Planet Burning*, (Allen Lane, London, 2006). J.H. Kunstler, *The Long Emergency: Surviving the Converging Catastrophes of the Twenty-First Century*, (Atlantic Books, London, 2005).

[72] W.Ophuls, *Requiem for Modern Politics: The Tragedy of the Enlightenment and the Challenge of the New Millennium*, (Westview Press, Boulder, Colorado, 1997).

[73] J.W. Smith (*et al*), *The Bankruptcy of Economics: Ecology, Economics and the Sustainability of the Earth*, (Macmillan, London, 1999).

[74] As above.

[75] J.M Gardiner, "A Perfect Moral Storm: Climate Change, Intergenerational Ethics and the Problem of Moral Corruption", *Environmental Values*, vol.15, 2006, pp. 397-413, cited p. 398.

[76] G.Hardin, "The Tragedy of the Commons", *Science*, vol.162, 1968, pp.1243-1248.

[77] J.M. Gardiner, "The Real Tragedy of the Commons", *Philosophy and Public Affairs*, vol. 30, no.4, 2001, pp. 387-416, and "The Pure Intergenerational Problem", *Monist*, vol.86, 2003, pp. 481-500.

[78] Gardiner cited note 74, p.404.

[79] B.Holden, *Democracy and Global Warming*, (Continuum, London and New York, 2002).

[80] D. Shearman and J.W.Smith, *The Climate Change Challenge and the Failure of Democracy*, (Praeger, Westport, 2007).

[81] J.W.Smith, *The Remorseless Working of Things*, (Kalgoorlie Press, Kalgoorlie,1992).

[82] J.W.Smith (*et al*), *Global Meltdown*, (Praeger, Westport, 1998).

## Bibliography


Adams, J., "Cost-Benefit Analysis: The Problem Not the Solution", *The Ecologist*, vol. 26, no.1, 1996, pp. 2-4.

*Agenda 21* (1972), report on the United Nations Conference on Environment and Development, Rio de Janeiro, June 3-14, at www.un.org/esa/sustdev/documents/agenda21/english/Agenda 21.pdf.

Allegro, J., *The Sacred Mushroom and the Cross*, (Hodder and Stoughton, London, 1970).

Allen, M., (*et al*), "Observational Constraints on Climate Sensitivity", in H.J. Schellnhuber (*et al* eds.), *Avoiding Dangerous Climate Change*, (Cambridge University Press, Cambridge, 2006), pp. 281-289.

Altizer, T. J. J. and Hamilton,W., *Radical Theology and the Death of God*, (Penguin Books, Harmondsworth, 1968).

Anderson, J. L., "Stone Age Minds at Work in the 21st Century Science: How Cognitive Psychology Can Inform Conservation Biology", *Conservation Biology in Practice*, vol. 2, 2001, pp. 18-25.

Appadurai, A., *The Social Life of Things: Commodities in Cultural Perspective*, (Cambridge University Press, Cambridge, 1986).

Appleyard, B., *Understanding the Present: Science and the Soul of Modern Man*, (Picador/Pan Books, London, 1993).

Araujo, M. B. and Rahbek, C., "How does Climate Change Affect Biodiversity?" *Science*, vol.313, no 5792, September 5 2006, pp.1396-1397, doi: 10.1126/science.1131758.

Arden Pope III, C., (*et al*), "Lung Cancer, Cardiopulmonary Mortality, and Long-Term Exposure to Fine Particulate Air Pollution", *Journal of the American Medical Association* vol.287, no.9, 2002, pp.1132-1141.

Arler, F., "Global Partnership, Climate Change and Complex Equality", *Environmental Values*, vol. 10, no.3, 2001, pp. 301-329.

Arnold, D. G. and Bustos, K., "Business Ethics and Global Climate Change", *Business and Professional Ethics Journal*, vol.24, nos. 1-2, 2005, pp. 103-130.

Assman, J., *Moses the Egyptian: The Memory of Egyptian Western Monotheism*, (Harvard University Press, Cambridge MA, 1998).

Athanasiou, T. and Baer, P., *Dead Heat: Global Justice and Global Warming*, (Seven Stories Press, New York, 2002).

Attarian, J., "Oil Depletion Revisited: Why the Peak is Probably Near", *The*

*Social Contract*, Winter 2004-2005, pp. 129-146.

Attfield, R., *Environmental Ethics: An Overview for the Twenty-First Century*, (Polity Press, Cambridge, 2003).

Avery, D. T., *Saving the Planet with Pesticides and Plastic*, (Hudson Institute, Indianapolis, 1995).

Baer, P. and T. Athanasiou, T., "Honesty About Dangerous Climate Change" at http://www.ecoequity.org/ceo/ceo_8_2.htm.

Bagozzi, R., (*et al.*), *The Social Psychology of Consumer Behaviour*, (Open University Press, Buckingham, 2002).

Bailey, R., (ed.), *The True State of the Planet*, (Free Press, New York, 1995).

Baker, R., "Burying the Problem', *The Age* (Melbourne), July 30 2005, p.3.

Baker, R., "Revealed: How Big Energy Won the Battle on Climate Change", *The Age*, July 30 2005, pp.1,2.

Ball, P., *Critical Mass: How One Thing Leads to Another*, (Farrar, Straus and Giroux, New York, 2004).

Ballard, D., "Using Learning Processes to Promote Change for Sustainable Development", *Action Research*, vol. 3, no.2, 2005.

Bandura, A., " Self-Efficacy: Toward a Unifying Theory of Behavior Change," *Psychological Review*, vol. 84, 1977, pp.191-215.

Bandura, A., "Self-Efficacy Mechanism in Human Agency", *American Psychologist*, vol. 37, 1982, pp. 122-147.

Banuri, T., (*et al*), *Climate Change 2001: Mitigation: A Report of Working Group III of the Intergovernmental Panel on Climate Change, Summary for Policymakers*, (Cambridge University Press, Cambridge, 2001), at http://www.ipcc.ch/pub/online.htm.

Banuri, T., *Summary for Policymakers. Climate Change 2001: Mitigation. A Report of Working Group III of the Intergovernmental Panel on Climate Change*, (Cambridge University Press, Cambridge, 2001), http://www.ipcc.ch/pub/online.htm.

Barnes, H. E., *The Twilight of Christianity*, (Richard R. Smith Inc., New York, 1931).

Barnett, J., "Adapting to Climate Change in Pacific Island Countries: The Problem of Uncertainty", *World Development*, vol. 29, 2001, pp. 977-993.

Barnett, R., *The Idea of Higher Education*, (SRHE and Open University Press, Buckingham, 1990).

Barry, B., *The Liberal Theory of Justice*, (Clarendon Press, Oxford, 1973).

Barton, S., "Problems in Preaching a Demythologized Christ I", *Interchange*, vol. 24, 1978, pp. 223-225.

Batra, R., *The Myth of Free Trade: A Plan for America's Economic Revival*, (Charles Scribner's Sons, New York, 1993).

Battersby, S., "Deep Trouble", *New Scientist* April 15 2006 at http://global.factiva.com/ha/default.aspx.

Beckerman, W. and Pasek, J., *Justice, Posterity, and the Environment*, (Oxford University Press, Oxford, 2001).

Beder, S., *Environmental Principles and Policies: An Interdisciplinary Approach*, (University of New South Wales Press, Sydney, 2006).

Beitz, C.,*Political Theory and International Relations*, (Princeton University Press, Princeton, New Jersey, 1979).

Benderly, B. L.," A Hippocratic Oath" for Scientists?", *Science Careers Forum*, March 2 2007 at http://sciencecareers.sciencemag.org/ career_ development/previous_issues/articles/2007_03_02/caredit_a0700030.

Bengtsson, L., "Geo-Engineering To Confine Climate Change: Is It At All Feasible?" *Climatic Change*, vol. 77, 2006, pp. 229-234.

Benton, M. J., *When Life Nearly Died: The Greatest Mass Extinction of All Time*, (Thomas and Hudson, London, 2003).

Berry, W., *The Unsettling of America*, (Sierra Club Books, San Francisco, 1986).

Bethell, T., *The Politically Incorrect Guide to Science*, (Regnery Publishing, Washington DC, 2005).

Betts, K. and Gilding, M., "The Growth Lobby and Australia's Immigration Policy", *People and Place*, vol. 14, no. 4, 2006, pp. 40-52.

Bindshadler, R., "Hitting the Ice Sheets Where It Hurts," *Science*, vol.311, no 5768, March 24 2006, pp. 1720-1721, doi: 10.1126/science.1125226.

Birch, C. and Cobb Jr., J. B., *The Liberation of Life: From Cell to the Community*, (Environmental Ethics Books, Denton, 1990).

Birch, C., *On Purpose*, (University of New South Wales University Press Kensington, New South Wales, 1990).

Black, R., "'Major Melt' for Alpine Glaciers", BBC News 4 April 2006 at http://news.bbc.co.uk/1/hi/sci/tech/4874224.stm.

Black, R., "Global Warming Risk 'Much Higher'", BBC News, May 23 2006 at http://news.bbc.co.uk/2/hi/science/nature/5006970.stm.

Blau, J., *The Visible Poor: Homelessness in the United States*, (Oxford University Press, New York, 1993).

Bloom, A., *The Closing of the American Mind: How Higher Education has Failed Democracy and Impoverished the Souls of Today's Students*, (Simon and Schuster, New York, 1987).

Bonner, H. B.,*Christianity and Conduct,* (Watts and Co, London, 1919).

Bony, S., (*et al*), "How Well Do We Understand and Evaluate Climate Change Feedback Processes?" *Journal of Climate*, vol. 19, no.15, August 2006, pp. 3445-3482.

Botting, D., *Humboldt and the Cosmos*, (Joseph, London, 1973).

Boyes, R., "Germany to Bury its Coal Industry," *The Australian*, February 2007, p. 10.

Bradbrook, A., (*et al.*,eds.), *The Law of Energy for Sustainable Development*, (Cambridge University Press, New York, 2005).

Bradley, R. S. and Jones, P. D., ""Little Ice Age" Summer Temperature Variations: Their Nature and Relevance to Recent Global Warming Trends," *Holocene*, vol.3, 1993, pp. 367-376.

Bradley, R. S., (*et al*), "Climate in Medieval Time", *Science*, doi: 10.1126/science. 1090372, 2003, pp. 404-405.

Brandt, R. B., "Morality and Its Critics", *American Philosophical Quarterly*, vol. 26, no.2, April 1989, pp. 89-100.

Brewer, M. B., "The Social Self: On Being the Same and Different at the Same Time", *Personality and Social Psychology Bulletin*, vol.17, 1991, pp. 475-482.

Briggs, S., "Worse to Come in Water Crisis", *The Advertiser* (Adelaide), November 4 2006, p. 5.

Bristow, P., "The Roots of Our Ecological Crisis", *Answers in Genesis (TJ Archive)*, vol. 15, no.1, April 2001, pp. 76-79 at http://www.answersingenesis.org/tj/v15/il/ecology.asp.

Brook, A. T., "What is Conservation Psychology?"*Population and Environmental Psychology Bulletin*, vol. 27, no.2, 2001, pp.1-2.

Brook, A. T.,*Effects of Contingencies of Self-Worth on Self-Regulation of Behavior*, (Unpublished doctoral dissertation, University of Michigan, Ann Arbor, 2005).

Broome, J., *Counting the Cost of Global Warming*, (White Horse Press, Isle of Harris, UK, 1992).

Broome J.,*Ethics Out of Economics*, (Cambridge University Press, Cambridge, 1999).

Broome, J., *Weighing Lives*, (Oxford University Press, Oxford, 2004).

Broome J., "Should We Value Population?" *Journal of Political Philosophy*, vol. 13, no.4, 2005, pp. 339-413, at pp. 411-413.

Brown, C., *International Relations Theory: New Normative Approaches*, (Harvester-Wheatsheaf, Hemel Hempstead,1992).

Brown, D. A., "Environmental Ethics and Public Policy", *Environmental Ethics*, vol.26, 2004, pp. 111-112.

Brown, L. R., *Plan B 2.0 :Rescuing a Planet Under Stress and a Civilization in Trouble*. (W.W. Norton, New York, 2006).

Brown, R. F., *The Birth of the Messiah*, (Doubleday, Gordon City, New York, 1977).

Bruce-Chwatt, L.J. and de Zulueta, J., *The Rise and Fall of Malaria in Europe:*

*A Historico-Epidemiological Study*, (Oxford University Press, Oxford, 1980).

Buchanan, M., *Ubiquity: Why Catastrophes Happen*, (Three Rivers Press, New York, 2002).

Budge, W. E. A., *Gods of the Egyptians*, (Dover Publications, New York, 1969).

Bultmann, R., *Jesus Christ and Mythology*, (SCM Press, London, 1960).

Bultmann, R., "New Testament and Mythology", in H.W. Bartsch (ed.), *Kerygma and Myth*, (Harper Torchbooks, New York, 1961).

Burns, W.C.G., "Potential Causes of Action for Climate Change Damages in International Fora: The Law of the Sea Convention", *McGill International Journal of Sustainable Development Law and Policy*, vol. 2, no. 1, March 2006, pp. 27-51.

Busby, C., "Poisoning in the Name of Progress", *The Ecologist*, vol. 29, no.7, 1999, pp. 395-401.

Byatt (*et al.*), I., "The Stern Review: A Dual Critique, Part II: Economic Aspects", *World Economics*, vol. 7, no. 4, October-December 2006, pp. 199-229.

Calderia, K. and Wickett, M. E., "Anthropogenic Carbon and ocean pH", *Nature*, vol.425, September 25 2003, p.365.

Caldicott, H., *Nuclear Power is Not the Answer to Global Warming or Anything Else*, (New Press, New York, 2006).

Callicott, J. Baird, "Conceptual Resources for Environmental Ethics in Asian Traditions of Thought: A Propaedeutic", *Philosophy East and West*, vol. 37, 1987, pp. 115-131.

Callicott, J. Baird, "Genesis Revisited: Murian Musings on the Lynn White Jr Debate", *Environmental History Review*, vol. 14, 1990, pp. 65-92.

Campbell, C., *The Romantic Ethic and the Spirit of Modern Consumerism*, (Basil Blackwell, Oxford, 1987).

Campbell, K., *Abstract Particulars*, (Blackwell, Oxford, 1990).

Campbell, T., (*et al*), *Protecting Human Rights: Instruments and Institutions*, (Oxford University Press, Oxford, 2003).

Camus, A., *The Myth of Sisyphus*, (Penguin, London, 2000).

Carens, J., "Aliens and Citizens: The Case for Open Borders", in W. Kymlicka (ed.), *The Rights of Minority Cultures*, (Oxford University Press, Oxford, 1995), pp. 331-349.

Carpenter, E., *Pagan and Christian Creed: Their Origin and Meaning*, (Kessinger Publishing, Kila MT, 1997).

Carson, R., *Silent Spring*, (Penguin Book, London, 1962).

Carter, R. M., (*et al.*), "The Stern Review: A Dual Critique, Part I: The Science", *World Economics*, vol. 7, no. 4, October-December 2006, pp. 167-198.

Catton, W. R., *Overshoot: The Ecological Basis of Revolutionary Change*, (University of Illinois Press, Urbana, 1982).

Center for Health and the Global Environment, Harvard Medical School, *Climate Change Futures: Health, Ecological and Economic Diemensions*, (Center for Health and Global Environment, Harvard Medical School, 2005).

Chaitin, G. J., *Information, Randomness and Incompleteness: Papers on Algorithmic Information Theory*, (World Scientific, Singapore, 1987).

Chan, J. C. L., "Comment on "Changes in Tropical Cyclone Number, Duration, and Intensity in a Warming Environment"," *Science*, vol. 311, no.5768, March 24 2006, doi:10. 1126/science. 1121522.

Chandler, R., *Understanding the New Age*, (Word, Waco,1988).

Chapman, J., *System Failure*, 2nd edition, (Demos. London, 2004).

Chapple, C. K. and Tucker, M. E. (eds.), *Hinduism and Ecology: The Intersection of Earth, Sky and Water*, (Harvard Divinity School, Cambridge, 2000).

Charlson, R. J., (*et al*) "Oceanic Phytoplankton, Atmospheric Sulphur, Cloud Albedo and Climate", *Nature*, vol.326, 1987, pp. 655-661.

Cheney, L. V., *Telling the Truth: Why Our Culture and Our Country Have Stopped Making Sense - And What We Can Do About It*, (Simon and Schuster, New York, 1995).

Christie-Murray, D., *A History of Heresy*, (Oxford University Press, Oxford and New York, 1989).

Church, J. A. and Gregory, J. M., "Changes in Sea Level", in J.T. Houghton (*et al* eds.), *Climate Change 2001: The Scientific Basis*, (Cambridge University Press, Cambridge, 2001), pp. 639-693.

Clark, S., "Saved by the Sun", *New Scientist*, September 16 2006, pp. 32-36.

Clayton, S. and Opotow, S., (eds.), *Identity and the Natural Environment*, (MIT Press, Cambridge MA, 2003).

Clayton , S. and Brook, A., "Can Psychology Help Save the World? A Model for Conservation Psychology", *Analyses of Social Issues and Public Policy*, vol.5, 2005, pp. 87-102.

Coady, T., (ed.), *Why Universities Matter*, (Allen and Unwin, Sydney 2000).

Coase R.H., "The Problem of Social Cost", *Journal of Law and Economics*, vol. 3, 1960, pp. 1-44.

Coase, R. H., *The Firm, the Market and the Law*, (University of Chicago Press, Chicago and London, 1988).

Cobb Jr., J. B., *Is it Too Late? A Theology of Ecology*, (Environmental Ethics Books, Denton, 1995).

Cohen, A. and Dascal, M., (eds.), *The Institution of Philosophy: A Discipline in Crisis?* (Open Court, La Salle, 1989).

Cohen, P., "Your Mission Is…", *New Scientist*, July 3 1999, pp. 38-41.

Colwell, R.R., "Global Warming and Infectious Diseases, *Science,* vol.274, 1996, pp. 2025-2031.

Colwell, R. R. and Patz, J. A., *Climate, Infectious Disease and Health*, (American Academy of Microbiology, Washington DC, 1998).

Comiso, J. C., "Abrupt Decline in Arctic Winter Sea Ice Cover", *Geophysical Research Letters*, vol. 33, 2006, L18504, doi:10.1029/2006GLO27341.

Consultative Group on International Agricultural Research. Inter-Center Working Group on Climate Change, *The Challenge of Climate Change: Research to Overcome its Impact on Food, Scarcity, Poverty, and National Resource Degradation in the Developing World*, (Consultative Group on International Agricultural Research, Inter-Center Working Group on Climate Change ,2002), at http://www.cgiar.org/pdf/climatechange.pdf.

Corsel, P. S., (*et al*), "Cost of Illness in the 1993 Waterborne Cryptosporidium Outbreak, Milwaukee, Wisconsin", *Emerging Infectious Diseases*, vol.9, no.4, April 2003, pp.426-431.

Courtillo, V., *Evolutionary Catastrophes: The Science of Mass Extinction*, (Cambridge University Press, Cambridge, 1999).

Cowan, T. and Parfit, D., "Against the Social Discount Rate", in P.Laslett and J. Fishkin (eds.), *Justice Between Age Groups and Generations*, (Yale University Press, New Haven, 1992), pp.144-161.

Coward, H. and Hurka, T., (eds.), *Ethics and Climate Change: The Greenhouse Effect*, (Wilfred Laurier University Press, Waterloo, Ontario, 1993).

Cox, H., *The Secular City*, (Pelican, London, 1968).

Cox, P. M., (*et al*), "Conditions for Sink-to Source Transitions and Runaway Feedbacks from Land Carbon Cycle", in H.J. Schellnhuber (et. al. eds.), *Avoiding Dangerous Climate Change*, (Cambridge University Press, Cambridge, 2006), pp. 155-161.

Craven, P., "Corporate Criticism Finds Popular Voice", *The Australian*, September 30 1998.

Crick F., *Of Molecules and Men*, (University of Washington Press, Seattle, 1966).

Cropper, M. L. and Oates, M. L., "Environmental Economics: A Survey", *Journal of Economic Literature*, vol. 30, 1992, pp. 674-740.

Crowley, F., *Degrees Galore: Australia's Academic Teller Machines*, (Published by Frank Crowley, 48 Clifton Drive, Port Macquarie, NSW 2444, Australia, 1998).

Crowley, T. J. and Lowery, T. S., "How Warm was the Medieval Warm Period? A Comment on 'Man-made Versus Natural Climate Change'", *Ambio*, vol.29, 2000, pp. 51-54.

Crystall, B., "The Big Clean-Up", *New Scientist*, September 3 2005, pp. 30-31.

Cullity, G., "Asking Too Much", *Monist*, vol. 86, 2003, pp. 402-418.
Cullity, G., *The Moral Demands of Affluence*, (Clarendon Press, Oxford, 2004).
Curley, E. M., *Descartes Against the Skeptics*, (Basil Blackwell, Oxford, 1978).
Curran, G., "Environment, Equality and Globalisation", in K. Horton and H. Patapan (eds.), *Globalisation and Equality*, (Routledge, London and New York, 2004), pp. 146-163.
Dahl, R., "A Changing Climate of Litigation", *Environmental Health Perspectives*, vol. 115, no.4, April 2007, pp. A 204-A 207.
Daly, H., *Steady-State Economics*, 2nd edition, (Earthscan, London, 1992).
Dancy, J., *Ethics Without Principles*, (Clarendon Press, Oxford, 2004).
Dansgaard, W., (*et al*), "North Atlantic Climate Oscilliations Revealed by Deep Greenland Ice Core", in F.E.Hansen and T.Takahashi (eds.), *Climate Processes and Climate Sensitivity*,(American Geophysical Union, Washington D.C., 1984), Geophysical Monograph No. 29, pp.288-298.
Darnton, A., "Strategic Thinking: Impact of Sustainable Development on Public Behaviour", (May 2004), at http://www.comminit.com/strategicthinking/st2006/thinking-1693.html..
Darnton, A., (*et al.*), *Promoting Pro-Environmental Behaviour: Existing Evidence to Inform Better Policy Making*, (October 2006), at http://www.defra.gov.uk/science/project_data/DocumentLibrary/SD14002 / SD14002_3712_FRP.pdf.
Daszak, P., (*et al.*), "Emerging Infectious Diseases of Wildlife - Threats to Biodiversity and Human Health", *Science*, vol. 287, 2000, pp. 443-449.
Davis, B., "A Greener Shade of Black", *New Scientist*, September 3 2005, pp. 38-40.
Dawkins, R., *The God Delusion*, (Bantam, London, 2006).
Day, J. M., (ed.), *Plato's Meno in Focus*, (Routledge, London, 1994).
Dayton, L., "Kyoto 'Too Late' to Stop Warming," *The Australian*, January 5 2006, p.3.
Dayton, L., "Record Ozone Hole Despite Cuts in CFCs", *The Weekend Australian,* October 21-22 2006, p.5.
Dayton, L., "Refuge from the Ice Next Time," *The Australian,* November 8 2006, p.24.
De Botton, A., *Status Anxiety*, (Pantheon Books, New York 2004).
De Freitas, C. R., "Are Concentrations of Carbon Dioxide in the Atmosphere Really Dangerous?" *Bulletin of Canadian Petroleum Geology*, vol.50, no.2, June 2002, pp.297-327.
De Shalit, A., *Why Posterity Matters: Environmental Policies and Future Generations*, (Routledge, London and New York, 1995).
Deffeyes, K. S., *Hubberts Peak: The Impending World Oil Shortage*, (Princeton

University Press, Princeton, 2001).
Defra (Department for Environment, Food and Rural Affairs), *Securing the Future - Developing UK Sustainable Development Strategy*, (The Stationery Office, London, 2005).
Dent, A., *Ecology and Faith: The Writings of Pope John Paul II*, (Arthur James, Berkhamsted, 1997).
Derrida, J., *Margins of Philosophy,* (University of Chicago Press, Chicago, 1982).
Deutch, J. M. and Moniz, G. J., "The Nuclear Option", *Scientific American*, September 2006, pp. 57-59.
Deutsch, M. and Gerard, H. G., "A Study of Normative and Informational Social Influence Upon Individual Judgment", *Journal of Abnormal Social Psychology*, vol.51, 1955, pp. 629-636.
Diamond, J., *Collapse: How Societies Choose to Fail or Survive*, (Allan Lane, London, 2005).
DiClemente, C. C., "Self Efficacy and Smoking Cessation Maintenance: A Preliminary Report", *Cognitive Therapy and Research*, vol.5, 1981, pp.175-187.
DiClemente, C. C., "Self Efficacy and the Addictive Behaviors", *Journal of Social and Clinical Psychology*, vol. 4, 1986, pp. 302-315.
DiClemente, C. C., (*et al.*), "The Process of Smoking Cessation: An Analysis of Precontemplation, Contemplation and Contemplation/Action", *Journal of Consulting and Clinical Psychology*, vol. 59, 1991, pp. 295-304.
Dierterlen, P., *Poverty: A Philosophical Approach*, (Rodopi Philosophical Studies/Amsterdam, New York, 2005).
Dimbleby, J., "The Coming War," *The Observer*, October 31 2004 at http://observer. guardian.co.uk/comment/story/0,6903,1340066,00.html.
Dittmar, H., *The Social Psychology of Material Possessions: To Have Is To Be*, (Harvester Wheatsheaf, Hemel, Hempstead, 1992).
Doana, T. W., *Bible Myths and Their Parallels in Other Religions,* (Kessinger Publishing, Kila MT, 1997).
Doherty, E., *The Jesus Puzzle: Did Christianity Begin with a Mythical Christ?* (Canadian Humanist Publications, Ottawa, 1999).
Doney, S. C.,"The Dangers of Ocean Acidification", *Scientific American*, vol.294, no.3, March 2006, pp. 38-45.
Douglas, M.,*Natural Symbols: Explorations in Cosmology*, (Barrie and Rockliff, London,1970).
Douglas, M. and Isherwood, B., *The World of Goods: Towards an Anthropology of Consumption*, (Routledge, London and New York, 1996).
Douthwaite, R., *The Growth Illusion,* (Resurgence, Bideford, Devon, 1992).
Downie, R. and MacNaughton, J., "Should Medical Students Read Plato?"

*Medical Journal of Australia*, vol. 170, 1999, pp. 125-127.
Driessen, P., *Eco-Imperialism: Green Power, Black Death*, (Merril Press, Bellevue, WA, 2003).
DTI, *Changing Patterns - UK Government Framework for Sustainable Consumption and Production*, (Department of Trade and Industry, London, 2003).
Duncan, R. C., "Big Jump in Ultimate Recovery Peak Would Ease, Not Reverse, Postproduction Decline" *Oil and Gas Journal*, July 19, 2004, pp.18-21.
Duncan, R. C., "The Olduvai Theory: Energy, Population, and Industrial Civilization", *The Social Contract*, Winter 2005-2006, pp.134-144.
Dupont, A., *East Asia Imperilled: Transnational Challenges to Security*, (Cambridge University Press, Cambridge, 2001).
Dupont, A. and Pearman, G., *Heating Up the Planet: Climate Change and Security*, (Lowy Institute Paper no 12, 2006) at http://www.lowyinstitute.org.
Dworkin, R., *Taking Rights Seriously*, (Duckworth, London, 1977).
Ebi, K.L. (*et al.*), *Climate Variability and Change and their Health Effects on Small Island States*, (Report on Regional Workshops and Conference Convened by WHO, WMO and UNEP).
Eckberg D. L. and Blocker, T. J., "Varieties of Religious Involvement and Environmental Concern", *Journal for the Scientific Study of Religion*, vol.28, 1989, pp. 509-517.
Economy, E. C., *The River Runs Black: The Environmental Challenge of China's Future*, (Cornell University Press, Ithaca and London, 2004).
Edwards, R., "Who Will Pay for a Nuclear Future?" *New Scientist*, June 10 2006, p.8.
Edwards, T., *Contradictions of Consumption: Concepts, Practices and Politics in Consumer Society*, (Open University Press, Milton Keynes, 2000).
Edwards, V., "Big Oil Says Reserves Are Plentiful", *The Australian*, September 12 2006, p.4.
Ekins, P., (ed.), *The Living Economy: A New Economics in the Making*, (Routledge and Kegan Paul, London, 1986).
Ekins, P., "Trade and Self-Reliance", *The Ecologist*, vol.19. September/October 1989, pp. 186-190.
Ekström, G., (*et al*), "Seasonality and Increasing Frequency of Greenland Glacier Earthquakes", *Science*, vol.311, March 24 2006, pp. 1756-1758, doi: 10.1126/science.1122112.
Elfstrom, G., *Ethics for a Shrinking World*, (MacMillan, London, 1990).
Elgin, D., *Voluntary Simplicity*, (William Morrow, New York, 1981).
Elliot, H., *Ethics for a Finite World: An Essay Concerning a Sustainable Future*,

(Fulcrum Publishing, Golden, Colorado, 2005).

Ellis, A., "Utilitarianism and International Ethics", in T.Nardin and D.R.Mapels (eds.), *Traditions of International Ethics*, (Cambridge University Press, Cambridge, 1992), pp. 158-179.

Elsner, J. B., "Evidence in Support of the Climate Change - Atlantic Hurricane Hypothesis", *Geophysical Research Letters*, vol.33, 2006, L167705, DOI: 10. 1029/2006GLO26869.

Emanuel, K., "Increasing Destructiveness of Tropical Cyclones over the Past 30 Years" *Nature*, vol.436, 2005, pp. 686-688.

Emanuel, K., "Emanuel Replies",*Nature*, vol.438, December 22/29, 2005, pp. E13, doi: 10. 1038/nature04427.

Enserink, M., "During a Hot Summer, Bluetongue Virus Invades Northern Europe", *Science*, vol. 313, 2006, pp. 1218-1219.

Epstein P. R., "Climate, Ecology, and Human Health", *Consequences*, vol. 3, 1997, pp.2-19.

Epstein, P.R., "Climate and Health", *Science* , vol.285, 1999, pp. 347-348.

Epstein, P.R., "Is Global Warming Harmful to Health?" *Scientific American*, vol. 283, no.2, 2000, pp. 50-57.

Erwin, D. H., *Extinction: How Life on Earth Nearly Ended 250 Million Years Ago*, (Princeton University Press, Princeton, 2006).

Essex C. and McKitrick, R., *Taken by Storm: The Troubled Science, Policy and Politics of Global Warming*, (Key Porter Books Toronto, 2002).

Etzioni, A., "Voluntary Simplicity: Characterization, Select Psychological Implications and Societal Consequences", in T. Jackson (ed.), *The Earthscan Reader in Sustainable Consumption*, (Earthscan, London, 2006), pp. 159-177.

Fanning, J. C., (*et al*), "Converting a Stomach to a Uterus: The Microscopic Structure of the Gastric Brooding Frog *Rheobatrachus Silus* ", *Gastroenterology*, vol. 82, 1982, pp. 62-70.

Featherstone, M., *Consumer Culture and Postmodernism*, (Sage, London,1991).

Feinberg, J., *Rights, Justice and the Bounds of Liberty*, (Princeton University Press, Princeton, New Jersey, 1980).

Fenton L. K.(et. al.), "Global Warming and Climate Forcing by Recent Albedo Changes on Mars", *Nature*, vol. 446, April 5, 2007, doi: 10.1038/nature05718.

Ferre, F., *Basic Modern Philosophy of Religion*, (George Allen and Unwin, London, 1968).

Feyerabend, P. K., *Against Method*, (New Left Books, London, 1975).

Feyerabend, P. K., *Science in a Free Society*, (New Left Books, London, 1978).

Fideler, D., *Jesus Christ, Sun of God: Ancient Cosmology and Early Christian*

*Symbolism*, (Quest Books, Wheaton IL, 1993).

Fincher, R. and Saunders, P., *Creating Unequal Futures? Rethinking Poverty, Inequality and Disadvantage*, (Allen and Unwin, Crows Nest NSW, 2001).

Fink, D. B., "Judaism and Ecology: A Theology of Creation", at http://environment.harvard.edu/religion/religion/judaism/index.html.

Finkelstein, I. and Silberman, N. A., *The Bible Unearthed: Archaeology's New Vision of Ancient Israel and the Origin of its Sacred Texts*, (Free Press, New York, 2001).

Fiorenza, E. S., *But She Said: Feminist Practices of Biblical Interpretation*,(Beacon Press, Boston, 1992).

Fisher, B. S., (*et al.*), *Technological Development and Economic Growth*. (Australian Bureau of Agricultural and Resource Economics, Canberra, 2006).

Flaherty, T., "Pilchards: Will We Ever Know?" *South Australian Regional Ripples*, vol. 6 no. 1, Autumn, 1999.

Flavin, C. and Gardner, G., "China, India, and the New World Order", in L.Starke (ed.), *State of the World 2006*, (Earthscan, London, 2006), pp. 3-23.

Fleming, D., "Energy and the Common Purpose: Descending the Energy Staircase with Tradeable Energy Quotas (TEQs)", at http://www.teqs.net/book/teqs.pdf.

Fletcher, G. P., *Loyalty: An Essay on the Morality of Relationships*, (Oxford University Press, Oxford, 1993).

Flynn, T., "Matthew vs. Luke", *Free Inquiry*, vol. 25, no.1, 2004, pp.34-35.

Folkestad T., (et al), "Evidence and Implications of Dangerous Climate Change in the Arctic", in H.J. Schellnhuber (*et al* eds.), *Avoiding Dangerous Climate Change*, (Cambridge University Press, Cambridge, 2006), pp. 215-218.

Folz, R., "Islam and Ecology Bibliography", at http://environment.harvard.edu/ eligion/religion/islam/bibliography.html.

Ford, D. F., *The Modern Theologians: A Introduction to Christian Theology in the Twentieth Century*, (Volume 1), (Basil Blackwell, Oxford, 1989).

Fox, M., *The Coming of the Cosmic Christ*, (Collins Dove, Melbourne, 1990).

Fox, W., *Towards a Transpersonal Ecology*, (State University of New York Press, New York, 1995).

Francis, L. P., "Global Systemic Problems and Interconnected Duties", *Environmental Ethics*, vol. 25, 2003, pp. 115-128.

Franklin, J., "The Sydney Philosophy Disturbance", *Quadrant*, April, 1999, pp. 16-21.

Freestone, D. and Streck, C., (eds.), *Legal Aspects of Implementing the Kyoto*

*Protocol Mechanisms: Making Kyoto Work*, (Oxford University Press, Oxford, 2005).

Frege, G., *Conceptual Notation and Related Articles*, translated and edited by T.W. Bynum, (Clarendon Press, Oxford, 1972).

Frei, H., *The Identity of Jesus Christ*, (Fortress, Philadelphia, 1975).

Frek T. and Gandy, P., *The Jesus Mysteries: Was the "Original Jesus" a Pagan God?* (Random House, New York, 1999).

Friedman, T. L., "The Power of Green". *International Herald Tribune*, April 5 2007.

Frior, J. and Jacobsen J. E., *The Crowded Greenhouse: Population, Climate Change, and Creating a Sustainable World*, (Yale University Press, New Haven, 2002).

Frye, N., *The Great Code: The Bible and Literature*, (Harcourt Brace Jovanovich Publishers, New York and London, 1981).

Fyfe, J C., "Southern Ocean Warming Due to Human Influence", *Geophysical Research Letters*, vol. 33, 2006, L19701, doi:10.1029/2006GLO27247.

Gaarder, J., *Sophie's World: A Novel About the History of Philosophy*, (Phoenix, London, 1996).

Gabriel, Y. and Lang, T., *The Unmanageable Consumer: Contemporary Consumption and Its Fragmentation*, (Sage, London, 1995).

Gadalla, M., *Historical Deception: The Untold Story of Ancient Egypt*, (Bastet Publishing, Eire, PA, 1996).

Gager, J. G., "The Gospels and Jesus: Some Doubts about Method", *The Journal of Religion*, vol.54, 1974, pp. 244-272.

Galtung, J., *Development, Environment and Technology: Towards a Technology for Self-Reliance*, (United Nations, New York, 1979).

Galtung, J., (*et al*), *Self-Reliance: A Strategy for Development*, (Bogle-L'Ouverture Publications Ltd, London, 1980).

Galtung, J., *Environment, Development and Military Activity*, (Norwegian Universities Press, Oslo, 1982).

Galtung, J., "Towards a New Economics: On the Theory and Practice of Self-Reliance", in P.Ekins (ed.), *The Living Economy*, (Routledge and Kegan Paul, London, 1986), pp. 97-109.

Gardiner, S. M., "The Real Tragedy of the Commons", *Philosophy and Public Affairs*, vol. 30, no.4, 2001, pp. 387-416.

Gardiner, S. M., "The Pure Intergenerational Problem", *Monist*, vol.86, 2003, pp. 481-500.

Gardiner, S. M., "Ethics and Global Climate Change", *Ethics*, vol.114, April 2004, pp. 555-600.

Gardiner, S. M., "The Global Warming Tragedy and the Dangerous Illusion of the Kyoto Protocol," *Ethics and International Affairs*, vol. 18, no.1, 2004, pp. 23-39.

Gardiner, S. M., "A Core Precautionary Principle", *Journal of Political Philosophy*, vol.14. no.1, 2006, pp. 33-60.

Gardiner, S. M., "A Perfect Moral Storm: Climate Change, Intergenerational Ethics and the Problem of Moral Corruption", *Environmental Values*, vol.15, 2006, pp. 397-413.

Gardiner, S. M., *Why Do Future Generations Need Protection?* E.D.F-Ecole Polytechnic, July 2006.

Gardner, G. and Stern, P., *Environmental Problems and Human Behavior*, 2nd edition, (Pearson, Boston MA, 2002).

Gaukroger, S., *Descartes: An Intellectual Biography*, (Clarendon Press, Oxford, 1995).

Gauthier, D., *Morals by Agreement*, (Clarendon Press, Oxford, 1986).

Gearty, C., *Can Human Rights Survive?* (Cambridge University Press, Cambridge, 2006).

Geller, E., "Evaluating Energy Conservation Programs: Is Verbal Report Enough?" *Journal of Consumer Research*, vol.8, 1981, pp. 331-335.

Gellner, E., "Reflections on Philosophy, Especially in America", in *The Devil in Contemporary Philosophy*, (Routledge, London, 1974), pp.37-38.

Georgescu-Roegen, N., *The Entropy Law and the Economic Process*, (Harvard University Press, Cambridge MA, 1971).

Gerlagh, R. and Papyrakis, E., "Are the Economic Costs of (Non-) Stabilizing the Atmosphere Prohibitive? A Comment", *Ecological Economics*, vol. 46, 2003, pp. 325-327.

Gerztenfeld, M., *Judaism, Environmentalism and the Environment: Mapping and Analysis*, (Jerusalem Institute for Israel Studies, Jerusalem,1998).

Gibbins, J., (*et al.*), "Scope for future CO2 Emission Reductions from Electricity Generation through the Deployment of Carbon Capture and Storage Technologies", in H.J.Schellnhuber (*et al.* eds.), *Avoiding Dangerous Climate Change*, (Cambridge University Press, Cambridge, 2006), pp. 379-383.

Gibbs, W. W., "Plan B for Energy" *Scientific American*, September 2006, pp. 78-87.

Giles, J., "US Posts Sensitive Climate Report for Public Comments", *Nature*, vol. 441,May 4 2006, pp.6-7, doi:10:1038/441006a.

Gini, A., *Why It's Hard to be Good*, (Routledge, New York and London, 2006).

Gladwell, M., *The Tipping Point: How Little Things Can Make a Big Difference*, (Back Bay Books, Los Angeles, 2002).

Glasby, G. P., "Abiogenic Origin of Hydrocarbons: An Historical Overview", *Resource Geology*, vol.56, no.1, 2006, pp.83-96.

Glen, W., *The Mass-Extinction Debates: How Science Works in a Crisis*, (Standford University Press, Stanford, 1994).

Gold, T., *The Deep Hot Biosphere*,(Copernicus Books, 1999).

Goodin, R. E., "What is So Special About Our Fellow Countrymen?" *Ethics*, vol. 98, 1988, pp. 663-686.

Goodman, R., "Taoism and Ecology", *Environmental Ethics*, vol.2, 1980, pp. 73-80.

Goodstein, D., *Out of Gas: The End of the Age of Oil*, (W.W. Norton, New York and London, 2004)

Gore, A., *Earth in the Balance: Ecology and the Human Spirit*, (Houghton Mifflin, Boston, 1992).

Gore, A., *An Inconvenient Truth: The Planetary Emergency of Global Warming and What We Can Do About It*, (Rodale, Emmaus PA, 2006).

Gore, A., *The Assault on Reason*, (Bloomsbury, London, 2007).

Gottlieb, R. S., *A Greener Faith: Religious Environmentalism and Our Planet's Future*, (Oxford University Press, Oxford, 2006).

Gowdy, J. M. and Olsen P. G., "Further Problems with Neoclassical Environmental Economics", *Environmental Ethics*, vol. 16, 1994, pp. 161-171.

Grabl, H., (*et al.* eds.), *World in Transition:Towards Sustainable Energy Systems*, (Earthscan, London, 2004).

Grant, L., *The Collapsing Bubble: Growth and Fossil Fuel*, (Seven Locks Press, Santa Ana CA, 2005).

Graves, K., *The World's Sixteen Crucified Saviours: Christianity before Christ*, (Kessinger Publishing, Kila MT, 1999).

Gray, J., *Liberalism*, (Open University Press, Milton Keynes, 1986).

Gray, J., *Enlightenment's Wake: Politics and Culture at the Close of the Modern Age*, (Routledge, London and New York, 1995).

Gray, J., *Straw Dogs: Thoughts on Humans and Other Animals*, (Granta Books, London, 2002).

Grayling, A. C., *Among the Dead Cities: The History and Moral Legacy of the WWII Bombing of Civilians in Germany and Japan*, (Walker and Company, New York, 2006).

Greeley, A., "Religion and Attitudes toward the Environment", *Journal for the Scientific Study of Religion*, vol.32, no. 1, March 1993, pp. 19-28.

Greenly, E., *The Historical Reality of Jesus*, (Watts and Co, London, 1927).

Greenwood, B., "Between Hope and a Hard Place", *Nature*, vol. 430, August 19

2004, pp. 926-927.
Gregory, J. M., (*et al*), "Climatology: Threatened Loss of the Greenland Ice-Sheet", *Nature*, vol. 428, April 8 2004, doi:10.1038/428616a.
Griffin, D. R., (ed.), *Spirituality and Society*, (State University of New York Press, Albany, 1988).
Griffin, D. R., (ed.), *The Reenchantment of Science*, (State University of New York Press, New York, 1988).
Griffin, D. R., (ed.), *Sacred Interconnections: Postmodern Spirituality, Political Economy and Art*, (State University of New York Press, Albany, 1990).
Grisez, G. G., *Beyond the New Theism: A Philosophy of Religion*, (University of Notre Dame Press, Notre Dame, 1975).
Groothuis, D., *Unmasking the New Age*, (InteVarsity, Downers Grove, 1986).
Grossman, D. A., "Warming up to a Not-So-Radical Idea: Tort-Based Climate Change Litigation", *Columbia Journal of Environmental Law*, vol. 28, 2003, pp.1-61.
Guelke, J. K., "Looking for Jesus in Christian Environmental Ethics", *Environmental Ethics*, vol.26, 2004, pp. 115-134.
Gupta, J., "The Climate Change Regime: De Jure and De Facto Commitment to Sustainable Development", in J.Paavola and I.Lowe (eds.), *Environmental Values in a Globalising World: Nature, Justice and Governance*, (Routledge, London and New York, 2005), pp. 159-178.
Haack, S., "Recent Obituaries of Epistemology", *American Philosophical Quarterly*, vol. 27, no.3, July 1990, pp. 199-212.
Hainzl S., (*et al*), "Evidence for Rainfall-Triggered Earthquake Activity", *Geophysical Research Letters*, vol.33, 2006, L19303. doi:10.1029/2006/GLO27642.
Hall, D. C. and Behl, R. J., "Integrating Economic Analysis and the Science of Climate Instability", *Ecological Economics*, vol. 57, 2006, pp. 442-465.
Hallam, A. and Wignall, P. B., *Mass Extinctions and Their Aftermath*, (Oxford University Press, Oxford, 1997).
Hallam, T., *Catastrophes and Lesser Calamities: The Causes of Mass Extinctions*, (Oxford University Press, Oxford, 2004).
Haller, S. F., *Apocalypse Soon? Wagering on Warnings of Global Catastrophe*, (McGill-Queen's University Press, Montreal and Kingston, 2002).
Hallin, P. O., "Environmental Concern and Environmental Behavior in Foley, a Small Town in Minnesota", *Environment and Behavior*, vol. 27, no. 4, 1995, pp. 558-578.
Halpern, D. and Bates, C., (with G. Beales and A.Heathfield), Prime Minister's Strategy Unit, Cabinet Office, *Personal Responsibility and Changing Behaviour: The State of Knowledge and Its Implications for Public Policy*,

(February, 2004) at http://www.number10.gov.uk/files/pdf/pr.pdf.

Hamilton, C. and Denniss, R., *Affluenza: When Too Much is Never Enough*, (Allen and Unwin, Crows Nest, New South Wales, 2005).

Hamilton, C. and Maddison, S., (eds.), *Silencing Dissent*, (Allen and Unwin, Crows Nest New South Wales, 2007).

Hamilton, C., *Scorcher: The Dirty Politics of Climate Change*, (Black Inc, Melbourne, 2007).

Hampson, F. O. and Reppy, J., (eds.), *Earthly Goods: Environmental Change and Social Justice*, (Cornel University Press, Ithaca and London, 1996).

Hampton, T., "Researchers Study Health Effects of Environmental Change", *Journal of the American Medical Association*, vol. 296, no.8, 2006, pp. 913-920.

Hanly, K., "The Problem of Social Cost: Coase's Economics versus Ethics", *Journal of Applied Philosophy*, vol. 9, 1992, pp. 77-83.

Hansen, J., "Is There Still Time to Avoid Disastrous Human-Made Climate Change?" Keynote Address to the Third Annual Climate Change Research Conference, Sacramento, California, September 13 2006 at http://www.climatechange.ca.gov/events/2006_conference/presentations/2006-09-13/2006-09-13_HANSEN. PDF.

Hansen, J., (*et al*), "Global Temperature Change", *Proceedings of the National Academy of Sciences,* vol. 103, no.39, September 26 2006, pp.14288-14293, published on-line www.pnas.org/cgi/doi/10.1073/pnas.0606291103.

Hardin, G., "The Tragedy of the Commons", *Science*, vol.162, 1968, pp.1243-1248.

Hardin, G., *Living Within Limits:Ecology, Economics, and Population Taboos*, (Oxford University Press, New York, 1993).

Hardin, G., *The Immigration Dilemma: Avoiding the Tragedy of the Commons*, (Federation for American Immigration Reform, Washington DC, 1995).

Hare, B., "Relationship Between Increases in Global Mean Temperature and Impacts on Ecosystems, Food Production, Water and Socio-Economic Systems", in H.J. Schellnhuber (*et al* eds.), *Avoiding Dangerous Climate Change*, (Cambridge University Press, Cambridge, 2006), pp. 177-185.

Hargrave, E., "What's Wrong? Who is to Blame?" *Environmental Ethics*, vol.25, 2003, pp. 3-4.

Harpur, T., The *Pagan Christ: Is Blind Faith Killing Christianity?* (Allen and Unwin, Sydney, 2004).

Harris, J. F., *Against Relativism: A Philosophical Defense of Method*, (Open Court, La Salle, 1993).

Harris, M. J., *Three Crucial Questions About Jesus*, (OM Publishing, Carlisle,

1994).
Harris, P. G., *International Equity and Global Environmental Politics: Power and Principles in U.S. Foreign Policy*, (Ashgate, Aldershot, 2001).
Harris, P. G., "Fairness, Responsibility, and Climate Change", *Ethics and International Affairs*, vol. 17, no.1, 2003, pp. 149-156.
Harris, S., *The End of Faith: Religion, Terror and the Future of Reason*, (W.W. Norton, New York, 2005).
Harte, J. (et al), "Global Warming and Soil Microclimate: Results from a Meadow-Warming Experiment", *Ecological Applications*, vol. 5, no.1, 1995, pp. 132-150.
Harte, J. (et. al.), "Biodiversity Conservation: Climate Change and Extinction Risk," *Nature*, vol. 430, July 1 2004, doi: 10. 1038/nature02718.
Hartmann, D. L., (*et al*), "Can Ozone Depletion and Global Warming Interact to Produce Rapid Climate Change?" *Proceedings of the National Academy of Sciences,* vol.97, no.4, February 15 2000, pp. 1412-1417.
Hartmann, T., *The Last Hours of Ancient Sunlight: The Fate of the World and What We Can Do Before It's Too Late*, (Three Rivers Press, New York, 2004).
Harvey, N., (*et al.*), "Geography and Environmental Studies in Australia: Symbiosis for Survival in the 21st Century", *Australian Geographical Studies*, vol. 40, no.1, March 2002, pp. 21-32.
Harvey, V. A., "New Testament Scholarship and Christian Belief", in R.J.Hoffman and Larue, G. A., (eds.), *Jesus in History and Myth*, (Prometheus Books, New York, 1986), pp.193-200.
Harwell, C.D., (*et al.*), "Diseases in the Ocean: Emerging Pathogens, Climate Links, and Anthropogenic Factors", *Science*, vol. 285, 1999, pp. 1505-1510.
Hassol, S. J. and Corell, R. W., "Arctic Climate Impact Assessment ",in H.J. Schellnhuber (*et al* eds.), *Avoiding Dangerous Climate Change*, (Cambridge University Press, Cambridge, 2006), pp. 205-213.
Hawken, P., (et. al.), *Natural Capitalism: Creating the Next Industrial Revolution*, (Little, Brown and Company, Boston, 1999).
Hawkins, D. G., (*et al.*), "What to Do About Coal", *Scientific American*, September, 2006, pp. 44-51.
Hay, J., (*et al.*), *Climate Variability and Change and Sea-level Rise in the Pacific Islands Region*, (South Pacific Regional Environment Programme, Samoa, 2003).
Hay, S. I., (*et al.*), "Urbanization, Malaria Transmission and Disease Burden in Africa", *Nature Reviews Microbiology*, vol. 3 no.1, 2005, pp. 81-90.
Hay, S.I., (*et al.*), "Climate Variability and Malaria Epidemics in the Highlands of

East Africa", *Trends in Parasitology*, vol. 21, no.2, 2005, pp. 52-53.

Hayden, H. C., *The Solar Fraud* (2nd edition), (Vales Lake, Pueblo West, 2004).

Hegerl, G. C., (*et al*), "Climate Sensitivity Constrained by Temperature Reconstructions over the Past Seven Centuries, *Nature*, vol. 440, April 20 2006, pp. 1029-1032, doi: 10.1038/nature04679.

Heidegger, M., *Being and Time*, (Basil Blackwell, Oxford, 1967).

Henderson, C., "Paradise Lost", *New Scientist*, August 5 2001, pp. 28-33.

Hennessy, K., (*et al*), *Climate Change Impacts on Fire-Weather in South East Australia*,(CSIRO, Canberra, 2006).

Hernàndez, O. and Monroe, M. C., "Thinking About Behavior", in B.A. Day and M.C. Monroe (eds.), *Environmental Education and Communication for a Sustainable World: A Handbook for International Practitioners*, (Academy for Educational Development, Washington DC, 2000), pp. 7-15.

Hewitt, G. M., "Genetic and Evolutionary Impacts of Climate Change," in T.E.Lovejoy and L.Hannah (eds.), *Climate Change and Biodiversity*, (Yale University Press, New Haven and London, 2006), pp.176-192.

Higgins, G., *Anacalypsis: An Inquiry into the Origin of Languages, Nations and Religions* (1833), (Kessinger Publishing, Kila MT, 1997).

Hill, J., (*et al.*), "Environment, Economic, and Energetic Costs and Benefits of Biodiesel and Ethanol Biofuels", *Proceedings of the National Academy of Sciences*, vol. 103, no.30, 2006, pp.11206-11210.

Hillman, M. with Fawcett, T., *How We Can Save the Planet*, (Penguin Books, London, 2004).

Hillman, M., "Personal Carbon Allowances", *British Medical Journal*, vol. 332, 2006, pp. 1387-1388.

Hodge, A., "Householders Cut Greenhouse Gases", *The Australian*, June 2 2006, p. 9.

Hodge, A., "Climate Pact has its Critics Fuming", *The Australian*, July 28 2005, p. 6.

Hoffert, M. J., (*et al.*), "Advanced Technology Paths to Global Climate Stability: Energy for a Greenhouse Planet", *Science*, vol. 298, 2002, pp. 981-987.

Hoffman, A. A. and Parsons, P. A., *Extreme Environmental Change and Evolution*, (Cambridge University Press, Cambridge, 1997).

Holden, B., *Democracy and Global Warming*, (Continuum, London and New York, 2002).

Homer-Dixon, T., *The Upside of Down: Catastrophe, Creativity and the Renewal of Civilization*, (Island Press, Washington DC, 2006).

Hood, R., "Global Warming", in R.G. Frey and C.H. Wellman, (eds.), *A Companion to Applied Ethics: Blackwell Companions to Philosophy*,

(Blackwell Publishing, Oxford, 2003), pp. 674-684.
Hooper, R., "Something in the Air", *New Scientist*, January 21 2006, pp. 40-43.
Horner, C. C., *The Politically Incorrect Guide to Global Warming and Environmentalism*, (Regnery Publishing Inc, Washington DC, 2007).
Hornung, E., *Conceptions of God in Ancient Egypt*, (Cornell University Press, Ithaca, 1982).
Horowitz, D., *The Professors: The 101 Most Dangerous Academics in America*, (Regnery Publishing, Washington DC, 2006).
Horton, J., *The End of Science: Facing the Limits of Knowledge in the Twilight of the Scientific Age*, (Addison-Wesley, Reading MA, 1996).
Houghton, J. T., (*et al* eds.), *Climate Change 2001: The Scientific Basis*,(Cambridge University Press, Cambridge, 2001).
Howat, I. M., (*et al*), "Rapid Changes in Ice Discharge from Greenland Outlet Glaciers," *Science*, vol. 315, no. 5818, March 16 2007, pp. 1559-1561.
Howson, C., *Hume's Problem: Induction and the Justification of Belief*, (Clarendon Press, Oxford, 2000).
Hoyos, C. D., (*et al*), "Deconvolution of the Factors Contributing to the Increase in Global Hurricane Intensity", *Science*, vol. 321, no. 5770, April 7 2006, pp. 94-97, doi: 10. 1126/science.1123560.
Hume, D., *A Treatise of Human Nature*, edited by L.A. Selby-Bigge,(Oxford University Press, Oxford, 1978).
Hunt, S. C., (*et al.*), "Cultivating Renewable Alternatives to Oil", in L.Starke (ed.), *State of the World 2006*, (Earthscan, London, 2006), pp. 61-77.
Hutton, D. and Connors, L., *A History of the Australian Environment Movement*, (Cambridge University Press, Melbourne, 1999).
Illing, D., "Hounded for Green Counsel", *The Australian*, June 2 1999.
IPCC, Working Group I, Fourth Assessment Report, *Climate Change 2007:* The Physical Science Basis (Summary for Policymakers), at http://www.ipcc.ch .
IPCC, Working Group II, Fourth Assessment Report, *Climate Change 2007: Climate Change Impacts, Adaptation and Vulnerability (Summary for Policymakers)* at http://www.ipcc.ch .
Irvine, A., "Latest Find from the Frege Archives", *Quadrant*, April 1999, pp. 53-55.
Jackson, T., *Motivating Sustainable Consumption - A Review of Evidence on Consumer Behaviour and Behavioural Change* (2005) at http://www.sd-research.org.uk/documents/MotivatingSCfinal.pdf.
Jackson, T., (ed.), *The Earthscan Reader in Sustainable Consumption*, (Earthscan, London, 2006).
Jacoby, R., The Last Intellectuals: American Culture in the Age of Academe,

(Basic Books, New York, 2000).

Jamieson, D., "Ethics, Public Policy, and Global Warming", *Science, Technology, and Human Values*, vol.17, no.2, 1992, pp. 139-153.

Jamieson, D., "Ethics and Intentional Climate Change", *Climatic Change*, vol. 33, 1996, pp. 323-336.

Jamieson, D., *Morality's Progress: Essays on Humans, Other Animals, and the Rest of Nature*, (Clarendon Press, Oxford, 2002).

Jamieson, D., "Ethics, Public Policy, and Global Warming", in A. Light and H. Rolston III (eds.), *Environmental Ethics: An Anthology*, (Blackwell Publishing, Oxford, 2003), pp. 371-379.

Jammer, M., *The Philosophy of Quantum Mechanics: The Interpretation of Quantum Mechanics in Historical Perspective*, (Wiley, New York, 1974).

Janz, N. K. and Becker, M. H., "The Health Belief Model: A Decade Later", *Health Education Quarterly*, vol. 11, 1984, pp. 1-47.

Jaspers, K., *The Idea of a University*, edited by K. Deutsch, (Owen, London, 1960).

Jencks, C., *The Homeless*, (Harvard University Press, Cambridge, Massachusetts, 2005).

Jha, A., "Energy Review Ignores Climate Change 'Tipping Point,' "*Guardian Unlimited,* September 4 2006 at http://www.guardian.co.uk/science/story/0,,/1864802,00.html.

Johnson, P., "Universities? We'd be Better Off Without Them", *The Australian*, September 18 1991, p. 21.

Jones C. D., (*et al.*), "Strong Carbon Cycle Feedbacks in a Climate Model with Interactive CO2 and Sulphate Aerosols", *Geophysical Research Letters,* vol. 30, May 9 2003, doi:10.1029/2003GLO16867, p. 1479.

Jones, B., *A Thinking Reed,* (Allen and Unwin, St. Leonards, New South Wales, 2006).

Jones, C., *Global Justice: Defending Cosmopolitanism*, (Oxford University Press, Oxford, 1999).

Jones, D., "The Ir/ Relevance of Environmental Ethics", *Environmental Ethics*, vol.75, 2003, pp. 223-224.

Joughlin, I., "Greenland Rumbles Louder as Glaciers Accelerate," *Science*, vol.311, March 24 2006, pp.1709, doi: 10.1126/science.1124496.

Kagan, S., *The Limits of Morality*, (Clarendon Press, Oxford, 1989).

Kahn, J. R. and Franceschi, D., "Beyond Kyoto:A Tax-Based System for the Global Reduction of Greenhouse Gas Emissions", *Ecological Economics*, vol. 58, 2006, pp. 778-787.

Kammen, D. M., "The Rise of Renewable Energy", *Scientific American*, September 2006, pp. 60-69.

Kant, I., *Foundations of the Metaphysics of Morals*, (Macmillan, London,1990).

Kaplan, R. D., "The Coming Anarchy: How Scarcity, Crime, Overpopulation, Tribalism, and Disease are Rapidly Destroying the Social Fabric of Our Planet", *The Atlantic Monthly*, February 1994, pp.44-76.

Karoly, D. J., (*et al*), "Detection of a Human Influence on North American Climate", *Science*, vol.302, no 5648, November 14 2003, pp.1200-1203, doi: 10.1126/science. 1089159.

Kavel, J., *The Enemy of Nature: The End of Capitalism or the End of the World?* (Zed Books, London, 2006).

Keating, K., *What Catholics Really Believe - Setting the Record Straight*, (Servant Publications, Ann Arbor, 1992).

Keen, S., "A Secular Apocalypse," *Bulletin of the Atomic Scientists*, January/February 2007, pp. 29-31.

Kekes, J., *The Nature of Philosophy*, (Basil Blackwell, Oxford, 1980).

Kennedy, D. and Hudson, B., "Ice and History", *Science*, vol.311, March 24 2006, pp. 1673, doi: 10.1126/science.1127485.

Kennedy,S., (*et al.*), *A Primer on the Macroeconomic Effects of an Influzena Pandemic*, Australian Treasury Working Paper Februray 2006.

Kenny, A.,*The Heritage of Wisdom: Essays in the History of Philosophy*, (Basil Blackwell, Oxford, 1987).

Kent, P. "Great Barrier Reef ' Already Doomed'", *The Advertiser,* November 6 2006, p. 8.

Kepple, F. and Röckmann, T., "Methane, Plants and Climate Change", *Scientific American*, vol. 296, no.2, February 2007.

Kerr, R. A., "A Worrying Trend of Less Ice, Higher Seas", *Science*, vol.311, no 5768, March 24 2006, pp. 1698-1701, doi: 10.1126/science.311.5768.1698.

Kerr, R. A., "Politicians Attack, But Evidence For Global Warming Doesn't Wilt", *Science*, vol.313, no. 5786, July 28 2006, p. 421:doi: 10.1126/science.313.5786.421.

Khazzoom, J. D., "Economic Implications of Mandated Efficiency Standards for Household Appliances", *Energy Journal,* vol. 1, 1980, pp. 21-39.

Kidner, D. W., "Why Psychology is Mute About the Environmental Crisis", *Environmental Ethics*, vol. 16, 1994, pp. 359-376.

Kimball, R., *Tenured Radicals: How Politics has Corrupted Our Higher Education*, (Harper and Row, New York, 1990).

King, L.P. and Szelenyi, S., *Theories of the New Class: Intellectuals and Power*, (University of Minnesota Press, Minneapolis, 2004).

Kininmonth, W., *Climate Change: A National Hazard,* (Multi-Science Publishing Co. Essex, 2004).

Kininmonth, W., " Don't be Gored into Going Along," *The Australian*, September 12 2006, p.12.

Kirkman, R., *Skeptical Environmentalism: The Limits of Philosophy and Science*, (Indiana University Press, Bloomington, 2002).

Kirp, D., *Shakespeare, Einstein and the Bottom Line: The Marketing of Higher Education*, (Harvard University Press, Cambridge, MA, 2003).

Klare, M.T.,*Resource Wars: The New Landscape of Global Conflict*, (Metropolitan Books, New York, 2001).

Knight, S., "Scientist Issues Grim Warning on Global Warming," *Timesonline*, June 15 2006 at http://www.timesonline.co.uk./article/0,, 2-2134760,00.html.

Koch, R. and Smith, C., *Suicide of the West*, (Continuum, London, 2006).

Kofbert, E., *Field Notes from a Catastrophe: Man, Nature and Climate Change*, (Bloomsbury Publishing, London, 2006).

Kollmus, A. and Agyeman, J., "Mind the Gap: Why Do People Act Environmentally and What are the Barriers to Pro-Environmental Behavior?" *Environmental Education Research*, vol. 8, no.3, 2002, pp.239-260.

Korten, D. C., *When Corporations Rule the World*, (Earthscan, London, 1995).

Kovel, J., *The Enemy of Nature*, (Zed Books, London, 2002).

Kruess,A. and Tschamtke, T., "Habit Fragmentation, Species Loss and Biological Control", *Science*, vol. 264, 1994, pp. 1581-1584.

Krutilla, J. V. and Fisher, A. C., *The Economics of Natural Environments*, (John Hopkins University Press, Baltimore MD, 1985).

Kuhn, A. B., *The Lost Light: An Interpretation of Ancient Scriptures* (1940), (Kessinger Publishing, Kila MT, 1997).

Kuhn, J., "Toward an Ecological Humanistic Psychology,"*Journal of Humanistic Psychology*, vol. 41, 2001, pp. 9-24.

Kump, L. R., (*et al*), "Massive Release of Hydrogen Sulfide to the Surface Ocean and Atmosphere During Intervals of Oceanic Anoxia", *Geology*, vol.33, no.5, May 2005, pp.397-400.

Kunstler, J. H., *The Long Emergency: Surviving the Converging Catastrophes of the Twenty-First Century*, (Atlantic Books, London, 2005).

Kuznets, S., *Economic Growth of Nations: Total Output and Production Structure*, (Belknap Press of Harvard University Press, Cambridge, Massachusetts, 1971).

Kverndokk, S., "Tradeable CO2 Emissions Permits: Initial Distribution as a Justice Problem", *Environmental Values*, vol.4, 1995, pp. 129-148.

LaHaye T. F. and Jenkins, J. B., *Left Behind: A Novel of the Earth's Last Days (Left Behind No.1)* (Tyndale House Publishers, Carol Stream IL, 1996).

Lai, K. L., "Conceptual Foundations for an Environmental Ethics: A Daoist Perspective", *Environmental Ethics*, vol.25, 2003, pp. 247-266.

Lakoff, G. and Johnson, M., *Metaphors We Live By*, (University of Chicago Press, Chicago, 1980).

Landsea, C. W., "Hurricanes and Global Warming, *Nature*, vol.438, 22/29 December, 2005, pp. E11-E13, doi: 10. 1038/nature04477.

Lashof, D. A., (*et al*), "Terrestrial Ecosystem Feedbacks to Global Climate Change", *Annual Review of Energy and the Environment*, vol. 22, 1997, pp. 75-118.

Lazare, D., "False Testament: Archaeology Refutes the Bible's Claim to History", *Harper's Magazine*, March 2002, pp. 39-47.

Leake, J. and Booth, R. "Iceland's Hot Rock to Power Europe", *The Australian*, May 14 2007, p. 13.

Lee, K., (*et al.*), "Global Change and Health – The Good, the Bad and the Evidence", *Global Change and Human Health*, vol.3, no.1, 2002, pp. 16-19.

Lee, K., "Global Relationships of Total Alkalinity with Salinity and Temperature in Surface Waters of the World's Oceans," *Geophysical Research Letters*, vol. 33, 2006, L. 19605, doi:10.1029/2006GLO27207.

Leeb, S. and Strathy, G., *The Coming Economic Collapse: How You can Thrive When Oil Costs $200 a Barrel*, (Warner Business Books, New York, 2006).

Leeuwin, J. W. S. and Smith, P., "Can Nuclear Power Provide Energy for the Future; Would it Solve the CO2 Emission Problem? (2003) at www.oprit.rug.nl/deenen/Technical.html.

Leggett, J., *The Empty Tank: Oil, Gas, Hot Air, and the Coming Global Financial Catastrophe,* (Random House, New York, 2005).

Lehrer, K., *Knowledge*, (Clarendon Press, Oxford, 1974).

Lehrer, K. and Wagner, C., *Rational Consensus in Science and Society*, (D. Reidel, Dordrecht, 1981).

Leidner, H., *The Fabrication of the Christ Myth*, (Surrey Books, Tampa Fl, 1999).

Lemons, J., "Atmospheric Carbon Dioxide: Environmental Ethics and Environmental Facts", *Environmental Ethics*, vol.5, 1983, pp. 21-32.

Lenton, T. M., "Millennial Timescale Carbon Cycle and Climate Change in an Efficient Earth System Model", *Climate Dynamics*, vol.26, 2006, pp. 687-711, doi:10.1007/s00382-0060-0109-9.

Leonard-Barton, D., "Living Lightly Can Mean Greater Independence, Richer Lives", *The Christian Science Monitor*, October 21 1980, p. 20.

Levin, M. G., *The Socratic Method: A Novel*, (Simon and Schuster, New York, 1987).

Levine, L. W., *The Opening of the American Mind: Canons, Culture, and History*, (Beacon Press, Boston, 1996).

Levins, R.T., (*et al*), "The Emergence of New Diseases", *American Scientist*, vol. 82, 1994, pp. 52-60.

Levison, I., "Law as Literature", *Texas Law Review*, vol.60, 1982, pp. 371-403.

Lewis, S. and Shanahon, D., "PM Pushes a New Kyoto", *The Australian*, November 1 2006, p.1.

Li, X., *Ethics, Human Rights and Culture: Beyond Relativism and Universalism*, (Palgrave Macmillan, New York, 2006).

Libet, B., (*et al* eds.), *The Volitional Brain: Towards a Neuroscience of Free Will*, (Imprint Academic, Thorverton, 1999).

Libet, B., *Mind Time: The Temporal Factor in Consciousness*, (Harvard University Press, Cambridge, 2004).

Lilley, I., "Whatever Happens, We'll Manage", *The Australian*, (Higher Education), May 17 2006, pp.44-45.

Lincoln, S. F., *Challenged Earth*, (Imperial College Press, London, 2006).

Linden, F., *The Winds of Change: Climate, Weather, and the Destruction of Civilizations*, (Simon and Schuster, New York, 2006).

Lindgren, E. and Gustafson, R., "Tick-borne Encephalitis in Sweden and Climate Change", *Lancet*, vol.358, 2001, pp. 16-87.

Lindsay, S. and Martens, W.J.M., "Malaria in the African Highlands: Past, Present and Future", *Bulletin of the World Health Organization*, vol. 78, 2000, pp. 33-45.

Lindzen, R. S., (*et al*), "Does the Earth have an Adaptive Infrared Iris?" *Bulletin of the American Meteorological Society*, vol.82, 2001, pp.417-432.

Lippman, W., *Public Opinion*, (Harcourt Brace and Company, New York, 1922).

Lomberg, B., "Stern Scare Blunted by the Figures", *The Australian*, November 6 2006, p. 16.

Lomborg B., *The Skeptical Environmentalist*, (Cambridge University Press, Cambridge, 2001).

Lomborg, B., "Stern Scare Blunted by the Figures", *The Australian*, November 6 2006, p.16.

Long, A. A., *The Cambridge Companion to Early Greek Philosophy*, (Cambridge University Press, Cambridge, 1999).

Lovejoy, T. E. and Hannah, L. L., (eds.), *Climate Change and Biodiversity*, (Yale University Press, New Haven and London, 2005).

Lovelock, J., *Gaia: A New Look at Life on Earth*, (Oxford University Press, Oxford, 1979).

Lovelock, J., *The Revenge of Gaia: Why the Earth is Fighting Back - and How We Can Still Save Humanity*, (Allen Lane, London, 2006).

Lovins, A., *Non-Nuclear Futures*, (Friends of the Earth, San Francisco, 1975).

Lowe, I., *A Big Fix: Radical Solutions for Australia's Environmental Crisis*, (Black Inc., Melbourne, 2005).

Lowe, J. A., (*et al*), "The Role of Sea-Level Rise and the Greenland Ice Sheet in Dangerous Climate Change: Implications for the Stabilisation of Climate", in H.J. Schellnhuber (*et al* eds.), *Avoiding Dangerous Climate Change*, (Cambridge University Press, Cambridge, 2006), pp. 29-36.

Lynas, M., "Shares in the Sky", *New Internationalist*, no. 357, June 2003 at http://www.newint.org/issue357/shares.htm.

Lyotard, J-F. and Thébaud,, J-L., *Just Gaming*, (University of Minnesota Press, Minneapolis, 1985).

Lyotard, J-F., *The Postmodern Condition: A Report on Knowledge*, (University of Minnesota Press, Minneapolis, 1984).

Lyster, R. and Bradbrook, A., *Energy Law and the Environment*, (Cambridge University Press, Melbourne, 2006).

Mabey, N., (*et al.*), *Argument in the Greenhouse: The International Economics of Controlling Global Warming*, (Routledge, London and New York, 1997).

MacCracken, M., "Climate Change Discussions in Washington: A Matter of Contending Perspectives", *Environmental Values*, vol. 15, 2006, pp. 381-395.

MacIntyre, A., *After Virtue*, 2nd edition, (Duckworth, London, 1985).

MacKenzie, D., "Bird Flu Outruns the Vaccines", *New Scientist*, November 4 2006, pp. 8-9.

Macquarrie, J., *In Search of Deity: An Essay in Dialectical Theism*, (SCM Press, London, 1984).

Macumbi, P., "Plague of My People," *Nature*, vol. 430. August 19 2004, p.925.

Maddy, P., *Realism in Mathematics*, (Clarendon Press, Oxford, 1990).

Makhijani, A., (*et al.* eds.), *Nuclear Wastelands: A Global Guide to Nuclear Weapons Production and Its Health and Environmental Effects*, (MIT Press, Cambridge, Massachusetts, 2000).

Malcolm, N., *Ludwig Wittgenstein: A Memoir*, (Oxford University Press, Oxford, 1958).

Mander, J. and Goldsmith, E., (eds.), *The Case Against the Global Economy and a Turn Toward the Local*, (Sierra Club Books, San Francisco,1996).

Mank, B.C., "Standing and Global Warming: Is Injury to All Injury to None?" *Environmental Law*, vol. 35, 2005, pp. 1-84.

Mann, M. E., (*et al*), "Global-Scale Temperature Patterns and Climate Forcing Over the Past Six Centuries," *Nature*, vol. 392, 1998, pp.779-787.

Mann, M. E., (*et al*), "Testing the Fidelity of Methods Used in Proxy-Based Reconstructions of Past Climate", *Journal of Climate*, vol. 18, 2005, pp.

4097-4107.

Manson, N. A., "Formulating the Precautionary Principle", *Environmental Ethics*, vol.24, 2002, pp. 263-274.

Marangudakis, M., "The Medieval Roots of Our Ecological Crisis", *Environmental Ethics*, vol.23, 2001, pp. 243-260.

Martens, P. and McMichael, A. J., *Environmental Change, Climate and Health*, (Cambridge University Press, Cambridge, 2002).

Martens, W.J.M., *Health and Climate Change: Modelling the Impacts of Global Warming and Ozone Depletion*, (Earthscan, London, 1998).

Martens, W.J.M., (*et al*), "Climate Change and Future Populations at Risk of Malaria", *Global Environmental Change*, vol. 9 (Supplement), 1999, S89-S107.

Martin, B., (eds., et. al.), *Intellectual Suppression*, (Angus and Robertson, North Ryde, New South Wales, 1986).

Martin, M., *The Case Against Christianity*, (Temple University Press, Philadelphia, 1991).

Martin-Schramm, J. B. and Stivers, R. L., *Christian Environmental Ethics: A Case Method Approach*, (Orbis Books, Maryknoll, New York, 2003).

Massey, G., *Ancient Egypt, The Light of the World: A Work of Reclamation and Restitution in Twelve Volumes*,(Kessinger Publishing, Kila MT, 2001).

Maxwell, N., *From Knowledge to Wisdom: A Revolution in the Aims and Methods of Science*, (Blackwell, Oxford, 1987).

McBride, S. and Wiseman, J., *Globalization and Its Discontents*, (Macmillan, London, 2000).

McCarthy, M., "Environment in Crisis: 'We Are Past the Point of No Return'," *The Independent,* January 16 2006 at http://news.independent.co.uk/environment/article338879.ece .

McCracken, G., *Culture and Consumption*, (Indiana University Press, Bloomington and Indianapolis, 1990).

McCulmon, J., "How Learned Books Nearly Perished", *The Australian*, May 6 1998, p. 40.

McGrath, A. E., *A Passion for Truth*, (Apollos, Leicester, 1996).

McGuinness, P. P., "Editorial: The Decline of Universities", *Quadrant*, April 1999, pp. 2-5.

McGuire, B., *A Guide to the End of the World: Everything You Never Wanted To Know*, (Oxford University Press, Oxford, 2003).

McGuire, B., *Surviving Armageddon: Solutions for a Threatened Planet*, (Oxford University Press, Oxford, 2005).

McIntyre, S. and McKitrick, R., "Corrections to Mann *et al* (1998) Proxy Data Base and Northern Hemisphere Average Temperature Series", *Energy and*

*Environment*, vol.14, 2003, pp. 751-771.
McIntyre, S. and McKitrick, R. "The M&M Critique of MBH98 Northern Hemisphere Climate Index: Update and Implications", *Energy and Environment*, vol.16, 2005, pp. 69-100.
McIntyre, S. and McKitrick, R., "Hockey Sticks, Principal Components, and Spurious Significance", *Geophysical Research Letters*, vol.32, 2005, L03710, doi: 10. 1029/2004GLO21750.
McKibben, B., "The Christian Paradox: How a Faithful Nation Gets Jesus Wrong", *Harper's Magazine*, August 2005, pp. 31-37.
McManus, J. F., (*et al*), "Collapse and Rapid Resumption of Atlantic Meridional Circulation Linked to Deglacial Climate Change", *Nature*, vol. 428, 2004, pp. 834-837.
McMichael, A. J., *Planetary Overload: Global Environmental Change and the Health of the Human Species,* (Cambridge University Press, Cambridge, 1993).
McMichael, A. J. and Githeko, A., (*et al*), "Human Health", Working Group II, Third Assessment Report. Intergovernmental Panel on Climate Change, *Climate Change 2001: Impacts, Adaptation, and Vulnerability*, (Cambridge University Press, Cambridge, 2001), pp. 453-485.
McMichael, A. J., *Human Frontiers, Environment and Disease: Past Patterns, Uncertain Futures*, (Cambridge University Press, Cambridge, 2001).
McMichael A.J., "The Biosphere, Health and Sustainable Development", *Science* , vol.297, 2002, p. 1093.
McMichael, A. J., "Population, Environment, Disease and Survival: Past Patterns, Uncertain Futures", *Lancet*, vol.359, 2002, pp. 1145-1148.
McMichael, A.J., (*et al.* eds.), *Climate Change and Human Health: Risks and Responses*, (World Health Organization, Geneva, 2003).
McMichael, A. J., (*et al.*), "Climate Change and Human Health: Present and Future Risks", *The Lancet*, Februrary 9 2006, doi:10.1016/S0140-6736(06) 68079-3.
McMurray, C. and Smith, R., *Diseases of Globalization: Socioeconomic Transitions and Health*, (Earthscan, London, 2001).
Meacher, M., "Why Plan for a Nuclear Future When World Uranium Supplies are Running Out?" *Guardian Weekly* July 21-27 2006, p. 17.
Mead, G., *Mind, Self and Society*, (University of Chicago Press, Chicago, 1934).
Meadows, D. H., (*et al*), *The Limits to Growth*, (Potomac Association, London, 1972).
Meinshausen, M., "What Does a 2° C Target Mean for Greenhouse Gas Concentrations? A Brief Analysis Based on Multi-Gas Emission Pathways and Several Climate Sensitivity Uncertainty Estimates", in H.J.

Schellnhuber (*et al* eds.), *Avoiding Dangerous Climate Change*, (Cambridge University Press, Cambridge, 2006), pp. 265-279.

Mele, A. R., *Free Will and Luck*, (Oxford University Press, Oxford, 2006).

Mendelson III, J. R., (*et al*), "Biodiversity: Confronting Amphibian Declines and Extinctions", *Science*, vol.313, no5783, 7 July 2006, doi: 10. 11261/science.1128396 at http://www.sciencemag.org/cgi/content/ full/313/5783/48.

Menestrel, M. L., (*et al*), "Processes and Consequences in Business Ethical Dilemmas: The Oil Industry and Climate Change", *Journal of Business Ethics*, vol. 41, 2002, pp. 251-266.

Menon, S. and Rotstayn, L. L., "The Radiative Influence of Aerosol Effects on Liquid-Phase Cumulus and Stratiform Clouds Based on Sensitivity Studies with Two Climate Models", *Climate Dynamics*, vol.27, no.4, September 2006, pp.345-356.

Merali, Z., "Return of the Atom", *New Scientist*, September 16, 2006, pp. 6-7.

Metz, B. and van Vuuren, D., "How, and at What Costs, Can Low-Level Stabilization be Achieved?- An Overview", in H.J. Schellnhuber (*et al.* eds.), *Avoiding Dangerous Climate Change*, (Cambridge University Press, Cambridge, 2006), pp. 337-345.

Meuvrel, O., "Cattle a Cause of Global Warming," *The Advertiser* (Adelaide), October 1 2005, p.69.

Meyer A., *Contraction and Convergence: The Global Solution to Climate Change*, (Green Books, Devon, 2000).

Meyer, A., "The United States has it Right on Climate Change in Theory", at http://www.opendemocracy.net/debates/article-6-129-2462.jsp.

Midgley, M., *Science as Salvation: A Modern Myth and its Meaning*, (Routledge, London and New York, 1992).

Millennium Ecosystem Assessment, *Ecosystems and Human Well Being: Health Synthesis*, (World Health Organization, Geneva, 2005).

Millennium Ecosystem Assessment, *Millennium Ecosystem Assessment Synthesis Report* March 23 2005 at http://www.millenniumassessment.org/ en/Products. EHWB.aspx.

Miller, D., (ed.), *Acknowledging Consumption: A Review of New Studies*, (Routledge, London and New York, 1995).

Miller, D., *On Nationality*, (Clarendon Press, Oxford, 1995).

Milliken, M., "World has a 10-year Window to Act on Climate Warming - NASA Expert", *Common Dreams News Center,* September 14 2006 at http://www. commondreams.org/headlines06/0914-01.html.

Mills, C. W., *Power, Politics, and People: The Collected Essays of C.Wright Mills*, edited by I.L. Horowitz, (Oxford University Press, New York,

1963).

Mimura, M., (*et al*), "Small Island States", IPCC Working Group II, Fourth Assessment Report, *Climate Change 2007: Climate Change Impacts, Adaptation and Vulnerability*, at http://www.ipcc.ch.

Mishan, E. J., "Pangloss on Pollution", *Swedish Journal of Economics*, vol.73, no. 1, 1971, pp. 113-120.

Mitchell, P. A., *The Making of the Modern Law of Defamation*, (Hart Publishing, Oxford, 2005).

Moellendorf, D., *Cosmopolitan Justice*, (Westview Press, Boulder, Colorado, 2002).

Monbiot, G., "How Much Reality Can You Take?" at.http:/www.monbiot.com/archives/2006109/21/how-much-reality-can-you-take/.

Monbiot, G., *Heat: How to Stop the Planet Burning*, (Allen Lane, London, 2006).

Montgomery, D. and Smith, A. E., "Price, Quantity and Technology Strategies for Climate Policy", in M.Schlesinger (*et al.* eds.), *Human-Induced Climate Change:An Interdisciplinary Assessment*, (Cambridge University Press, Cambridge, 2007).

Monto, A. S., "The Threat of an Avian Influenza Pandemic", *New England Journal of Medicine*, vol. 352, no 4, 2005, pp. 323-325.

Moonaw, W. R., (*et al.*), "Technological and Economic Potential of Greenhouse Gas Emissions Reduction", in B.Metz (*et al.*eds.), *Climate Change 2001: Mitigation. Contribution of Working Group III to the Third Assessment Report of the IPCC*, (Cambridge University Press, Cambridge, 2001).

Morgan, R. and Barton, J., *Biblical Interpretation*, (Oxford University Press, Oxford, 1988).

Morris, S., "Mini Tax on Carbon to Prepare for Future", *The Australian*, April 5, 2007, p.1.

Moser, P. K., *Empirical Justification*, (D. Reidel, Dordrecht, 1985).

Multisectorial Initiative on Potent Industrial Greenhouse Gases, *MIPSGGS Newsletter*, September 2006 at www.mipiggs.org.

Murray C., *Human Accomplishment: The Pursuit of Excellence in the Arts and Sciences, 800 BC to 1950*, (HarperCollins, New York, 2003).

Muth, R. M. and Hendee, J. C., "Technology Transfer and Human Behavior", *Journal of Forestry*, 1980, pp. 141-144.

Nails, D., (ed.), *Naturalistic Epistemology*, (D.Reidel, Dordrecht, 1987).

Nash, R., (ed.), *Liberation Theology*, (Mott Media, Milford, Michigan, 1984).

National Emissions Trading Taskforce, *Possible Design for a National Greenhouse Gas Emission Trading Scheme* (2006) at.www.emissionstrading.net.au.

National Research Council of the National Academies (NRC), Committee on Surface Temperature Reconstructions for the Last 2,000 Years, Board on Atmospheric Sciences and Climate Division on Earth and Life Studies, *Surface Temperature Reconstructions for the Last 2,000 years* (2006) at http://newton.nap.edu/pdf/0309102251/pdf_image.

National Research Council, *Abrupt Climate Change: Inevitable Surprises*, (National Academy Press, Washington DC, 2002).

Nelson, R. H., *Economics as Religion: From Samuelson to Chicago and Beyond,* (Pennsylvania State University Press, University Park, Pennsylvania, 2001).

Nesje, A., (*et al*), "Were Abrupt Late Glacial and Early-Holocene Climate Changes in Northwest Europe Linked to Freshwater Outbursts to the North Atlantic and Arctic Oceans?" *The Holocene*, vol.14, 2004, pp. 299-310.

Neumayer, E., "In Defence of Historical Accountability for Greenhouse Gas Emissions", *Ecological Economics*, vol. 33, 2000, pp. 185-192.

Newman, J. H., *The Idea of a University*, introduction to G.N. Shuster, (Image Books, Garden City, New York, 1959).

Newton-Smith, W. H., *The Rationality of Science*, (Routledge and Kegan Paul, Boston, 1981).

Nicholls, R. J. and Lowe, J. A., "Climate Stabilisation and Impacts of Sea-Level Rise", in H.J. Schellnhuber (*et al* eds.), *Avoiding Dangerous Climate Change*, (Cambridge University Press, Cambridge, 2006), pp. 195-202.

Nickel, J. W., *Making Sense of Human Rights*, (University of California Press, London, 1987).

Nietzsche, F., *Schopenhauer as Educator*, translated by J.W. Hillesheim and M.R. Simpson, (South Bend, Indiana, Gateway, 1965).

Nordhaus W., "The *Stern Review* on the Economics of Climate Change" (2006) at http://nordhaus.econ.yale.edu/SternReviewDz.pdf.

North R.D., *Life on a Modern Planet: A Manifesto for Progress*, (Manchester University Press, Manchester, 1995).

Nyong, A. and Niang-Diop, I., "Impacts of Climate Change in the Tropics: The African Experience",in H.J. Schellnhuber (*et al* eds.), *Avoiding Dangerous Climate Change*, (Cambridge University Press, Cambridge, 2006), pp. 235-241.

O'Brien, A., "Carpenter Warms to Geothermal Energy", *The Weekend Australian*, November 18-19 2006, p.2.

O'Hear, A., *After Progress: Finding the Old Way Forward*, (Bloomsbury, London, 1999).

O'Keefe, B., "Warming ' Can't be Blamed for Storms'", *The Australian*,

February 21, 2006, p.6.
O'Neill, O., *Faces of Hunger: An Essay on Poverty, Justice and Development*, (Allen and Unwin, London, 1986).
O'Neill, O., *Towards Justice and Virtue: A Constructive Account of Practical Reasoning*, (Cambridge University Press, Cambridge, 1996).
Ogden, J., "High Hopes for Hydrogen", *Scientific American*, September 2006, pp. 70-77.
Ophuls, W., *Requiem for Modern Politics: The Tragedy of the Enlightenment and the Challenge of the New Millennium*, (Westview Press, Boulder, CO, 1998).
Orchison, K., "450 years' Electricity in Hot Rocks", *The Weekend Australian*, September 9-10 2006, (Power Generation Supplement), p.2.
Oreskes, N., "The Scientific Consensus on Climate Change," *Science*, vol. 306, December 3 2004, p.1686.
Ornstein, R. and Ehrlich, P., *New World New Mind: Moving Toward Conscious Evolution*, (Doubleday, New York, 1989).
Orr, D. W., "Moderization and the Ecological Perspective", in D.W. Orr and M.S. Soroos (eds.), *The Global Predicament*, (University of North Carolina Press, Chapel Hill, 1979), pp. 75-89.
Orr, D. W., "Slow Knowledge", *Conservation Biology*, vol. 10, no.3, June 1996, pp. 669-702.
Orr, D. W., "Speed", *Conservation Biology*, vol. 12, no.1, February 1998, p. 4-7.
Orr, D. W., *Earth in Mind: On Education, Environment and the Human Prospect*, (Island Press, Washington DC, 1994).
Orr, J. C., (*et al*), "Anthropogenic Ocean Acidification Over the Twenty-First Century and Its Impact on Calcifying Organisms", *Nature*, vol.437, September 29 2005, pp. 681-686.
Osfeld, R. S. (*et al*), "Conservation Medicine: The Birth of Another Crisis Discipline", in A.A. Aguirre (*et al* eds.), *Conservation Medicine: Ecological Health in Practice* ,(Oxford University Press, Oxford, 2002), pp. 17-26.
Osofsky, H. and Burns, W.C.G., *Adjudicating Climate Change: Sub-National, National and Supra-National Approaches*, (Cambridge University Press, Cambridge, 2007).
Padilla, E., "Climate Change, Economic Analysis and Sustainable Development", *Environmental Values*, vol. 13, 2004, pp. 523-544.
Paehlke, R., *Environmentalism and the Future of Progressive Politics* (Yale University Press, New Haven, 1989).
Page, E., "Intergenerational Justice and Climate Change", *Political Studies*, vol. 47, 1999, pp. 53-66.

Pagels, E.,*The Origin of Satan*, (Vintage Books, New York, 1996).

Pakrass, M., (*et al*), "Conservation Medicine: An Emerging Field", in P. Raven and T.Williams (eds.), *Nature and Human Society: The Quest of a Sustainable World*, (National Academy Press, Washington DC, 1999) pp. 551-556.

Papineau, D., "Is Epistemology Dead?" *Proceedings of the Aristotelian Society*, vol. 82, 1982, pp. 129-142.

Parfit, D., *Reasons and Persons*, (Clarendon Press, Oxford, 1984).

Paterson, M., *Global Warming and Global Politics*, (Routledge, London and New York, 1996).

Patz J. A., (*et al.*), "Global Climate Change and Emerging Infectious Diseases", *Journal of the American Medical Association*, vol. 275, 1996, pp. 217-233.

Patz, J. A., "A Human Disease Indicator for the Effects of Recent Global Climate Change", *Proceedings of the National Academy of Sciences*, vol. 99, no. 20, 2002, pp. 12506-12508.

Pearce, D., *Economic Values and the Natural World*, (Earthscan, London, 1993).

Pearce, F., "Dark Future Looms for Arctic Tundra", *New Scientist*, January 21 2006, p.15.

Pearce, F., "If We Don't Stop Burning Oil…", *New Scientist*, February 18 2006, p.10.

Pearce, F., "The Parched Planet", *New Scientist*, February 25 2006 pp.31-36.

Pearce, F., "Grudge Match", *New Scientist,* March 18 2006, pp. 40-43.

Pearce, F., "Ecopolis Now", *New Scientist*, June 17 2006, pp. 36-42.

Pearce, F.," Kyoto Promises are Nothing but Hot Air," *New Scientist*, June 24 2006, p.10.

Pearce, F., "One Degree and We're Done For", *New Scientist*, September 30 2006, p.8.

Pearce F., "State of Denial", *New Scientist*, November 4 2006, pp.18-21.

Pearce, F., *When the Rivers Run Dry: Water - The Defining Crisis of the Twenty-First Century*, (Beacon Press, Boston, 2006).

Pearce, F., *With Speed and Violence: Why Scientists Fear Tipping Points in Climate Change*, (Beacon Press, Boston, 2007).

Peerenboom, R. P., "Beyond Naturalism: A Reconstruction of Daoist Environmental Ethics", *Environmental Ethics*, vol.13, 1991, pp. 3-22.

Peñalver, E. M., "Acts of God or Toxic Torts? Applying Tort Principles to the Problem of Climate Change", *Natural Resources Journal*, vol. 38, Fall 1998, pp. 563-601.

Pepper, S. C., *World Hypotheses: A Study in Evidence*, (University of California Press, Berkeley, 1957).

Peters, T., *The Comic Self*, (Harper, San Francisco, 1991).

Pew Center on Global Climate Change, *Beyond Kyoto: Advancing the International Effort Against Climate Change* at.http:/www.pewclimate.org/docUplands/Long%2DTerm%20Target%2E.pdf.

Pielke, Jr., R. A., "Are There Trends in Hurricane Destruction?" *Nature*, vol.438, December 22/29, 2005, pp. E11, doi: 10. 1038/nature04426.

Pinguelli-Rosa, L. and Munasinghe, M., (eds.), *Ethics, Equity and International Negotiations on Climate Change,* (Edward Elgar, Cheltenham, 2003).

Pirsig, R. M., *Zen and the Art of Motorcycle Maintenance: An Inquiry into Values*, (Perennial Classics, New York, 1999).

Plimer, I., "All Hot and Bothered", *The Independent Weekly* (Adelaide), July 8-14 2006, p.12.

Po-keung, I., "Taoism and the Foundations of Environmental Ethics", *Environmental Ethics*, vol.5, 1983, p. 335.

Porritt, J., *Capitalism as if the World Matters,* (Earthscan, London, 2006).

Posner, R. A., *Public Intellectuals: A Study of Decline*, (Harvard University Press, Cambridge MA, 2001).

Postel S., "Self-guarding Freshwater Ecosystems", in L.Starke (ed.), *State of the World 2006*, (Earthscan, London, 2006), pp.41-60.

Pounds, J. A., "Widespread Amphibian Extinctions from Epidemic Disease Driven by Global Warming," *Nature*, vol. 439, January 12 2006, pp. 161-167, doi: 10.1038/nature04246.

Prager, M. H. and Shertzer, K W., "Remembering the Future": A Commentary on "Intergenerational Discounting: A New Intuitive Approach"", *Ecological Economics*, vol. 60, 2006, pp. 24-26.

Preston, B.L., (*et al*), *Climate Change in the Asia/Pacific Region*, (CSIRO, Melbourne, 2006) at http://www.csiro.au/resources/pfkd.html.

Price, C.,*Time, Discounting and Value*, (Blackwell, Oxford, 1993).

Primavesi, A., *Gaia's Gift*, (Routledge, London, 2003).

Prime Minister Strategy Unit, *Personal Responsibility and Changing Behavior: The State of Knowledge and Its Implications for Public Policy*, (February, 2004) at http://www.number10.gov.uk/files/pdf/pr.pdf.

Prochaska, J. O. and DiClemente, C. C., "Stages and Processes of Self-Change of Smoking: Toward an Integrative Model of Change", *Journal of Consulting and Clinical Psychology*, vol. 51, 1983, pp. 390-395.

Prochaska, J. O., "Strong and Weak Principles for Progressing from Precontemplation to Action on the Basis of Twelve Problem Behaviors", *Health Psychology*, vol.13, 1994, pp. 47-51.

Rahmstorf, S., (*et al.*), "Recent Climate Observations Compared to Projections,"

*Science*, February 1 2007, doi: 10.11261/science.1136843.

Rapley, C., "The Antarctic Ice Sheet and Sea Level Rise", in H.J. Schellnhuber (*et al* eds.), *Avoiding Dangerous Climate Change*, (Cambridge University Press, Cambridge, 2006), pp. 25-27.

Rapport, D. J., "Ecosystem Health: An Emerging Integrative Science", in D.J.Rapport (*et al* eds.), *Evaluating and Monitoring the Health of Large Scale Ecosystems*, (Springer, Heidelberg, 1995), pp. 5-34.

Rawls, J., *A Theory of Justice*, Oxford University Press, Oxford, 1973).

Readings, B., *The University in Ruins*, (Harvard University Press, Cambridge, Massachusetts, 1996).

Reed, E. S., *Encountering the World: Toward an Ecological Psychology*, (Oxford University Press, New York, 1996).

Reiter, P., "Climate Change and Mosquito-Borne Diseases", *Environmental Health Perspectives*, vol. 109, March 2001, pp. 141-161.

Report of the Secretary-General, *We the Peoples: The Role of the United Nations in the Twenty-First Century* at http://unpan1.un.org/intradoc/groups/public/documents/UN/UNPAN000923.pdf.

Report of the Senate Environment, Communications, Information Technology and the Arts Reference Committee, *The Heat is On: Australia's Greenhouse Future*, (Parliament of the Commonwealth of Australia, Canberra, November, 2006).

Revel, J-F., *Anti-Americanism*, (Encounter Books, New York, 2004).

Richards, G., "Academic Loses E-Mail After Criticising Uni", *The Age*, April 21, 1999.

Rignot, E. and Kanagaratnam, P., "Changes in Velocity Structure of the Greenland Ice Sheet", *Science*, vol.311, no 5763, February 17 2006, pp. 986-990, doi: 10.1126/science.1121381.

Roberts, I., "Review of M.Hillman with T.Fawcett, *How We Can Save the Planet*", *British Medical Journal*, vol. 332, June 10 2006, p. 1398.

Roberts, P., *The End of Oil: On the Edge of a Perilous New World*, (Houghton Miflin, New York, 2004).

Robertson, J. M., *The Historical Jesus: A Survey of Positions*, (Watts and Co, London, 1916).

Robinson, J. A. T., *Honest to God*, (SCM Press, London, 1963).

Rogers D.J. and Randolph, S. F., "The Global Spread of Malaria in a Future, Warmer World", *Science*, vol. 289, 2000, pp. 1763-1765.

Rolston III, H., "Can the East Help the West to Value Nature?" *Philosophy East and West*, vol. 37, 1987, pp. 172-190.

Romm, J. J., *The Hype About Hydrogen: Fact and Fiction in the Race to Save the Climate*, (Island Press, Washington DC, 2005).

Ross, L. D. and Nisbett, R. F., *The Person and the Situation*, (McGraw-Hill, New York, 1991).

Ross, L. D., "The Intuitive Psychologist and his Shortcomings: Distortions in the Attribution Process", in L. Berkowitz (ed.), *Advances in Experimental Social Psychology*, vol. 10, (Academic Press, New York, 1977).

Rossi, R. H., *Down and Out in America: The Origins of Homelessness*, (University of Chicago Press, Chicago,1991).

Rozak, T., (*et al* eds.), *Ecopsychology: Restoring the Earth, Healing the Mind*, (Sierra Club Books, San Francisco, 1995).

Rucker, R., *Infinity and the Mind: The Science and Philosophy of the Infinite*, Harvester Press, Sussex, 1982).

Runge, C. F. and Senauer, B., "How Biofuels Could Starve the Poor", *Foreign Affairs*, May/June 2007 at http://www.foreignaffairs.org.

Running, S. W., "Is Global Warming Causing More, Larger Wild Fires?" *Science*, vol.311, no. 5789,August 18 2006, pp. 927-928, doi: 10.1126/science.1130370.

Rushkoff, D., *Nothing Sacred: The Truth About Judaism*, (Crown Publishing, New York, 2003).

Russell, I. M., (ed.), *Feminist Interpretation of the Bible*, (Basil Blackwell, Oxford, 1985).

Rutherford, S., (*et al*), "Proxy - Based Northern Hemisphere Surface Reconstructions: Sensitivity to Method, Predictor Network, Target Season, and Target Domain", *Journal of Climate*, vol.18, 2005, pp. 2308-2329.

Ryckmans, P., *The View from the Bridge: The 1996 Boyer Lectures*, (ABC Books, Sydney, 1996).

Sabin, M. (ed.) *The Earth has a Soul: The Nature Writings of C.G.Jung*, (North Atlantic Books, Berkley, California, 2002).

Sackett, D., (*et al*), *Evidence-based Medicine*, (Churchill Livingstone, New York, 1997).

Sagoff, M., "Some Problems with Environmental Economics", *Environmental Ethics*, vol. 10, 1988, pp. 55-74.

Sagoff, M.,*The Economy of the Earth*, (Cambridge University Press, Cambridge, 1988).

Sajó, A., (ed.) *Human Rights with Modesty: The Problem of Universalism*, (Martinus Nijhoff, Leiden and Boston, 2004).

Sajó, A., (ed.), *Abuse: The Dark Side of Fundamental Rights*, (Eleven International Publishing Co, Amsterdam, 2006).

Salameh M., "How Realistic are OPEC's Proven Oil Reserves?" *Petroleum Review*, August 2004.

Salliot, J., "Kyoto Plan No Good, Minister Argues", *Globe and Mail*, April 8

2006, p.A 5.

Samsam Bakhtiari, A. M., "World Oil Production Capacity Model Suggests Output Peak by 2006-2007", *Oil and Gas Journal*, April 26, 2004, pp.18-20.

Sandin, P., "The Precautionary Principle and the Concept of Precaution", *Environmental Values*, vol.13, 2004, pp.461-475.

Sandrea, I., "Deep Water Oil Discovery Rate May Have Peaked: Production Peak May Follow in 10 years', *Oil and Gas Journal*, vol. 102, no.28, July 26 2004, p.18.

Santer, B. D., (*et al*), "Forced and Unforced Ocean Temperature Changes in Atlantic and Pacific Tropical Cyclogenesis Regions", *Proceedings of the National Academy of Sciences*, vol.103, no.38, September 19 2006, pp. 13905-13910, doi:10. 1073/pnas.0602861103.

Santmire, H. P.,*Nature Reborn: The Ecological and Cosmic Promise of Christian Theology*, (Fortress Press, Minneapolis, 2000).

Sauer-Thompson, G. and Smith, J. W., *The Unreasonable Silence of the World: Universal Reason and the Wreck of the Enlightenment Project*, (Ashgate, Aldershot, 1997).

Saul, J. R., *The Unconscious Civilisation*, (Penguin Books, Ringwood, Victoria, 1997).

Saul, J. R., *The Collapse of Globalism and the Reinvention of the World*, (Viking/Penguin, London, 2005).

Saunders, C. D., (*et al.*), "Using Psychology to Save Biodiversity and Human Well Being", *Conservation Biology*, vol. 20, no. 3, 2006, pp.702-705.

Saunders, C. D., "The Emerging Field of Conservation Psychology," *Human Ecology Review*, vol.10, no.2, 2003, pp.137-149.

Saunders, C. D.,"Growing Green Kids", *Chicago Wilderness Magazine*, Fall 2004, at http://chicagowildernessmag.org/issues/fall2004/greenkids.html.

Saunders, C. D., (*et al.*), "Using Psychology to Save Biodiversity and Human Well Being", *Conservation Biology*, vol. 20, no. 3, 2006, pp.702-705.

Sayre, K. M., "An Alternative View of Environmental Ethics", *Environmental Ethics*, vol. 13, 1991, pp. 195-213.

Scheffer, M., (*et al*), "Positive Feedback Between Global Warming and Atmospheric CO2 Concentrations Inferred from Past Climate Change", *Geophysical Research Letters*, vol.33, 2006, L10702, doi:10.1029/2005/GLO25044.

Schellnhuber, H., (*et al* eds.), *Avoiding Dangerous Climate Change*, (Cambridge University Press, Cambridge, 2006).

Schiermeier, Q., "Cleaner Skies Leave Global Warming Forecasts Uncertain", *Nature*, vol. 435, May 12 2005, p. 135.

Schiermeier, Q., "Trouble Brews Over Contested Trend in Hurricanes", *Nature*, vol. 435,June 23 2005, pp.1008-1009, doi: 10.1038/4351008b.

Schlesinger, M. E., (*et al*), "Assessing the Risk of a Collapse of the Atlantic Thermohaline Circulation", in H.J. Schellnhuber (*et al* eds.), *Avoiding Dangerous Climate Change*, (Cambridge University Press, Cambridge, 2006), pp. 37-47.

Schneider, D. P., (*et al*)," Antarctic Temperatures Over The Past Two Centuries from Ice Cores", *Geophysical Research Letters*, vol.33, 2006, L16707. doi:10.1029/2006/GLO27057.

Schor, J. B., *The Overworked American: The Unexpected Decline of Leisure*, (Basic Books, New York,1991).

Schroll, M. A., "Remembering Ecopsychology's Origins: A Chronicle of Meetings, Conversations, and Significant Publications", at http://www.ecopsychology.org/journal/ezine/ep_origins.html.

Schwartz, P. and Randall, D., "An Abrupt Climate Change Scenario and Its Implications for United States National Security", October 2003 at http://www.environmentaldefense.org/documents/3566_AbruptClimateChange.pdf.

Schwartz, S. P., "Why It Is Impossible to be Moral", *American Philosophical Quarterly*, vol. 36, no.4, October 1999, pp. 351-360.

Schweitzer, A., *Quest for the Historical Jesus: A Critical Study of its Progress from Reimarus to Wrede*, (3rd edition), (Adam and Charles Black, London, 1956).

Semeniuk, I., "Can E.O. Wilson Really Save the World?" *New Scientist,* September 30 2006, pp. 54-56.

Shanahan, D. and Warren, M., "Howard Defiant in Face of Global Warming Warning", *The Australian*, October 31 2006, p. 1.

Shearman, D. and Sauer-Thompson, G., *Green or Gone: Health, Ecology, Plagues, Greed and Our Future*, (Wakefield Press, Adelaide, 1997).

Shearman, D. and Butler, C., Submission by Doctors for the Environment Australia, *Uranium Mining, Processing and Nuclear Energy Review* (July 2006) at www.dea.org.au.

Shearman, D. and Smith, J. W., *The Climate Change Challenge and the Failure of Democracy*, (Praeger Press, Westport, 2007).

Shue, H., "Avoidable Necessity: Global Warming, International Fairness, and Alternative Energy," in I.Shapiro and J.W. DeCew (eds.), *Theory and Practice, Nomas XXXVII,* (New York University Press, New York and London, 1995), pp. 239-265.

Shue, H., "Global Environment and International Inequality", *International Affairs*, vol. 75, 1999, pp. 531-545.

Shue, H., "Climate" in D. Jamieson (ed.), *A Companion to Environmental Philosophy: Blackwell Companions to Philosophy*, (Blackwell Publishers, Oxford, 2001), pp. 449-459.

Shue, H., "Legacy of Danger: The *Kyoto Protocol* and Future Generations", in K. Horton and H. Patapan (eds.), *Globalisation and Equality*, (Routledge, London and New York, 2004), pp. 164-178.

Simmons, M., *Twilight in the Desert: The Coming Saudi Oil Shock and the World Economy*, (John Wiley and Sons, New York, 2005).

Simms, A., "To the Rescue", *New Scientist*, February 2006, p. 50.

Simms, A., *Ecological Debt: The Health of the Planet and the Wealth of Nations*,(Pluto Press, London, 2005).

Simon, J. L., *The Ultimate Resource*, (Princeton University Press, Princeton, 1981).

Singer, P., "Famine, Affluence and Morality", *Philosophy and Public Affairs*, vol. 1, 1972, pp. 229-244.

Singer, P., *One World*: *The Ethics of Globalization*, (Yale University Press, New Haven and London, 2002).

Singer, S. F. and Avery, D. T., *Unstoppable Global Warming Every 1,500 Years*, (Rowman and Littlefield, Lanham, 2007).

Skinner, B. F., "Why We Are Not Acting to Save the World", in *Upon Further Reflection*, (Prentice-Hall, Englewood Cliffs, New Jersey,1987).

Slott, J. M., "Coastline Responses to Changing Storm Patterns", *Geophysical Research Letters*, vol.33, 2006, L18404. doi:10.1029/2006/GLO27445.

Small, H., (ed.), *The Public Intellectual*, (Blackwell Publishing, Oxford, 2002).

Smith, C. E. and Smith, J. W., "Economics, Ecology, and Entropy: The Second Law of Thermodynamics and the Limits to Growth", *Population and Environment*, vol.17, no.4, March 1996, pp.309-321.

Smith, D., "How Will it All End?" in B. Cartledge (ed), *Health and the Environment*, (Oxford University Press, Oxford, 1994).

Smith, D. M., (*et al.*), "Improved Surface Temperature Prediction for the Coming Decade from a Global Climate Model," *Science*, vol. 317, August 10 2007, pp. 796 - 799.

Smith, J. W., "Formal Logic: A Degenerating Research Programme in Crisis", *Cogito*, vol. 2, no.3, September 1984, pp.1-8.

Smith, J. W., *Reductionism and Cultural Being: A Philosophical Critique of Sociobiological Reductionism and Physicalist Scientific Unificationism*, (Martinus Nijhoff, The Hague, 1984).

Smith, J. W., *Reason, Science and Paradox*, (Croom Helm, London, 1985).

Smith, J. W., *The Progress and Rationality of Philosophy as a Cognitive Enterprise*, (Avebury, Aldershot, 1988).

Smith, J. W., *The High Tech Fix: Sustainable Ecology or Technocratic Megaprojects for the 21st Century,* (Avebury, Aldershot, 1991).

Smith, J.W., (ed.), *Immigration, Population and Sustainable Environments: The Limits to Australia's Growth,* (Flinders Press, Bedford Park, 1991).

Smith, J. W.,*The Remorseless Working of Things*, (Kalgoorlie Press, Kalgoorlie,1992).

Smith, J. W. (et. al.), *Is the End Nigh? Internationalism, Global Chaos and the Destruction of Earth,* (Avebury, Aldershot, 1995).

Smith J. W. and Sauer-Thompson, G., *Beyond Economics: Sustainability, Globalization and National Self-Sufficiency,* (Avebury, Aldershot, 1996).

Smith J. W. and G. Sauer-Thompson, G., *The Unreasonable Silence of the World: Universal Reason and the Wreck of the Enlightenment Project,* (Avebury, Aldershot, 1997).

Smith, J. W., (et al), *Healing a Wounded World: Economics, Ecology and Health for a Sustainable Life*, (Praeger, Westport and London, 1997).

Smith, J. W. and Sauer-Thompson, G., "Civilization's Wake: Ecology, Economics and the Roots of Environmental Destruction and Neglect", *Population and Environment*, vol.19, no.6, July 1998, pp.541-575.

Smith, J. W. (et. al.), *Global Meltdown: Immigration, Multiculturalism and National Breakdown in the New World Disorder*, (Praeger, Westport, Connecticut and London, 1998).

Smith J.W.,*(et al.), The Bankruptcy of Economics: Ecology, Economics and the Sustainability of the Earth,* (Macmillan, London, 1999).

Smith, J. W.(et. al.), *Global Anarchy in the Third Millennium? Race, Place and Power at the End of the Modern Age,* (Macmillan, London and St. Martin's Press, New York, 2000).

Smith, J. and Shearman D., *Climate Change Litigation: Analysing the Law, Scientific Evidence and Impacts on the Environment, Health and Property,* (Presidian Legal Publications, Adelaide, 2006).

Smith, M., *Jesus the Magician,* (Victor Gollancz, London, 1978).

Smith, P., *Killing the Spirit: Higher Education in America*, (Viking, New York, 1990).

Smith, P. F., "Contraction and Convergence: Myth and Reality," *British Medical Journal*, vol. 332, June 24 2006, p. 1509.

Smith, T., "The Case Against Free Market Environmentalism", *Journal of Agricultural and Environmental Ethics*, vol. 8, no. 2, 1995, pp. 126-144.

Smith, T., "Response to Narveson", *Journal of Agricultural and Environmental Ethics*, vol. 8, no. 2, 1995, pp. 157-158.

Snow, R., "The Invisible Victims", *Nature*, vol.430, August 19 2004, pp. 934-935.

Socolow, R. H. and Pacala, S. W., "A Plan to Keep Carbon in Check", *Scientific American,* September 2006, pp. 28-35.

Soden, B. J., (*et al.*), "The Radiative Signature of Upper Tropospheric Moistening", *Science*, vol. 310, 2005, pp. 841-844.

Sosa, E., "Nature Unmirrored, Epistemology Naturalized," *Synthese,* vol. 55, 1983, pp. 49-72.

Soule, E., "Assessing the Precautionary Principle", *Public Affairs Quarterly*, vol. 14, no.4, October 2000, pp. 309-328.

Soule, M. E., "What is Conservation Biology?" *BioScience*, vol.35, 1985, pp. 727-734.

Soule, M. E., *Conservation Biology: The Science of Scarcity and Diversity*, (Sinauer Associates, Sunderland, MA,1986).

Spong, J. S., *Rescuing the Bible from Fundamentalism*, (HarperSanFrancisco, New York, 1991).

Spong, J. S., *Born of a Woman: A Bishop Rethinks the Birth of Jesus,* (HarperSanFrancisco, New York, 1992).

Spong, J. S., *Why Christianity Must Change or Die*, (HarperSanFrancisco, New York, 1998).

Starkey, R. and Anderson, K., *Domestic Tradable Quotas: A Policy Instrument of Reducing Greenhouse Gas Emissions from Energy Use*, Tyndall Centre Technical Report, No. 39 at www.tyndall.ac.uk/publications/tech_reports/ tech_reports.shtml.

Stern Review, *The Economics of Climate Change*, October 31 2006 at www.sternreview.org.uk.

Stern, P. and Dietz, T., "The Value Basis of Environmental Concern", *Journal of Social Issues*, vol. 50, 1994, pp. 65-84.

Stern, P. C., "Psychology and the Science of Human-Environment Interactions", *American Psychologist*, vol. 55, 2000, pp. 523-530.

Stern, P., "Toward a Coherent Theory of Environmentally Significant Behavior", *Journal of Social Issues*, vol. 56, no.3, 2000, pp.407-424.

Stick, J., "Can Nihilism be Pragmatic?" *Harvard Law Review*, vol.100, 1986, pp. 332-401.

Stick, S., "Enviromental Tobacco Smoke-Physicians Must Avoid Fanning the Flames", *Australian and New Zealand Journal of Medicine*, vol. 30, 2000, pp. 436-439.

Stone, R., "Nuclear Trafficking: 'A Real and Dangerous Threat'" *Science*, vol. 292, 2001, pp. 1632-1636.

Stott, R., "Contraction and Convergence: Healthy Response to Climate Change", *British Medical Journal*, vol.332, June 10 2006, pp. 1385-1387.

Strauss, A. L., "The Legal Option: Suing the United States in International

Forums For Global Warming Emissions", *Environmental Law Institute*, vol. 33, 2003, pp.10185-10191 at http://www.eli.org.

Strauss, D. F., *The Old Faith and the New: A Confession*, (Asher and Co, London, 1973).

Strauss, D. F.,*The Life of Jesus Critically Examined*, (SCM Press, London, 1973).

Stringer, P., (ed.), *Confronting Social Issues: Applications of Social Psychology*, (Academic Press, London,1982).

Sumner, L W., *The Moral Foundation of Rights*, (Clarendon Press, Oxford, 1987).

Suppiah, R., (et. al.) *Climate Change Under Enhanced Greenhouse Conditions in South Australia*, (Climate Impacts and Risk Groups, CSIRO Marine and Atmosphere Research, June 2006).

Sutton, A. J., (*et al*), *Methods for Meta-Analysis in Medical Research*, (Wiley, Chichester, 2000).

Svensmark, H. and Calder, N., *The Chilling Stars: A New Theory of Climate Change*, (Icon Books, Cambridge, 2007).

Tabor, G. M., (*et al*), "Conservation Biology and the Health Sciences: Defining the Research Priorities of Conservation Medicine", in M.E. Soule (et. al., eds.), *Research Priorities in Conservation Biology*, 2nd edition, (Island Press, Washington DC,2001), pp. 155-173.

Tabor, G. M., "Defining Conservation Medicine", in A.A. Aguirre (*et al* eds.), *Conservation Medicine: Ecological Health in Practice,* (Oxford University Press, Oxford, 2002), pp. 8-16.

Tarasov, L. and Peltier, W. R., "Arctic Freshwater Forcing of the Younger Dryas Cdd Reversal", *Nature*, vol. 435, 2005, pp. 662-665.

Tatum, A., "Benefits of Humanities Highlighted", *Australian Doctor*, March 19 1999.

Taylor, A. E., So*crates*, (Hyperion Press, Westport, 1979).

Teichmann, R., *Abstract Entities*, (Macmillan, Houndmills,1992).

Tester, J. W., (*et al.*), *The Future of Geothermal Energy: Impact of Enhanced Geothermal Systems (EGS) on the United States in the 21st Century* (2007) at http://geothermal.inel.gov/publications/future_of_geothermal_energy.pdf.

Thacker, P. D., "The Many Travails of Ben Santer", *Environmental Science and Technology* (On-line), August 9 2006 at http://pubs.acs.org/subscribe/journals/esthag-w/2006/aug/policy/pt-santer.html.

Thiering, B., *Jesus and the Riddle of the Dead Sea Scrolls*, (Harper, San Francisco, 1992).

Thompson, J., *Justice and World Order: A Philosophical Inquiry,* (Routledge, London, 1992).

Thompson, L. G., (et al), "Abrupt Tropical Climate Change: Past and Present", *Proceedings of the National Academy of Sciences*, vol. 103, no.28, July 11 2006, pp. 10536-10543, doi: 10.1073/pnas.0603900103.

Thorpe, A., "A Fake Fight", *New Scientist*, March 17 2007, p. 24.

Tickell, C. Sir, "Religion and the Environment," Lecture delivered to *The Earth Our Destiny* Conference, Portsmouth Cathedral, November 30 2002 at http://www.crispintickell.com/page18.html.

Tiles, M., "Philosophy of Mathematics", in N. Bunnin and E.P. Tsui-James (eds.), *The Blackwell Companion to Philosophy*, (Blackwell Publishing, Oxford, 2003), pp.345-374.

Tobin, G. A., (*et al*), *The Uncivil University: Politics and Propaganda in American Education*, (Institute for Jewish and Community Research, San Francisco, 2005).

Topolski, J., (ed.), *Historography: Between Modernism and Postmodernism*, (Rodopi, Amsterdam, 1994).

Torn, M. S. and Harte, J., "Missing Feedbacks, Asymmetric Uncertainties, and the Underestimation of Future Warming", *Geophysical Research Letters*, vol.33, 2006, L10703, doi:10.1029/2005/GLO225540.

Trainer, F.E., *Abandon Affluence!* (Zed Books, London, 1985).

Trainer, F. E., *Developed to Death: Rethinking Third World Development*, (Green Print, London, 1989).

Trainer, F. E., *The Conserver Society: Alternatives for Sustainability*, (Zed Books, London, 1995).

Trainer, T., *The Nature of Morality*, (Avebury, Aldershot, 1991).

Trainer, T., "Development: The Radical Alternative View", at http://socialwork.arts.unsw.edu.au/tsw/D99.Development.RadView.html.

Trainer, T., "Social Responsibility: The Most Important, and Neglected, Problem of All?" at http://socialwork.arts.unsw.edu.au/tsw/D98.SocialResponsibility.html.

Trainer, T., "The Stern Review: Critical Notes on its Abatement Optimism", Unpublished manuscript, November 2006. Contact: T.Trainer@unsw.edu.au.

Trainer, T., *Renewable Energy Cannot Sustain Consumer Society*, (Springer, Dordrecht, 2007).

Trubek, D. M., "Where the Action is: Critical Legal Studies and Empiricism", *Stanford Law Review*, vol. 36, 1984, pp. 575-622.

Turley, C., (*et al*), "Reviewing the Impact of Increased Atmospheric CO2 on Ocean pH and the Marine Ecosystem", in H.J.Schellnhuber (*et al*), *Avoiding Dangerous Climate Change*, (Cambridge University Press, Cambridge, 2006), pp. 65-70.

Turner, D. and Hartzell, L.," The Lack of Clarity in the Precautionary Principle", *Environmental Values*, vol.13, 2004, pp.449-460.

Turner, J., "Significant Warming of the Antarctic Winter Troposphere", *Science*, vol.311, no 5769, March 31 2006, pp. 1914-1917, doi: 10.1126/science.1121652.

Tyler, M. J., (*et al*), "Inhibition of Gastric Acid Secretion in the Gastric Brooding Frog *Rheobatrachus Silus*", *Science*, vol. 220, 1983, pp. 609-610.

Tyler, M. J., *Australian Frogs*, (Viking O'Neil, Penguin Books, Melbourne, 1989).

Tytler, R., *Re-Imagining Science Education*, ( Australian Council for Education Research, Camberwell, Victoria, 2007).

UK Department for Environment, Food and Rural Affairs (Defra), *Triggering Widespread Adoption of Sustainable Behaviour*, (Behaviour Change: A Series of Practical Guides for Policy Makers and Practitioners, Number 4, Summer 2006) at http//www.defra.gov.uk/scienceproject_data/Documentlibrary/SD_14006/SD14006_3804_INF.pdf.

UK Meteorological Office, *The Greenhouse Effect and Climate Change: A Briefing from the Hadley Centre*, (UK Hadley Centre for Climate Prediction and Research, Devon, 1999).

UK Secretary of State for Environment, Food and Rural Affairs, *The UK Government Sustainable Development Strategy, Securing the Future: Delivering UK Sustainable Development Strategy*, (March, 2005) at http://www.sustainable-development.gov.uk/publications/uk-strategy/index.htm.

UK Sustainable Consumption Roundtable, *I Will If You Will: Towards Sustainable Consumption*, (Sustainable Development Commission and National Consumer Council, May 2006), at http://www.sd-commission.org.uk/publications/downloads/I-Will-If-You-Will. pdf.

United Nations Environment Programme, *GEO 2000*, Press Release, Nairobi, September 15 1999.

United Nations Environment Programme, *Global Environment Outlook 3*, (Earthscan, London, 2002).

Van Liedekerke, L., "John Rawls and Derek Parfit's Critique of the Discount Rate", *Ethical Perspectives*, vol. 11, 2004, pp. 72-83.

Van Lieshout, M., (*et al.*), "Climate Change and Malaria: Analysis of the SRES Climate Change and Socio-Economic Scenarios", *Global Environmental Change*, vol. 14, 2004, pp. 87-99.

Vandentorren, S., (*et al*), "August 2003 Heat Wave in France: Risk Factors for Death of Elderly People Living at Home", *European Journal of Public*

*Health*, vol.16, no. 6, 2006, pp. 583-591.

Vanderheiden, S., "Knowledge, Uncertainty, and Responsibility: Responding to Climate Change", *Public Affairs Quarterly*, vol. 18, no.2, 2004, pp. 141-158.

Veblen, T., *The Higher Learning in America: A Memorandum on the Conduct of Universities by Business Men*, (Sagamore Press, New York, 1957).

Velicona, I. and Wahr, J., "Measurements of Time-Variable Gravity Show Mass Loss in Antartica", *Science*, vol.311, no 5768, March 24 2006, pp. 1754-1756, doi: 10.1126/science.1123785.

Verheyen, R., *Climate Change Damage and International Law: Prevention Duties and State Responsibility*, (Martinus Nijhoff Publishers, Leiden, 2005).

Vernadsky, V. I., *The Biosphere*, (Springer-Verlag, New York, 1998).

Victor, D. C., *The Collapse of the Kyoto Protocol and the Struggle to Stop Global Warming*, (Princeton University Press, Princeton, 2001).

Vidal, J., "Nuclear Plants Bloom. Is the Reviled N-Power the Answer to Global Warming?" *The Guardian*, August 12 2004 at http://www.guardian.co.uk/life/feature/story/0,,1280884,00.html.

Vining, J. and Ebreo, A., "Emerging Theoretical and Methodological Perspectives on Conservation Behavior", in R.Bechtel and A.Churchman (eds.), *Handbook of Environmental Psychology*, (John Wiley, New York, 2002), pp. 541-558.

Vogel, G., "Picking Up the Pieces After Hwang", *Science*, vol. 312, April 28 2006, pp. 516-517.

Von Hobe, M. (*et. al.*), "Severe Ozone Depletion in Cold Arctic Winter 2004-2005", *Geophysical Research Letters*, vol.33, 2006, L17815, doi:10.1029/2006GLO26945.

Von Storch, H. E. and Zorita, F., "Comment on 'Hockey Sticks, Principal Components, and Spurious Significance, by S.M. McIntrye and R.L. McKitrick", *Geophysical Research Letters*, vol.32, 2005, L20701, doi: 10. 1029/2005GLO22753.

Wahlquist, A., "More Heat, Less Rain on Way for States," *The Australian*, November 6 2006, p. 4.

Walker, M. D., (et. al.), "Plant Community Responses to Experimental Warming Across the Tundra Biome", *Proceedings of the National Academy of Sciences*, vol. 103, no.5, January 31 2006, pp. 1342-1346, doi: 10.1073/pnas.0503198103.

Wallace, A. G., "Educating Tomorrow's Doctors: The Thing that Really Matters is that We Care", *Academic Medicine*, vol. 72, no 4, April 1997, pp. 253-258.

Wallace, J. M., "What Science Can and Cannot Tell Us About Greenhouse Warming," *Bridges*, vol. 7, nos. 1-2, 2000, pp.1-15.

Walzer, M., *Spheres of Justice: A Defence of Pluralism and Equality*, (Basil Blackwell, Oxford, 1983).

Ward, P. D., "Impact from the Deep", *Scientific American*, vol.295, no.4, October 2006, pp. 42-49.

Warren, M., "Carbon Trading Market to Open 'As Early as 2011'", *The Australian*, May 29 2007.

Warren, M., "Science Tempers Fears on Climate," *The Weekend Australian*, September 2-3 2006, p.1.

Warren, M., "The Clean Green Dream:", *The Weekend Australian*, November 4-5 2006, p. 19.

Warren, R., "Impacts of Global Climate Change at Different Annual Mean Global Temperature Increases", in H.J. Schellnhuber (*et al* eds.), *Avoiding Dangerous Climate Change*, (Cambridge University Press, Cambridge, 2006), pp. 93-131.

Watson, R. T., (*et al.*), *Climate Change 2001: Synthesis Report. A Contribution of Working Groups I, II, and III to the Third Assessment Report of the IPCC*, (Cambridge University Press, Cambridge, 2001), at http://www.ipcc.ch/pub/online.htm.

Watson, R.T and McMichael, A.J., "Global Climate Change - The Latest Assessment: Does Global Warming Warrant a Health Warning?" *Global Change and Human Health*, vol. 2, 2001, pp. 64-75.

Wegman, E. J., (*et al*), *Ad Hoc Committee Report on the ' Hockey Stick' Global Climate Reconstruction* at http://energy.commerce.house.gov/108/home/07142006_Wegman_Report.pdf.

Westerling, A. L., (*et al*), "Warming and Early Spring Increase Western US Forest Wildfire Activity", *Science*, vol.311, no. 5789,August 18 2006, pp. 940-943, doi: 10.1126/science.1128834.

Wheless, J., *Forgery in Christianity: A Documented Record of the Foundations of the Christian Religion* (1930), (Kessinger Publishing, Kila MT, 1997).

White, G. E., "The Inevitability of Critical Legal Studies", *Stanford Law Review*, vol. 36, 1984, pp. 649-672.

White, L., "The Roots of Our Ecological Crisis", *Science*, vol. 155, 1967, pp. 1203-1207.

White, S., "The Nuclear Power Option - Expensive, Ineffective and Unnecessary", *The Sydney Morning Herald*, July 13 2005, at http://www.smh.com.au/news/Opinion/The_nuclear_power_option_expensive_ineffective_and_unneccessary/2005/06/12/1118514925517.html.

Whitney, E., "Lyn White, Ecotheology, and History", *Environmental Ethics*, vol.15, 1993, pp. 151-169.

Wigley, T. M. L. and Schlesinger, M. E., "Analytical Solution for the Effect of Increasing CO2 on Global Mean Temperature", *Nature*, vol. 315, 1985, pp. 649-652.

Wild, M., (*et al*), "From Dimming to Brightening: Decadal Changes in Solar Radiation at Earth's Surface", *Science*, vol. 308, May 6 2005, pp. 847-850.

Williams, B., "Debate Over Peak Oil Issue Boiling Over, With Major Implications for Industry, Society," *Oil and Gas Journal*, vol.101, no.27, July 14 2003, pp.18-19.

Williams, B., *Ethics and the Limits of Philosophy*, (Harvard University Press, Cambridge, 1985).

Williams, D. and Baverstock, K., "Chernobyl and the Future: Too Soon for a Final Diagnosis", *Nature*, vol. 440, 2006, pp. 993-994.

Williamson, T. and Radford, A., "Building, Global Warming and Ethics", in W. Fox (ed.), *Ethics and the Built Environment*, (Routledge, London and New York, 2000), pp. 57-72.

Wilson, E. O., *Biophilia*, (Harvard University Press, Cambridge, MA, 1984).

Wilson, E. O., *Consilience: The Unity of Knowledge*, (Knopf, New York, 1998).

Wilson, E. O., *The Creation: An Appeal to Save Life on Earth*, (WW Norton, New York, 2006).

Wilson, J. K., *The Myth of Political Correctness: The Conservative Attack on Higher Education*, (Duke University Press, Durham, 1995).

Wilson, N., "BP, Rio in Clean Coal Plan", *The Australian*, May 22 2007, pp. 1-4.

Wilson, P., "Blair Puts Carbon Targets Into Law", *The Australian*, March 15 2007, p. 8.

Winter D. D. and Koger, S. M., *The Psychology of Environmental Problems*, 2nd edition, (Lawrence Erlbaum Assoc., Mahwah, New Jersey, 2004).

Witze, A., "Tempers Flare at Hurricane Meeting," *Nature*, vol. 441,May 4 2006, doi: 10.1038/441011a.

Wolf, C., "Intergenerational Justice", in R.G. Frey and C.H. Wellman, (eds.), *A Companion to Applied Ethics: Blackwell Companions to Philosophy*, (Blackwell Publishing, Oxford, 2003), pp. 279-294.

Wolterstorff, N., *On Universals: An Essay on Ontology*, (University of Chicago Press, Chicago, 1970).

World Health Organization (WHO), *Heat Waves: Risks and Responses*, (Health and Global Environmental Change Series No2, WHO Regional Office for Europe, Denmark, 2004).

World Meteorological Organization, *Antarctic Ozone Bulletin*, No.3 2006 at http://www.wmo.ch/web/arep/06/ant-bulletin-3-2006.pdf..

Wunsch, C., "Abrupt Climate Change: An Alternative View," *Quarterly Research*, vol. 65, 2006, pp. 191-203.

WWF Freshwater Program, *Rich Countries, Poor Water* (2006) at http://wwf.org.au/publications/rich-countries-poor-water/.

Yaffe, M. D. (ed.), *Judaism and Environmental Ethics: A Reader*, (Lexington Books, Lanham, 2001).

Yencken, D., (*et al*), *Environment, Education and Society in the Asia-Pacific: Local Traditions and Global Discourses*, (Routledge, London, 2000).

Youngquist, W., *GeoDestinies: The Inevitable Control of Earth Resources over Nations and Individuals*, (National Book Company, Portland OR, 1997).

Zakaria, F., "The Case for a Global Carbon Tax", *Newsweek* April 16, vol.149, no. 16 2007.

Zimmerman, G. L., (*et al.*), "A 'Stages of Change' Approach to Helping Patients Change Behavior", *American Family Physician*, March 1 2000 at http://www.aafp.org/afp/20000301/1409.html.

Zwart, P. J., *About Time: A Philosophical Inquiry into the Origin of Nature and Time*, (North Holland Publishing Co, Amsterdam, 1976).

# Index